Valve Amplifiers

Valve Amplifiers

Second edition

Morgan Jones

Newnes

OXFORD AUCKLAND BOSTON JOHANNESBURG MELBOURNE NEW DELHI

Newnes
An imprint of Butterworth-Heinemann
Linacre House, Jordan Hill, Oxford OX2 8DP
225 Wildwood Avenue, Woburn, MA 01801-2041
A division of Reed Educational and Professional Publishing Ltd

A member of the Reed Elsevier plc group

First published 1995
Reprinted 1996, 1997, 1998
Second edition 1999

British Library Cataloguing in Publication Data
A catalogue record for this book is available from the British Library

Library of Congress Cataloguing in Publication Data
A catalogue record for this book is available from the Library of Congress

ISBN 0 7506 4425 7

Composition by Genesis Typesetting, Rochester, Kent
Cover photograph by Morgan Jones and Sean Lancastle
Printed and bound in Great Britain by
Biddles Ltd, Guildford and King's Lynn

Contents

Preface to first edition

Twenty-five years ago the author bought his first valve amplifier; it cost him £3, and represented many weeks' pocket money. Whilst his pocket money has increased, so have his aspirations, and the DIY need was born.

Although there were many sources of information on circuit design, the electronics works gave scant regard to audio design, whilst the hi-fi books barely scratched the surface of the theory. The author therefore spent much time in libraries trying to link this information together to form a basis for audio design. This book is the result of those years of effort and aims to present thermionic theory in an accessible form without getting too bogged down in maths. Primarily, it is a book for practical people armed with a scientific calculator, a power drill and a soldering iron.

The author started a BSc in Acoustical Engineering, but left after a year to join BBC Engineering as a Technical Assistant, where he received excellent training in electronics and rose to the giddy heights of a Senior Engineer before being made redundant by BBC cuts. During that time, he became a member of the Audio Engineering Society, and designed and constructed many valve pre-amplifiers and power amplifiers, loudspeakers, pick-up arms and a pair of electrostatic headphones.

Note for the CD generation: *Pick-up arms are exquisitely engineered mechanical contrivances that support a lump of rock on the end of a scaffold pole being scraped by the wiggly groove in a 300 mm (flattish) vinyl disc known as an LP. The resulting sound is often very good.*

Preface to second edition

The author is deeply grateful to all the readers who bought the first edition of *Valve Amplifiers*, and feels that it is only right and proper that his royalties were invested in better test equipment. However, intensive testing, together with helpful feedback from readers, revealed a number of howlers in the first edition, for which the author can only apologize humbly.

The author's continued investigations in search of the 'perfect' valve RIAA pre-amplifier have led to a thorough investigation of practical components. The Beast direct-drive electrostatic loudspeaker amplifier required a push–pull output, and can easily be modified for a balanced input, whilst most 1 bit DACs inherently produce a balanced output (digital phase splitter!), so balanced operation throughout looks more and more desirable.

Dedication

The author would like to dedicate this book to the dwindling band of BBC engineers, particularly at BBC Southampton, and also to those at BBC Wood Norton, of which he has many colourful memories.

Acknowledgements

Many people gave various forms of encouragement, some of it unintentional, but the author would particularly like to thank Mark Bartrum and Dave Mansell for their heroic proofreading, but also for their perceptive questions, which opened many new lines of enquiry. Photographs were taken by Tony Wade, in the course of which the author was colourfully educated in the art of 'hi-fi' photography, and absorbed such terms as 'reciprocity failure' and 'it'll be alright on the print', for which he is eternally grateful.

Many commercial enterprises were most helpful with information or loaned equipment, and their names have been mentioned in the text, but particular credit is due to Brian Sowter of Sowter Transformers for his patience in listening to the author's bizarre requests.

Morgan Jones

1

Circuit analysis

In order to look at the interesting business of designing and building valve amplifiers, we will need some knowledge of electronics fundamentals. Unfortunately, fundamentals are not terribly interesting, and to cover them fully would consume the entire book. Ruthless pruning has therefore been necessary in order to condense what is needed into one chapter.

It is thus with deep sorrow that the author has had to forsake complex numbers and vectors, whilst the omission of differential calculus is a particularly poignant loss. All that is left is ordinary algebra, and although there are *lots* of equations, they are timid, miserable creatures and quite defenceless.

If you are comfortable with basic electronic terms and techniques, then please feel free to skip directly to Chapter 2, where valves appear.

Mathematical symbols

Unavoidably, a number of mathematical symbols are used, some of which you may have forgotten (or not even previously met):

$a = b$ a equals b
$a \approx b$ a is approximately equal to b
$a \propto b$ a is proportional to b
$a \neq b$ a is not equal to b
$a > b$ a is greater than b
$a < b$ a is less than b
$a \geq b$ a is greater than, or equal to, b
$a \leq b$ a is less than, or equal to, b

As with the $=$ and $\neq$ symbols, the four preceding symbols can have a slash through them to negate their meaning ($a \nless b$, a is not less than b).

$\sqrt{a}$	the number which when multiplied by itself is equal to a (square root)
a^n	a multiplied by itself n times. $a^4 = a \times a \times a \times a$ (a to the power n)
$\pm$	plus or minus
∞	infinity
$^\circ$	degrees – either of temperature °C, or of an angle (360° in a circle)
$\|\|$	parallel – either parallel lines, or an electrical parallel connection

Electrons and definitions

Electrons are *charged* particles. Charged objects are attracted to other charged particles, or objects. A practical demonstration of this is to take a balloon, rub it briskly against a jumper, and then place the face that was rubbed against a wall. Let go. The balloon remains stuck to the wall. This is because we have charged the balloon, and so there is an attractive force between it and the wall. (Although the wall was initially uncharged, placing the balloon on the wall induced a charge.)

Charged objects come in two flavours: negative and positive. Unlike charges attract and like charges repel. Electrons are negative and, if we could find some, positrons would be positive.

If we don't have ready access to positrons, how can we have a positively charged object? Suppose we had an object that was negatively charged, because it had 2000 electrons clustered on its surface. If we had another, similar, object that only had 1000 electrons on its surface, then we would say that the first object was more negatively charged than the second, but as we can't count how many electrons we have, we might just as easily have said that the second object was more positively charged than the first. It's just a matter of which way you look at it.

To charge our balloon, we had to do some work, and use energy. We had to overcome friction when rubbing the balloon against the woollen jumper. In the process, electrons were moved from one surface to the other. Therefore one object (the balloon) has acquired an excess of electrons and is negatively charged, whilst the woollen jumper has lost the same number of electrons and is positively charged.

The balloon would therefore stick to the jumper. Or would it? Certainly it will be attracted to the jumper, but what happens when we place the two in contact? The balloon does not stick. This is because the fibres of the jumper were able to touch all of the charged area on the balloon, and the electrons were so attracted to the jumper that they moved back onto the jumper, thus neutralizing the charge.

At this point we can discard vague talk of balloons and jumpers because we have just observed electron flow.

An electron is very small, and doesn't have much of a charge, so we need a more practical unit for defining charge. That practical unit is the *coulomb* (C). We could now say that 1 coulomb of charge had flowed between one point and another, which would be equivalent to saying that approximately 6 180 000 000 000 000 000 electrons had passed, but much handier.

Simply being able to say that a large number of electrons has flowed past a given point is not in itself very helpful. We might say that a billion cars have travelled down a particular section of motorway since it was built, but if you were planning a journey down that motorway, you would want to know what was the flow of cars *per hour* through that section.

Similarly in electronics, we are not concerned with the total flow of electrons since the dawn of time, but we do want to know about electron flow at any given instant. So we could define the flow as the number of coulombs of charge that flowed past a point in one second. This is still rather long winded and we will abbreviate yet further.

We will call the flow of electrons a *current*, and, as the coulomb/second is unwieldy, a new unit will be defined, the *ampere* (A). Because the ampere is such a useful unit, the definition linking current and charge is usually stated in the following form:

One coulomb is the charge moved by one ampere flowing for one second.

$$\text{charge (coulombs)} = \text{current (amperes)} \times \text{time (seconds)}$$

This is still rather unwieldy, so symbols are assigned to the various units: charge has the symbol Q, current I, and time t:

$$Q = It$$

This is a very useful equation, and we will meet it again when we look at capacitors (which store charge).

Meanwhile, current has been flowing, but why did it flow? If we are going to move electrons from one place to another, we need a force to cause this movement. This force is known as the *electromotive force* (EMF). Current will continue to flow whilst this force is applied, and will flow from a higher potential to a lower potential.

If two points are at the same potential, no current will flow between them. What is important is the *potential difference* (pd).

A potential difference causes a current to flow between two points. As this is a new property, we will need a unit, a symbol and a definition to describe it. We mentioned work being done in charging the balloon, and in its very precise and physical sense, this is how we can define potential difference, but first we must define *work*:

One joule of work is done if a force of one newton moves one metre from its point of application.

This very physical interpretation of work can be understood easily once we realize that it means that one joule of work would be done by moving one kilogramme a distance of one metre in one second. Since charge is directly related to the mass of electrons moved, the physical definition of work can be modified to define the force that causes the movement of charge.

Unsurprisingly, because it causes the motion of electrons, the force is called the electromotive force (EMF), and it is measured in *volts*:

If one joule of work is done moving one coulomb of charge, then the system is said to have a potential difference of one volt (V).

$$\text{work done (joules)} = \text{charge (coulombs)} \times \text{potential difference (volts)}$$

$$W = QV$$

The concept of work is important because work can only be done by the expenditure of energy, which is therefore also expressed in joules:

$$\text{work done (joules)} = \text{energy expended (joules)}$$

$$W = E$$

In our specialized sense, doing work means moving charge (electrons) to make currents flow.

Batteries and lightbulbs

If we want to make a current flow, we need a *circuit*. A circuit is exactly that; a loop or path through which a current can flow, from its starting point all the way round the circuit, to *return* to its starting point. Break the circuit and the current ceases to flow.

The simplest circuit that we might imagine is a battery connected to a lightbulb via a switch. We *open* the switch to stop the current flow (open circuit), and close it to light the bulb. Meanwhile our helpful friend (who has been watching all this) leans over and drops a thick piece of copper across the battery terminals, causing a *short* circuit.

The bulb goes out. Why?

Ohm's law

To answer the last question we need some property that defines how much current will flow and where. That property is *resistance*, so we will need another definition, units, and a symbol:

If a potential difference of one volt is applied across a resistance, resulting in a current of one ampere, then the resistance has a value of one ohm (Ω).

$$\text{potential difference (volts)} = \text{current (amperes)} \times \text{resistance (ohms)}$$

$$V = IR$$

This is actually a statement of Ohm's law, rather than a strict definition of resistance, but we don't need to worry too much about that.

We can rearrange the previous equation to make I or R the subject:

$$I = \frac{V}{R}$$

$$R = \frac{V}{I}$$

These are incredibly powerful equations and should be committed to memory.

The circuit in Fig. 1.1 is switched on, and a current of 0.25 A flows. What is the resistance of the bulb?

$$R = \frac{V}{I} = \frac{240}{0.25} = 960\,\Omega$$

Now this might seem like a trivial example, since we could easily have measured the resistance of the bulb to 3½ significant figures using our shiny, new, digital multimeter. But could we? The hot resistance of a lightbulb is

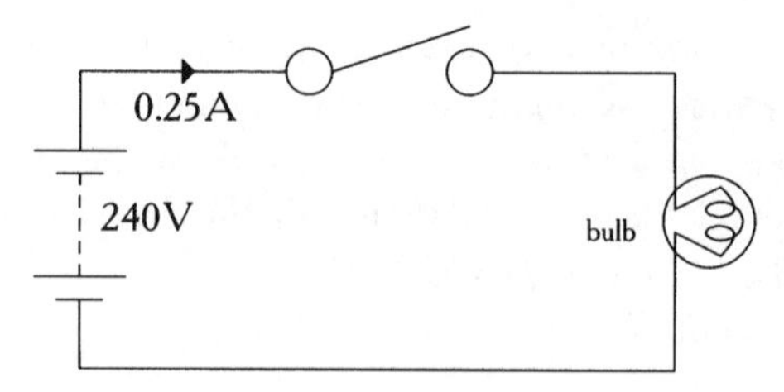

Fig. 1.1 *Use of Ohm's law to determine the resistance of a hot lightbulb*

very different to its cold resistance; in the example above, the cold resistance was measured as being 80 Ω.

We could now work the other way, and ask what current will flow through an 80 Ω resistor connected to 240 V?

$$I = \frac{V}{R} = \frac{240}{80} = 3\,\mathrm{A}$$

Incidentally, this is why lightbulbs are most likely to fail at switch-on. The high initial current that flows before the filament has warmed up and changed its resistance, stresses the weakest parts of the filament, they become so hot that they vaporize, and the bulb blows.

Power

In the previous example, we looked at a lightbulb, and rated it by the current that flowed through it when connected to a 240 V battery. But we all know that lightbulbs are rated in *watts*, so there must be some connection between the two:

One watt of power is expended if one joule of work is done in one second.

$$\text{power (watts)} = \frac{\text{work done (joules)}}{\text{time taken (seconds)}}$$

$$P = \frac{W}{t}$$

This may not seem to be the most useful of definitions, and indeed it isn't, but by combining it with some earlier equations:

$$W = QV$$

So:

$$P = \frac{QV}{t}$$

But:

$$Q = It$$

So:

$$P = \frac{IVt}{t}$$

We obtain:

$$P = IV$$

This is a fundamental equation of equal importance to Ohm's law. Substituting the Ohm's law equations into this yields:

$$P = \frac{V^2}{R} = I^2 R$$

We can now use these equations to calculate the power rating of our lightbulb. Since it drew 0.25 A when fed from 240 V, and had a hot resistance of 960 Ω, we can use any of the three equations.

Using:

$$P = \frac{V^2}{R} = \frac{240^2}{960} = 60\,\text{W}$$

It will probably not have escaped your notice that this lightbulb looks suspiciously like an AC mains lightbulb, and that the battery was rather large. We will come back to this later.

Kirchhoff's laws

There are two of these: a current law and a voltage law. They are both very simple and at the same time very powerful.

The current law states:

The algebraic sum of the currents flowing into, and out of, a node is equal to zero.

$$0 = I_1 + I_2 + I_3 + \ldots$$

What it says in a more relaxed form is that what goes in, comes out. If we have 10 A going into a junction, then that current must also leave that junction – it might not all come out on one wire, but it will all come out. A sort of conservation of current, if you like. See Fig. 1.2.

Current flowing into the node: $I_1 = 10A$
Currents leaving the node: $I_2 = 4A$, $I_3 = 6A$
Total current leaving the node: $I_{total} = 10A$

$I_1 = 10A$
$I_2 = 4A$
$I_3 = 6A$

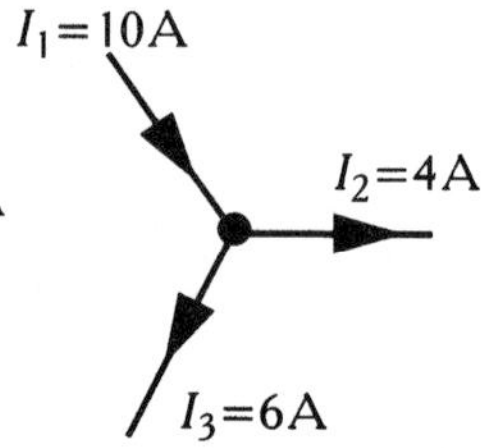

Fig. 1.2 *Currents at a node (Kirchhoff's current law)*

Current flowing into the node:	$I_1 = 10\,A$
Currents leaving the node:	$I_2 = 4\,A$
	$I_3 = 6\,A$
Total current leaving the node:	$I_{total} = 10\,A$

From the point of view of the node, the currents leaving the node are flowing in the opposite direction to the current flowing into the node, so we *must* give them a minus sign before plugging them into the equation:

$$
\begin{aligned}
0 &= I_1 + I_2 + I_3 \\
&= 10\text{A} + (-4\text{A}) + (-6\text{A}) \\
&= 10 - 4 - 6
\end{aligned}
$$

This may have seemed pedantic, since it was obvious from the diagram that the incoming currents equalled the outgoing currents, but you may need to find a current when you don't even know the direction in which it is flowing. Using this convention forces the correct answer!

It is vital to make sure that your signs are correct.

The voltage law states:

The algebraic sum of the EMFs and potential differences acting around any loop is equal to zero.

This law draws a very definite distinction between EMFs and potential differences. EMFs are *sources* of electrical energy (such as batteries), whereas potential differences are the *voltages dropped* across components. Another way of stating the law is to say that the algebraic sum of the EMFs must equal the algebraic sum of the potential drops around the loop. Again, you could consider this to be a conservation of voltage. See Fig. 1.3.

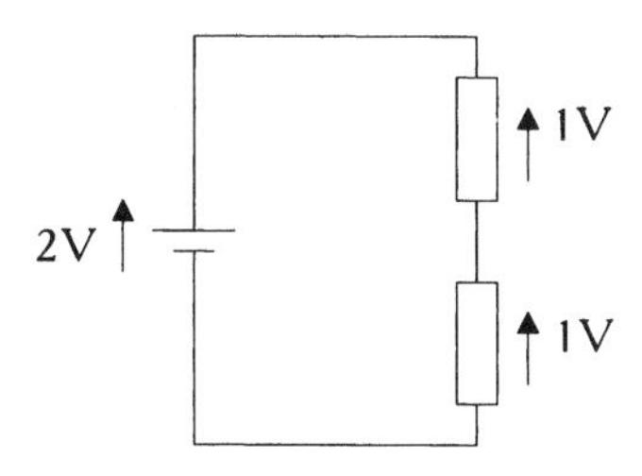

Fig. 1.3 *Summation of potentials within a loop (Kirchhoff's voltage law)*

Resistors in series and parallel

If we had a *network* of resistors, we might want to know what the total resistance was between terminals A and B. See Fig. 1.4.

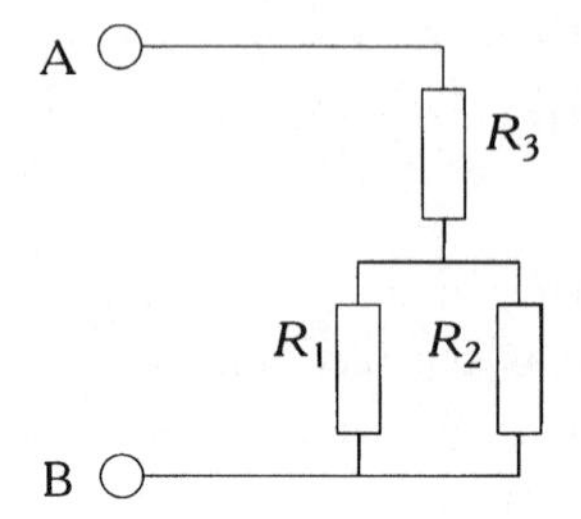

Fig. 1.4 *Series/parallel resistor network*

We have three resistors: R_1 is in *parallel* with R_2, and that *combination* is in *series* with R_3.

As with all problems, the thing to do is to break it down into its simplest parts. If we had some means of determining the value of resistors in series, we could use it to calculate the value of R_3 in series with the combination of R_1 and R_2, but as we don't yet know the value of the parallel combination, we must find this *first*. This question of order is most important and we will return to it later.

If the two resistors (or any other component, for that matter) are in parallel, then they must have the same voltage drop across them. Ohm's law might therefore be a useful starting point:

$$I = \frac{V}{R}$$

Using Kirchhoff's current law, we can state that:

$$I_{\text{total}} = I_{\text{R1}} + I_{\text{R1}} + \ldots$$

So:

$$\frac{V}{R_{\text{parallel}}} = \frac{V}{R_1} + \frac{V}{R_2} + \ldots$$

Dividing by V:

$$\frac{1}{R_{\text{parallel}}} = \frac{1}{R_1} + \frac{1}{R_2} \ldots$$

The reciprocal of the total parallel resistance is equal to the sum of the reciprocals of the individual resistors.

For the special case of only two resistors, we can derive the equation below:

$$R_{\text{parallel}} = \frac{R_1 R_2}{R_1 + R_2}$$

This is often known as 'product over sum', and whilst it is useful for mental arithmetic, it is slow to use on a calculator (more keystrokes).

Now that we have cracked the parallel problem, we need to crack the series problem.

First, we will simplify the circuit. We can now calculate the total resistance of the parallel combination and replace it with one resistor of that value – an *equivalent* resistor. See Fig. 1.5.

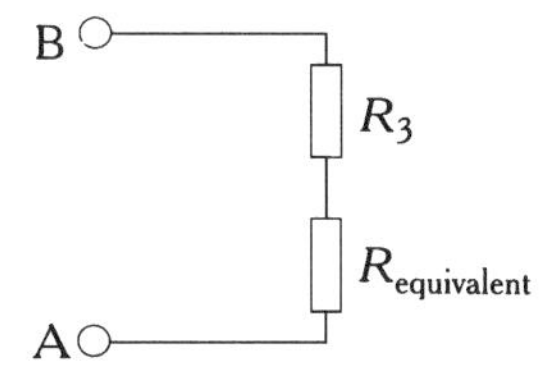

Fig. 1.5 *Simplification of Fig. 1.4 using an equivalent resistor*

Using the voltage law, the sum of the potentials across the resistors must be equal to the driving EMF:

$$V_{\text{total}} = V_{\text{R1}} + V_{\text{R2}} + \ldots$$

Using Ohm's law:

$$V_{\text{total}} = IR_1 + IR_2 + \ldots$$

But if we are trying to create an equivalent resistor, whose value is equal to the combination, we could say:

$$IR_{total} = IR_1 + IR_2 + \ldots$$

Hence:

$$R_{series} = R_1 + R_2 + \ldots$$

The total resistance of a combination of resistors is equal to the sum of their individual resistances.

Using the parallel and series equations we are now able to calculate the total resistance of *any* network. See Fig. 1.6.

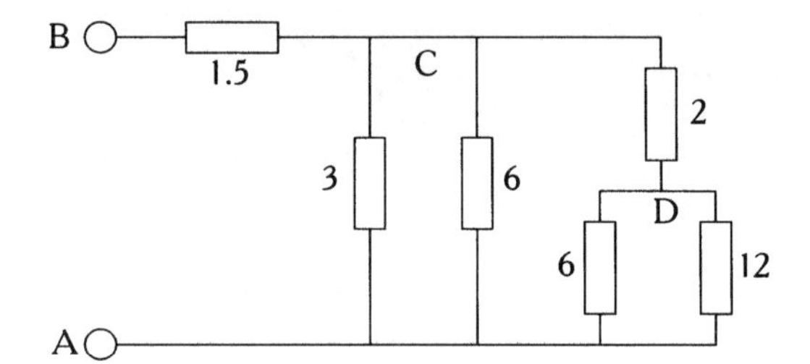

Fig. 1.6

Now this may look horrendous, but it is not a problem if we attack it logically. The hardest part of the problem is not wielding the equations or numbers, but where to start.

We want to know the resistance looking into the terminals A and B, but we don't have any rules for finding this directly, so we must look for a point where we can apply our rules. We can only apply one rule at a time, so we look for a combination of components that is made up *only* of series *or* parallel components.

In this example, we find that between node A and node D, there are *only* parallel components. We can calculate the value of an equivalent resistor and substitute it back into the circuit:

$$R_{parallel} = \frac{\text{product}}{\text{sum}}$$

$$= \frac{6 \times 12}{6 + 12}$$

$$= 4\,\Omega$$

We redraw the circuit. See Fig. 1.7.

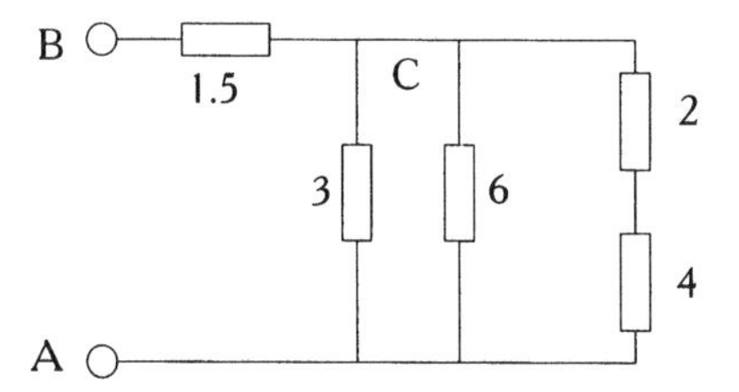

Fig. 1.7

Looking again, we see now that the *only* combinations made up of series *or* parallel components are between node A and node C, but we have a choice – either the series combination of the 2 Ω and 4 Ω, or the parallel combination of the 3 Ω and 6 Ω. The one to go for is the series combination. This is because it will result in a single resistor that will then be in parallel with the 3 Ω and 6 Ω resistors. We can cope with the three resistors in parallel later:

$$R_{\text{series}} = R_1 + R_2$$
$$= 4 + 2$$
$$= 6\,\Omega$$

We redraw the circuit. See Fig. 1.8.

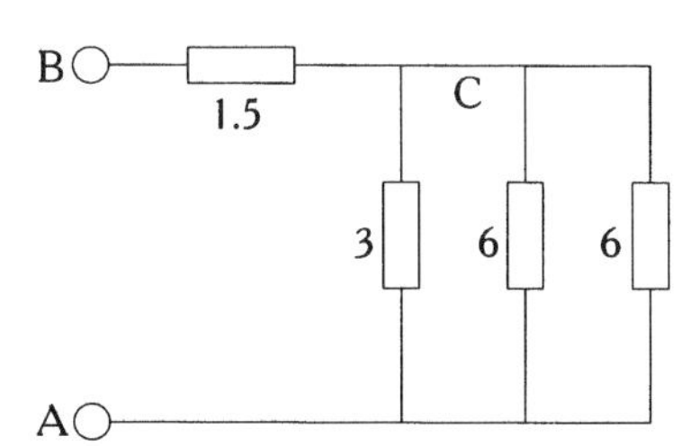

Fig. 1.8

We now see that we have three resistors in parallel:

$$\frac{1}{R_{\text{parallel}}} = \frac{1}{R_1} + \frac{1}{R_2} + \frac{1}{R_3}$$

$$= \frac{1}{3} + \frac{1}{6} + \frac{1}{6}$$

$$= \frac{2}{3}$$

Hence:

$$R_{\text{parallel}} = \frac{3}{2} = 1.5\,\Omega$$

We have now reduced the circuit to two 1.5 Ω resistors in series, and so the total resistance is 3 Ω.

This took a little time, but it demonstrated some useful points that will enable you to analyse networks much faster the second time around:

- The critical stage is choosing the starting point.
- The starting point is generally as far away from the terminals as it is possible to be.
- The starting point is made up of a combination of *only* series *or* parallel components.
- Analysis tends to proceed outwards from the starting point towards the terminals.
- Redrawing the circuit helps. You may even need to redraw the original circuit if it doesn't make sense to you. Redrawing as analysis progresses reduces confusion and errors – do it!

Potential dividers

Figure 1.9 shows a potential divider. This could be made up of two discrete resistors or it could be the moving wiper of a volume control. As before, we will suppose that a current I flows through the two resistors. We want to know the ratio of the output voltage to the input voltage. See Fig. 1.9.

$$V_{\text{out}} = IR_2$$

$$V_{\text{in}} = IR_1 + IR_2$$

$$= I(R_1 + R_2)$$

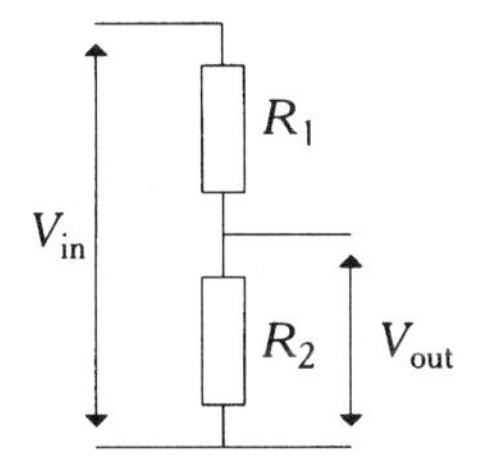

Fig. 1.9 *Potential divider*

Hence:

$$\frac{V_{out}}{V_{in}} = \frac{IR_2}{I(R_1 + R_2)}$$

$$= \frac{R_2}{R_1 + R_2}$$

This is a very important result and, used intelligently, can solve virtually anything.

Equivalent circuits

So far we have looked at networks of resistors and calculated equivalent *resistances*. Now we will extend the idea to equivalent *circuits*. This is a tremendously powerful concept for circuit analysis.

It should be noted that this is not the only method, but it is usually the quickest, and kills 99% of all known problems. Other methods include Kirchhoff's laws combined with a lot of simultaneous equations and the superposition theorem. These methods may be found in standard texts, but they tend to be cumbersome and so we will not discuss them here.

The Thévenin equivalent circuit

When we looked at the potential divider, we were able to calculate the ratio of output voltage to input voltage. If we were now to connect a battery across the input terminals, we could calculate the output voltage. Using our earlier tools, we could also calculate the total resistance looking into the output terminals. As before, we could then redraw the circuit, and the result is known as the *Thévenin equivalent circuit*. If two black boxes were made, one containing the original circuit and the other the Thévenin equivalent, you would not be able to tell from the output terminals which was which. The concept is simple to use and can break down complex networks quickly and efficiently. See Fig. 1.10.

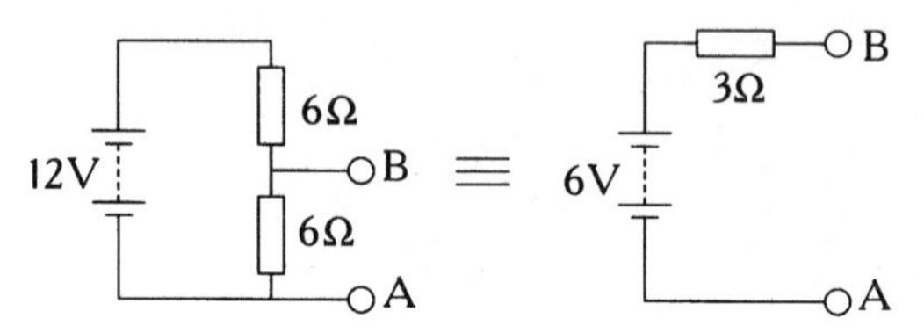

Fig. 1.10 *A 'black box' network, and its Thévenin equivalent circuit*

This is a simple example to demonstrate the concept. First, we will find the equivalent resistance, often known as the *output resistance*. Now, in the world of equivalent circuits, batteries are perfect voltage sources, they have zero *internal resistance* and look like a short circuit when we consider their resistance. Therefore, we can ignore the battery, or replace it with a piece of wire whilst we calculate the resistance of the total circuit:

$$R_{output} = \frac{R_1 R_2}{R_1 + R_2}$$

$$= 3\,\Omega$$

Next, we need to find the output voltage. We will use the potential divider equation:

$$\frac{V_{out}}{V_{in}} = \frac{R_2}{R_1 + R_2}$$

$$= \frac{1}{2}$$

So:

$$V_{out} = \frac{1}{2} \cdot V_{in}$$

$$= 6\,\text{V}$$

Now for a much more complex example: this will use all the previous techniques and really demonstrate the power of Thévenin. See Fig. 1.11.

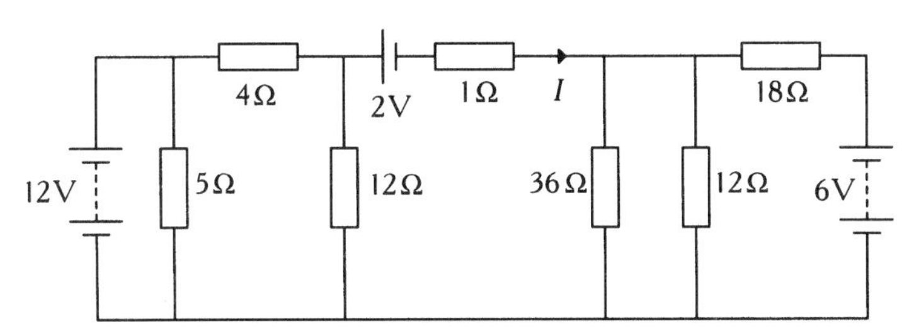

Fig. 1.11

For some obscure reason, we want to know the current flowing in the 1 Ω resistor. The first thing to do is to redraw the circuit. Before we do this we can observe that the 5 Ω resistor in parallel with the 12 V battery is irrelevant. Yes, it will draw current from the battery, but it doesn't affect the operation of the rest of the circuit. A short circuit in parallel with 5 Ω is *still* a short circuit, so we will throw it away. See Fig. 1.12.

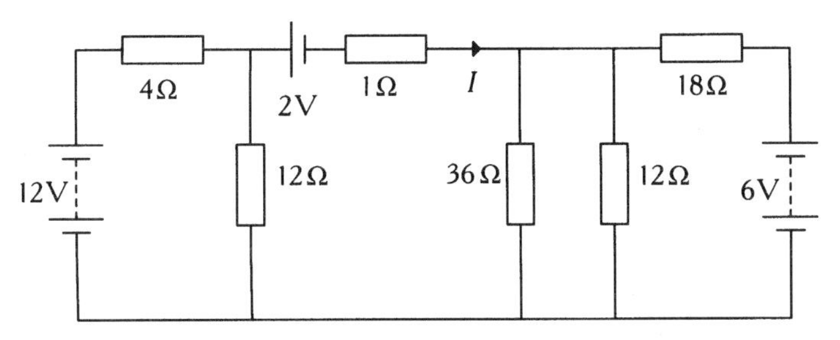

Fig. 1.12

Despite our best efforts, this is still a complex circuit, so we need to break it down into modules that we recognize. Looking at the left-hand side, we see a battery with a 4 Ω and 12 Ω resistor which looks suspiciously like the simple problem that we saw earlier, so let's break the circuit there and make an equivalent circuit. See Fig. 1.13.

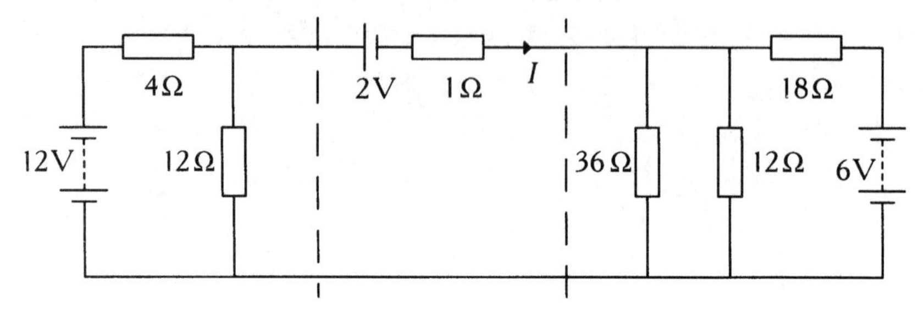

Fig. 1.13

Using the potential divider rule:

$$V_{out} = \frac{R_2}{R_1 + R_2} \cdot V_{in}$$

$$= 9V$$

Using 'product over sum':

$$R_{out} = \frac{R_1 R_2}{R_1 + R_2}$$

$$= 3\,\Omega$$

Looking at the right-hand side we can perform a similar operation to the right of the dotted line.

First, we find the resistance of the parallel combination of the 36 Ω and 12 Ω resistors, which is 9 Ω. We now have a potential divider, whose output resistance is 6 Ω, and the Thévenin voltage is 2 V.

Now we redraw the circuit. See Fig. 1.14.

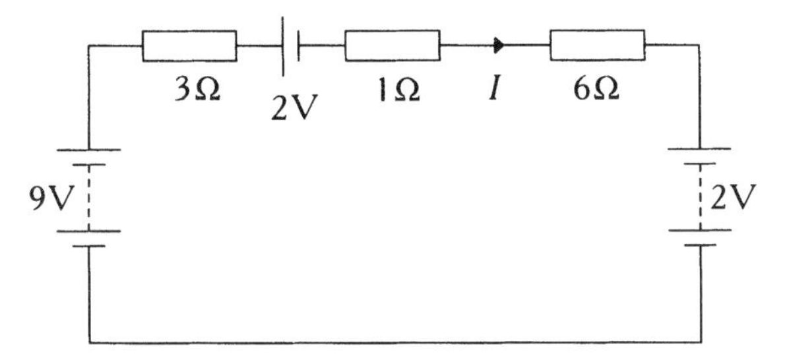

Fig. 1.14

We can make a few observations at this point. First, we have three batteries in series, why not combine them into one battery? There is no reason why we should not do this provided that we take note of their polarities. Similarly, we can combine some or all of the resistors. See Fig. 1.15.

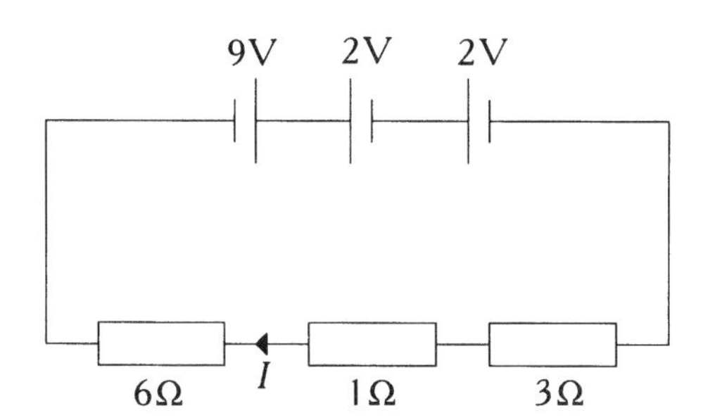

Fig. 1.15

The problem now is trivial, and a simple application of Ohm's law will solve it. We have a total resistance of 10 Ω, and a 5 V battery, so the current must be ½ A.

Useful points to note:

- Look for components that are irrelevant, such as resistors directly across battery terminals.
- Look for potential dividers on the outputs of batteries and 'Thévenize' them. Keep on doing so until you meet the next battery.
- Work from battery terminals outwards.
- Keep calm, and try to work neatly – it will save mistakes later.

Although it is possible to solve almost all problems using a Thévenin equivalent circuit, sometimes a Norton equivalent is more convenient.

The Norton equivalent circuit

The Thévenin equivalent circuit was a perfect voltage source in *series* with a resistance. The *Norton* equivalent circuit is a perfect *current* source in *parallel* with a resistance. See Fig. 1.16.

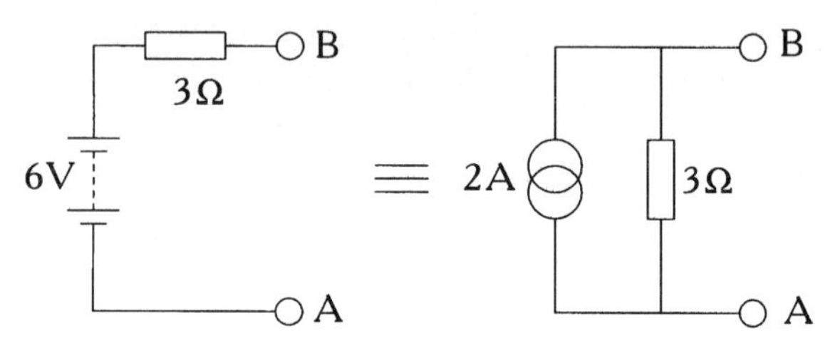

Fig. 1.16 *The Norton equivalent circuit*

We can easily convert from a Norton source to a Thévenin source, or vice versa, because the resistor has the same value in both cases. We can easily find the value of the current source by short circuiting the output of the Thévenin source and calculating the resulting current – this is the Norton current.

To convert from a Norton source to a Thévenin source, we leave the source open circuit and calculate the voltage developed across the Norton resistor – this is the Thévenin voltage.

For the vast majority of problems, the Thévenin equivalent will be quicker, mostly because we become used to thinking in terms of voltages that can easily be measured by a meter, or viewed on an oscilloscope. Occasionally, a problem will arise that is intractable using Thévenin, and converting to a Norton equivalent causes the problem to solve itself. These problems usually involve the summation of a number of currents, when the only other solution would be to resort to Kirchhoff and simultaneous equations. An example of a problem best solved by Norton occurs in Chapter 5, where the values of feedback resistors for a power amplifier are determined.

Units and multipliers

All the calculations up to this point have been arranged to use convenient values of voltage, current and resistance. In the real world, we will not be so fortunate, and to avoid having to use scientific notation, which takes longer to write, and is virtually unpronounceable, we will prefix our units with multipliers.

Prefix	Abbreviation	Multiplies by
atto	a	10^{-18}
femto	f	10^{-15}
pico	p	10^{-12}
nano	n	10^{-9}
micro	μ	10^{-6}
milli	m	10^{-3}
kilo	k	10^{3}
mega	M	10^{6}
giga	G	10^{9}
tera	T	10^{12}

Note that the *case* of the prefix is important; there is a large difference between 1 mΩ and 1 MΩ. Electronics uses a very wide range of values, and whilst you will only rarely see atto, femto, giga and tera used, the other prefixes are in common use.

Electronics engineers commonly abbreviate further, and you will often hear a 22 pF (picofarad) capacitor referred to as 22 'puff', whilst the 'ohm' is commonly dropped for resistors, and 470 kΩ (kilo-ohm) would be pronounced as 'four-seventy-kay'.

A rather more awkward abbreviation that arose before high resolution printers became available (early printers and typewriters couldn't print 'μ'), is the abbreviation of μ (micro) to m. This abbreviation still persists, particularly in American text, and you will occasionally see a 10 mF capacitor specified, although the context makes it clear that what is actually meant is 10 μF. For this reason, true 10 mF capacitors are invariably specified as 10 000 μF.

Unless an equation states otherwise, you may assume that it uses the base physical units, so an equation involving capacitance and time constants would expect you to express capacitance in farads and time in seconds. Thus 75 μs = 75×10^{-6} seconds, and the value of capacitance resulting from an equation might be 2.2×10^{-10} F = 220 pF.

The decibel

The human ear spans a vast dynamic range from the near silence heard in an empty recording studio to the deafening noise of a nearby pneumatic drill. If we were to plot this range linearly on a graph, the quieter sounds would hardly be seen, whereas the difference between the noise of the drill and a jet engine would be given a disproportionate amount of room on the graph. What we need is a graph that gives an equal weighting to *relative* changes in level of both quiet and loud sounds. By definition, this implies a logarithmic scale on the graph, but electronics engineers went one better, and invented a logarithmic ratio known as the *decibel* (dB) which was promptly hijacked by the acoustical engineers. (The fundamental unit is the bel, but this is inconveniently large, so the decibel is more commonly used, and the capital B is usually dropped.)

The dB is *not* an absolute quantity. It is a *ratio*, and it has one formula for use with currents and voltages, and another for powers:

$$\text{dB} = 20 \log_{10}\left(\frac{V_1}{V_2}\right) = 20 \log_{10}\left(\frac{I_1}{I_2}\right) = 10 \log_{10}\left(\frac{P_1}{P_2}\right)$$

The reason for this is that $P \propto V^2$ or I^2, and with logarithms, multiplying the logarithm by 2 is the same as squaring the original number. Using a different formula to calculate dBs when using powers ensures that the resulting dBs are equivalent, irrespective of whether they were derived from powers or voltages.

This might seem like a lot of complication when all we wanted to do was to describe the difference in two signal levels, but the dB will be found to be a very handy unit.

Useful common dB values are:

dB	V_1/V_2	P_1/P_2
0	1	1
3	$\sqrt{2}$	2
6	2	4
20	10	100

Because dBs are derived from logarithms, they obey all the rules of logarithms, and adding dBs is the same as multiplying the ratios that generated them. Note that dBs can be negative, implying loss, or a drop in level.

For example, if we had two cascaded amplifiers, one with a voltage gain of ½ and the other with a voltage gain of 10, then by *multiplying* the individual gains, the combined voltage gain would be 5, but we could also find this in dB, by saying that one amplifier had –6 dB of gain whilst the other had 20 dB, and *adding* the gains in dB to give a total gain of 14 dB.

When *designing* amplifiers, we will not often use the above example, as we will find it more convenient to use absolute voltages, but we will frequently use the dB to describe filter responses.

Alternating current

All of the techniques used so far have used *direct current* (DC), where the current is constant and flows in one direction only. Listening to DC is not very interesting, so we now need to look at *alternating currents* (AC).

All of the previous techniques of circuit analysis can be applied equally well to AC signals.

The sine wave

The *sine* wave is the simplest possible alternating signal. The equation that generates this signal is:

$$v = V_{peak} \sin(\omega t + \theta)$$

where: v = the instantaneous value at time t

V_{peak} = the peak value

ω = angular frequency in radians/second ($\omega = 2\pi f$)

t = the time in seconds

θ = a constant phase angle

The above concepts are shown on the diagram. See Fig. 1.17.

There is a convention in use in this equation. *Upper* case letters denote DC, or constant values, whereas *lower* case letters denote the instantaneous AC, or changing, value. It is a form of shorthand to avoid having to specify separately that a quantity is AC or DC. It would be nice to say that this convention is rigidly applied, but it is often neglected, and the context of the symbols usually makes it clear whether the quantity is AC or DC.

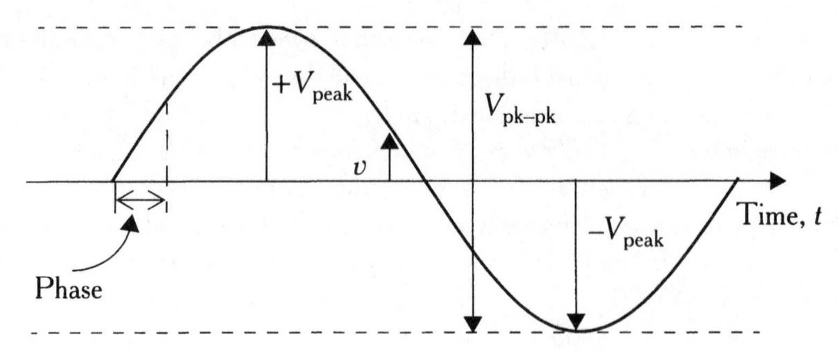

Fig. 1.17

In electronics, the word 'peak' (pk) has a very precise meaning and, when used to describe an AC waveform, it means the voltage from zero volts to the peak voltage reached, either positive or negative. Peak to peak (pk–pk), means the voltage from positive peak to negative peak and, for a symmetrical waveform, $V_{pk-pk} = 2V_{pk}$. Although electronics engineers habitually use ω to describe frequency, they do so only because calculus requires that they work in radians. Since $\omega = 2\pi f$, we can rewrite the equation as:

$$v = V_{peak} \sin(2\pi ft + \theta)$$

If we now inspect this equation, we see that apart from time t, there are other constants that we could vary before we allow time to change and determine the waveform. We can change V_{peak}, and this will change the *amplitude* of the sine wave, or we can change f, and this will change the *frequency*. The inverse of frequency is *period*, which is the time taken for one full cycle of the waveform to occur:

$$\text{period } T = \frac{1}{f}$$

If we listen to a sound that is a sine wave, and change the amplitude, this will make the sound louder or softer, whereas varying frequency will change the pitch. If we vary θ (phase), it will sound the same if we are listening to it, and unless we have an external reference, the sine wave will look exactly the same viewed on an oscilloscope. Phase becomes significant if we compare one sine wave with another sine wave *of the same frequency*, or a harmonic of that frequency. Attempting to compare phase between sine waves of unrelated frequencies is meaningless.

Now that we have described sine waves, we will look at them as they would appear on the screen of an oscilloscope. See Fig. 1.18.

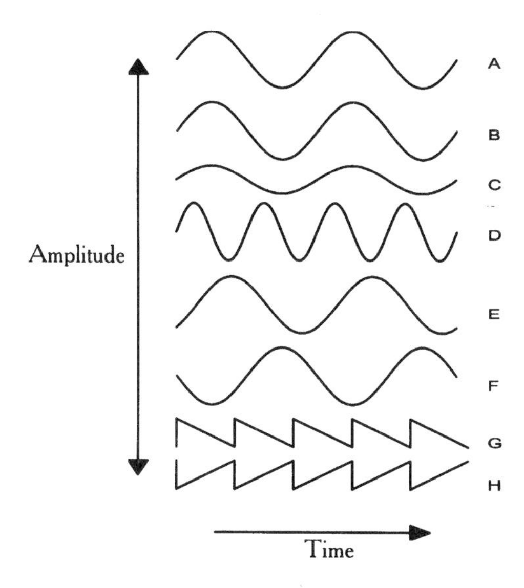

Fig. 1.18

Sine waves A and B are identical in amplitude, frequency and phase. Sine wave C has lower amplitude, but frequency and phase are the same. Sine wave D has the same amplitude, but a higher frequency. Sine wave E has identical amplitude and frequency to A and B, but the phase θ has been changed.

Sine wave F has been *inverted*. Although, for a sine wave, we cannot see the difference between a 180° phase change and an inversion, for asymmetric waveforms, there is a distinct difference. We should therefore be very careful when we say that two waveforms are 180° out of phase with one another, that we do not actually mean that one is inverted with respect to the other.

The *sawtooth* waveforms G and H are inverted, and it can be seen that this is completely different to a 180° phase change. (Strictly, the term 'phase splitter' is incorrect for this reason, but a more technically correct description would be clumsy to use.)

The transformer

When the electric light was introduced as an alternative to the gas mantle, there was a great debate as to whether the distribution system should be AC or DC. The outcome was settled by the enormous advantage of the *transformer* which could step up, or step down, the voltage of an AC supply. The DC supply could not be manipulated in this way, and evolution took its course.

A transformer is essentially a pair of electrically insulated windings that are magnetically coupled to each other, usually on an iron core. They vary from fingernail size to the size of a large house, depending on power rating and operating frequency, with high frequency transformers being smaller. The symbol for a transformer is modified depending on the core material. Solid lines indicate a laminated iron core, dotted lines denote a dust core, whilst an air core has no lines. See Fig. 1.19.

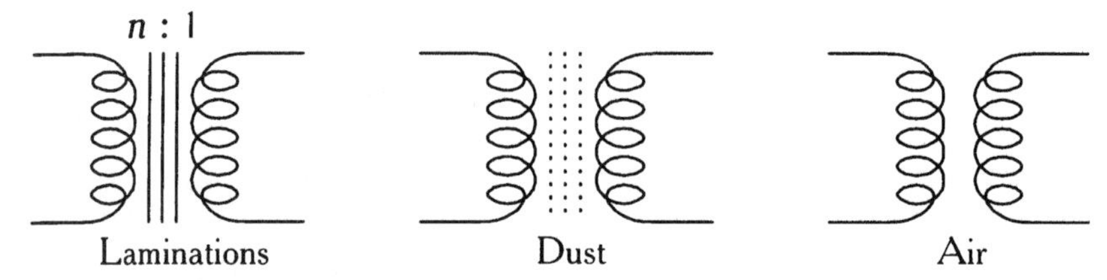

Fig. 1.19 *Transformer symbols*

The perfect transformer changes one AC voltage to another, more convenient, voltage with no losses whatsoever; all of the power at the input is transferred to the output:

$$P_{in} = P_{out}$$

Having made this statement, we can now derive some useful equations:

$$V_{in} \cdot I_{in} = V_{out} \cdot I_{out}$$

Rearranging:

$$\frac{V_{in}}{V_{out}} = \frac{I_{out}}{I_{in}} = n$$

The new constant n is very important, and is the *turns ratio* between the input and output windings of the transformer. Habitually, when we talk about transformers, the input winding is known as the *primary*, and the output winding is the *secondary*. Occasionally, a transformer may have another winding, known as a *tertiary* winding, but it is more common to refer to multiple primaries and secondaries.

When the perfect transformer steps voltage down, perhaps from 240 V to 12 V, the current ratio is stepped up, and for each ampere of primary current, 20 A must be drawn by the secondary. This implies that the resistance of the load on the secondary is different to that seen looking into the primary. If we substitute Ohm's law into the conservation of power equation:

$$\frac{V^2_{\text{primary}}}{R_{\text{primary}}} = \frac{V^2_{\text{secondary}}}{R_{\text{secondary}}}$$

$$\frac{R_{\text{secondary}}}{R_{\text{primary}}} = \left(\frac{V_{\text{primary}}}{V_{\text{secondary}}}\right)^2 = n^2$$

The transformer changes resistances by the *square* of the turns ratio. This will become very significant when we use audio transformers, and we need to match the resistance of a loudspeaker to the output valves.

As an example, an output transformer with a primary to secondary turns ratio of 31.6:1 would allow the output valves to see the 8 Ω loudspeaker as an 8 kΩ load, whereas the loudspeaker sees the Thévenin output resistance of the output valves stepped down by an identical amount.

The concept of looking into a device in one direction, and seeing one thing, whilst looking in the opposite direction, and seeing another, is very powerful, and we shall use it frequently when we investigate simple amplifier stages.

Practical transformers are not perfect, and we will investigate their imperfections in greater detail in Chapter 3.

Capacitors, inductors and reactance

Previously, when we analysed circuits, they were composed purely of resistors and voltage or current sources.

We now need to introduce two new components: *capacitors* and *inductors*. Capacitors have the symbol C and the unit of capacitance is the *farad* (F). 1 F is an extremely large capacitance and more common values range from a few pF to tens of thousands of μF. Inductors have the symbol L and the unit of inductance is the *henry* (H). The henry is quite a large unit and common values range from a few μH to tens of H.

The simplest capacitor is made of a pair of separated plates, whereas an inductor is a coil of wire, and this physical construction is reflected in their graphic symbols. See Fig. 1.20.

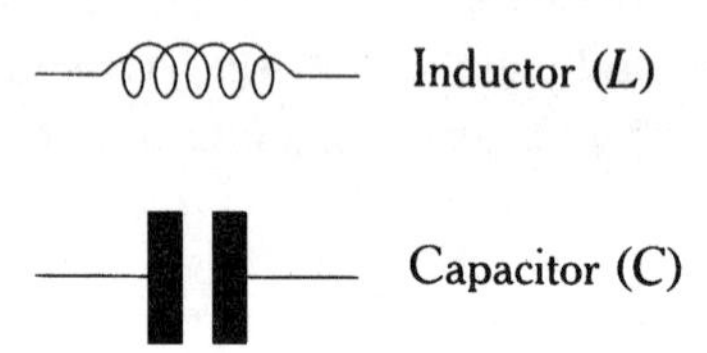

Fig. 1.20 *Inductor and capacitor symbols*

Resistors have resistance, whereas capacitors and inductors have *reactance*. Reactance is the AC equivalent of resistance and is given the symbol X. We will often have circuits where there is a combination of inductors and capacitors, so it is normal to add a subscript to denote which reactance is which:

$$X_C = \frac{1}{2\pi fC}$$

$$X_L = 2\pi fL$$

Looking at these equations, we see that reactance changes with frequency, as well as with the value of the component. We can plot this relationship as a graph. See Fig. 1.21.

The inductor has a reactance of zero at zero frequency. More intuitively, it is a short circuit at DC. As we increase frequency, its reactance rises.

The capacitor has infinite reactance at zero frequency. It is open circuit at DC. As frequency rises, reactance falls.

A circuit that is made up of only one capacitor, or one inductor, is not very interesting, and we might want to describe the behaviour of a circuit made up of resistance and reactance, such as a moving-coil loudspeaker. See Fig. 1.22.

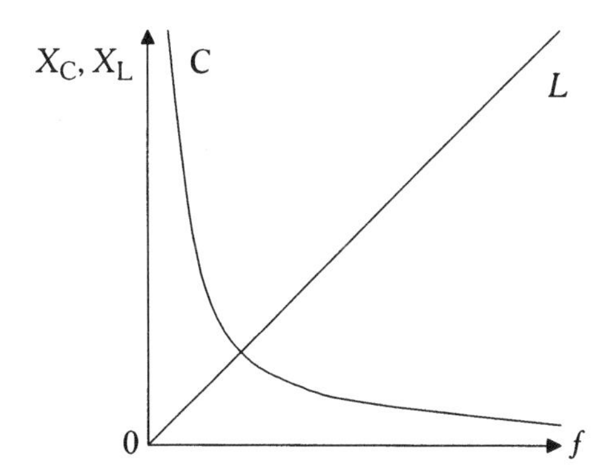

Fig. 1.21 *Reactance of inductor and capacitor against frequency*

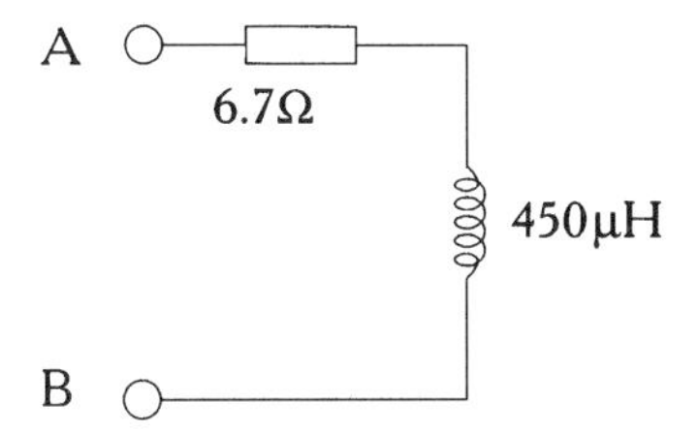

Fig. 1.22

We have a combination of resistance and inductance, but at the terminals A, B we will see neither a pure inductance, nor a pure resistance, but a combination of the two factors which is known as *impedance*.

In a traditional electronics book we would now lurch into the world of vectors, phasors and complex number algebra. Whilst fundamental AC theory is essential for electronics engineers who have to pass examinations, we cannot justify the mental trauma needed to cover the topic in depth, and so we will simply pick out useful results that are relevant to our highly specialized field of interest.

Filters

We mentioned that reactance varies with frequency. This property can be used to make a *filter* that allows some frequencies to pass unchecked, whilst others are attenuated. See Fig. 1.23.

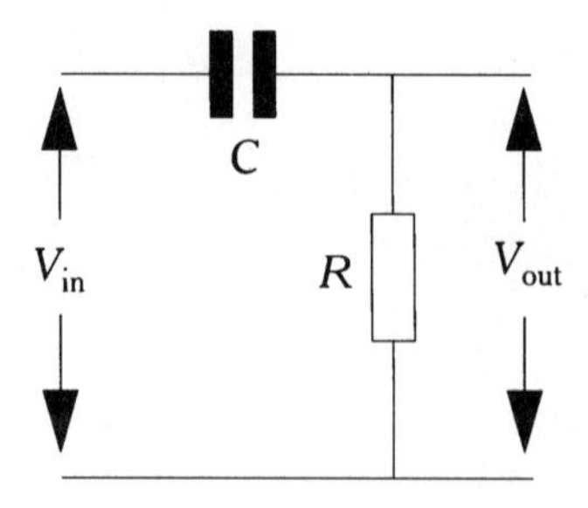

Fig. 1.23 *CR high-pass filter*

All filters are based on potential dividers. In this filter, the upper leg of the potential divider is a capacitor, whereas the lower leg is a resistor. We stated earlier that a capacitor is an open circuit at DC. This filter will therefore have infinite attenuation at DC – it *blocks* DC. At infinite frequency, the capacitor is a short circuit, and the filter passes the signal with no attenuation, so the filter is known as a *high-pass* filter. See Fig. 1.24.

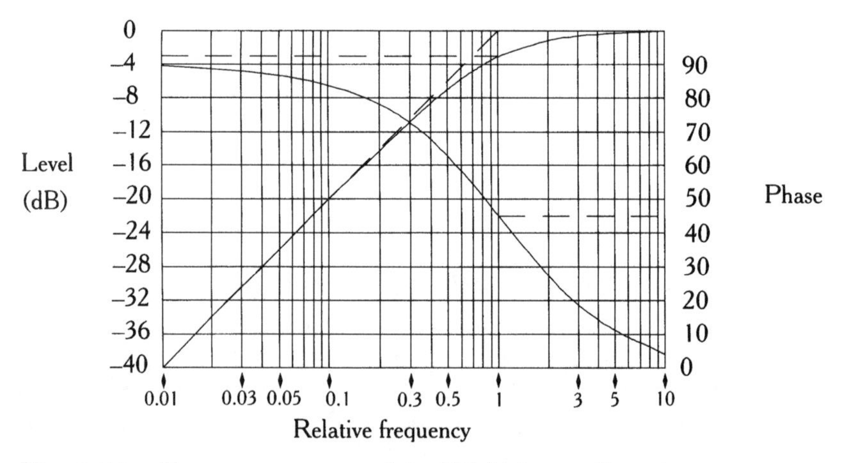

Fig. 1.24 *Frequency response of the CR high-pass filter*

Frequency is plotted on a logarithmic scale in order to encompass the wide range of values without cramping. When we needed a logarithmic unit for amplitude ratios, we invented the dB, but a logarithmic unit for frequency already existed, so engineers stole the *octave* from the musicians. An octave is simply a halving or doubling of frequency, and corresponds to eight 'white keys' on a piano keyboard.

The curve has three distinct regions: the *stop-band*, *cut-off* and the *pass-band*.

The stop-band is the region where signals are stopped, or attenuated. In this filter, the attenuation is inversely proportional to frequency, and we can see that at a sufficiently low frequency, the shape of the curve in this region becomes a straight line. If we were to measure the slope of this line, we would find that it tends towards 6 dB/octave.

Note that the *phase* of the output signal changes with frequency, with a maximum change of 90° when the curve finally reaches 6 dB/octave.

This slope is very significant, and all filters with only one reactive element will have an ultimate slope of 6 dB/octave. As we add more reactive elements, we can achieve a higher slope, and for this reason, filters are often referred to by their *order*, which is the number of reactive elements contributing to the slope. A third order filter would have three reactive elements, and its ultimate slope would therefore be 18 dB/octave.

Although the curve reaches an ultimate slope, the behaviour at cut-off is of interest, not least because it allows us to say at what frequency the filter begins to take effect. On the diagram, a line was drawn to determine the ultimate slope. If this is extended until it intersects with a similar line continued from the pass-band attenuation, the point of intersection is the filter cut-off frequency. (You will occasionally see idealized filter responses drawn in this way, but this does not imply that filter response actually changes abruptly from pass-band to stop-band.)

If we now drop a line down to the frequency axis from the cut-off point, it will pass through the curve, and the filter response at this point will be 3 dB down on the pass-band value. The cut-off frequency is therefore also known as $f_{-3\text{dB}}$, or the −3 dB point, and at this point the phase curve is at its steepest, with a phase change of 45°.

Second order filters, and above, have considerable freedom in the way that the transition from pass-band to stop-band is made, and so the *class* of filter is often mentioned in conjunction with names like Bessel, Butterworth and Chebychev, in honour of their originators.

Although we initially investigated a high-pass *CR* filter, there are other combinations that can be made using one reactive component and a resistor. See Fig. 1.25.

We now have a pair of high-pass filters, and a pair of *low-pass* filters; the low-pass filters have the same slope, and cut-off frequency can be found from the graph in the same way.

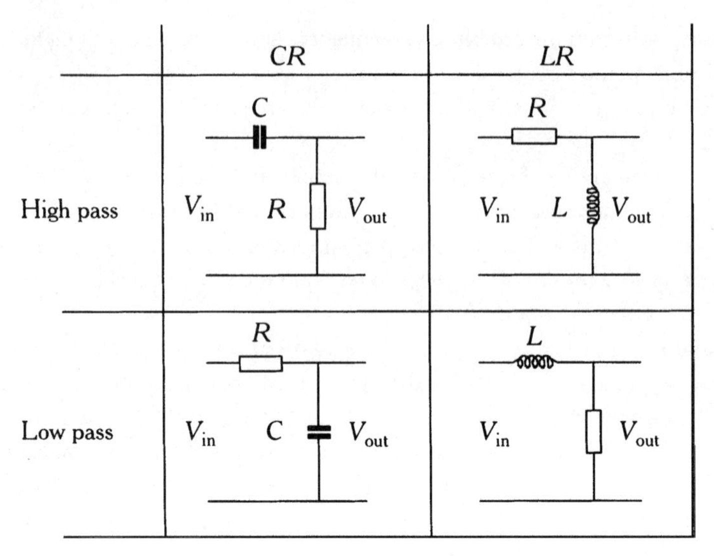

Fig. 1.25

Now that we are familiar with the shape of the curves of these simple filters, it would be far more convenient to refer to them by their cut-off frequency and slope (the word 'ultimate' is commonly neglected). For these simple filters, the equation for cut-off frequency is the same whether the filter is high pass or low pass. For a *CR* filter:

$$f_{-3\text{dB}} = \frac{1}{2\pi CR}$$

And for an *LR* filter:

$$f_{-3\text{dB}} = \frac{R}{2\pi L}$$

Time constants

In audio, simple filters or *equalization* networks are often described in terms of their *time constants*. These have a very specialized meaning which we will touch upon later, but in this context, they are simply used as a shorthand form of describing a first order filter that allows component values to be calculated quickly.

For a CR network, the time constant τ (tau) is:

$$\tau = CR$$

For an LR network:

$$\tau = \frac{L}{R}$$

Because it is a *time* constant, the units of τ are seconds, but audio time constants are habitually given in μs. We can easily calculate the cut-off frequency of the filter from its time constant τ:

$$f = \frac{1}{2\pi\tau}$$

Note that τ is quite distinct from period, which is given the symbol T.

Examples of time constants in audio are: equalization for cassette tape – 120 μs or 70 μs; RIAA equalization for LP records – 3180, 318, 75 μs; UK FM broadcast radio de-emphasis – 50 μs.

A 120 μs HF de-emphasis needs a low-pass filter, usually CR, and all we do is to choose a pair of component values whose product equals 120 μs; 1 nF and 120 kΩ would do nicely. See Fig. 1.26.

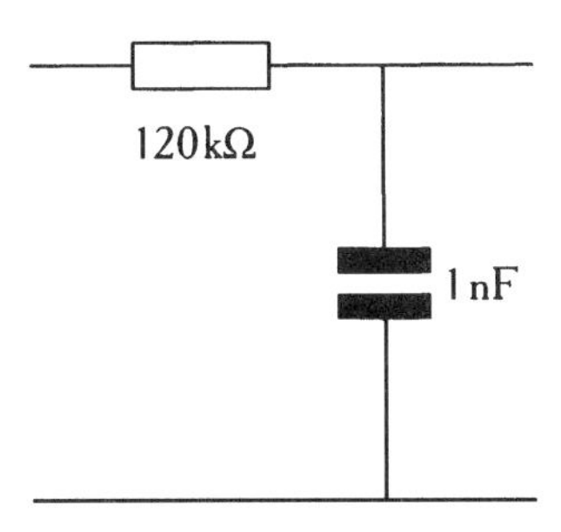

Fig. 1.26 *A 120 μs de-emphasis network*

Resonance

So far, we have made filters using only one reactive component, but if we make a network using a capacitor *and* an inductor, we will find that we have a *resonant* circuit. Resonance occurs everywhere in the natural world, from the sound of a tuning fork, to the bucking and twisting of the Tacoma Narrows bridge. (A bridge that finally collapsed during a storm because the wind excited a structural resonance.) A resonant electronic circuit is shown in Fig. 1.27.

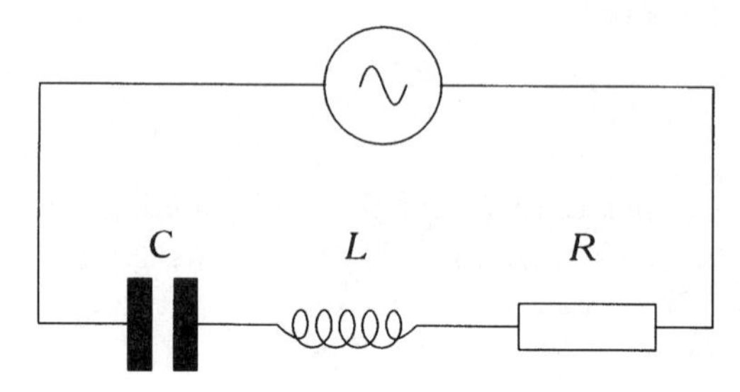

Fig. 1.27 *Series resonant circuit*

If we were to sweep the frequency of the source, whilst measuring the current drawn, we would find that at the resonant frequency, the current would rise to a maximum which would be determined purely by the resistance of the resistor. The circuit would appear as if the other components were not there. We could then plot a graph of current against frequency. See Fig. 1.28.

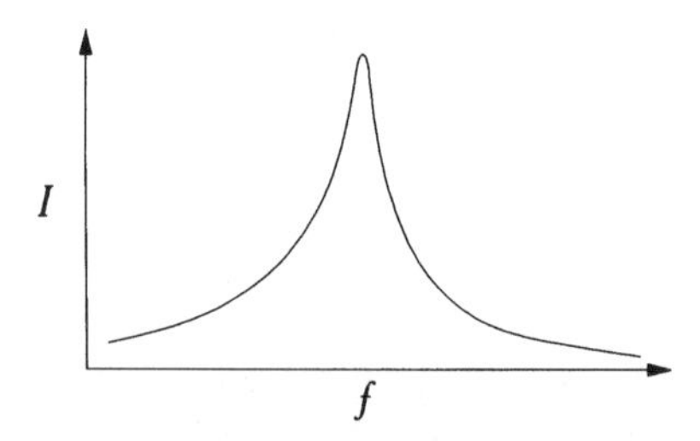

Fig. 1.28 *Current against frequency for series resonant circuit*

The sharpness and height of this peak is determined by the Q or *magnification factor* of the circuit:

$$Q = \frac{1}{R}\sqrt{\frac{L}{C}}$$

This shows us that a small resistance can cause a high Q, and this will be very significant later. The frequency of resonance is:

$$f = \frac{1}{2\pi\sqrt{LC}}$$

Our first resonant circuit was a *series* resonant circuit, but *parallel* resonance, where the *total* current falls to a minimum at resonance, is also possible. See Fig. 1.29.

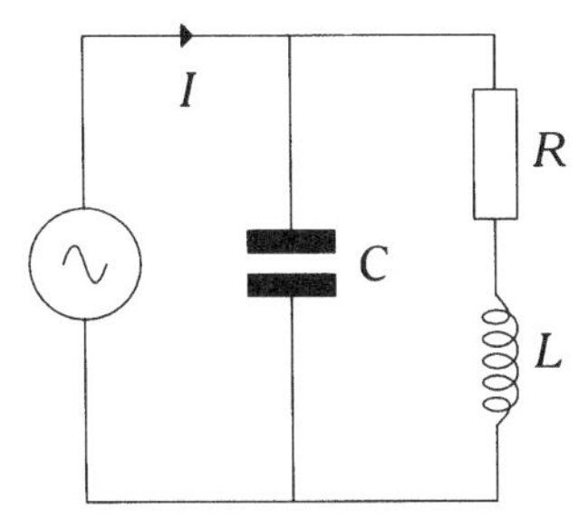

Fig. 1.29 *Parallel resonant circuit*

Provided that Q > 5, the above equations will be reasonably accurate for parallel resonance. We will not often worry greatly about the accuracy of resonant calculations, since we do not want resonances in audio and we will do our best to remove them.

RMS and power

We mentioned power earlier, when we investigated the flow of current through a lightbulb using a 240 V battery. Mains electricity is AC and is 240 V_{AC} 50 Hz in England, but how do we specify the 240 V?

If we had a valve heater filament, it would be most useful if it could operate equally well from AC or DC. As far as the valve is concerned, AC electricity will heat the filament equally well, just so long as we apply the correct voltage.

The RMS voltage of any waveform is the DC voltage that would have the same heating effect as the original waveform.

RMS is short for *R*oot of the *M*ean of the Squares, which refers to the method of calculating the value. Fortunately, the ratios of V_{RMS} to V_{peak} have been calculated for the common waveforms, and in audio design, we are mostly concerned with the sine wave, for which:

$$V_{peak} = \sqrt{2} \cdot V_{RMS}$$

All AC voltages are given in V_{RMS} unless specified otherwise, so a heater designed to operate at $6.3\,V_{AC}$ would work equally well connected to $6.3\,V_{DC}$.

We have only mentioned RMS voltages, but we can equally well have RMS currents, in which case:

$$P = V_{RMS} \cdot I_{RMS}$$

There is no such thing as an RMS watt!
Please refer to the definition of RMS.

The square wave

Up until now, all of our dealings have been with sine waves, which are pure tones. When we listen to music, we do not hear pure tones, instead we hear a *fundamental* with various proportions of *harmonics* whose frequencies are arithmetically related to the frequency of the fundamental. We are able to distinguish between one instrument and another because of the differing proportions of the harmonics, and because of the *transient* at the beginning of each note.

A useful waveform for *quickly* testing amplifiers would have many harmonics and a transient component. The square wave has precisely these properties, in that it is composed of a fundamental frequency plus odd harmonics whose amplitudes steadily decrease with frequency. See Fig. 1.30.

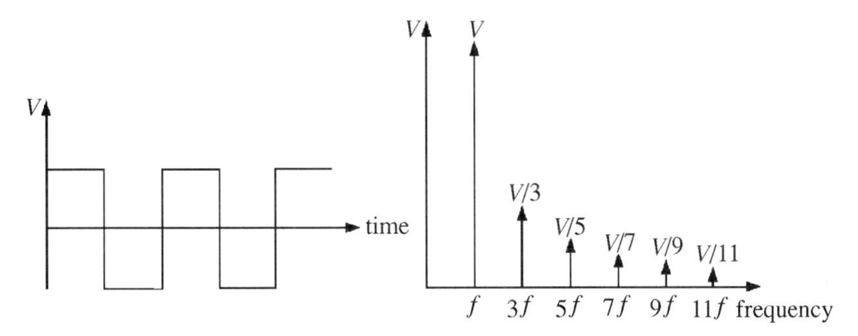

Fig. 1.30 *Square wave viewed in time and frequency*

A square wave is thus an *infinite series* of harmonics, all of which must be summed from the fundamental to the infinite harmonic. We can express this argument mathematically as a *Fourier* series, where f is the fundamental frequency:

$$\text{square wave} = \frac{4}{\pi} \sum_{n=1}^{\infty} \frac{\sin[(2n - 1) \cdot 2\pi ft]}{2n - 1}$$

This is a shorthand formula, but it is much easier to understand the distribution of harmonics if we express them as below:

$$\text{square wave} \propto 1(f) + \frac{1}{3}(3f) + \frac{1}{5}(5f) + \frac{1}{7}(7f) + \frac{1}{9}(9f) + \ldots$$

We can now see that the harmonics die away very gradually, and that a 1 kHz square wave has significant harmonics well beyond 20 kHz. What is not explicitly stated by these formulae is that the relative phase of these components is critical. The square wave thus not only tests amplitude response, but also phase response.

Square waves and transients

We briefly mentioned earlier that the square wave contained a transient component. One way of viewing a square wave is to treat it as a DC level whose polarity is inverted at regular intervals. At the instant of inversion, the voltage therefore has to change instantaneously from its negative level to its positive level. This abrupt change at the leading edge of the square wave is the transient, and because it occurs so quickly, it must contain a high proportion of high frequency components. Although we already knew that the square wave contained these high frequencies, it is only at the leading edge that they all sum constructively, and so any change in high frequency response will be seen at this leading edge.

We have considered the square wave in terms of frequency, now we will consider it as a series of transients in *time*, and investigate its effect on the behaviour of *CR* and *LR* networks.

The best way of understanding this topic is with a mixture of intuitive reasoning coupled to a few graphs. Equations *are* available, but we very rarely need to use them.

We will start by looking at the voltage across a capacitor when a voltage *step* is applied via a series resistor. See Fig. 1.31.

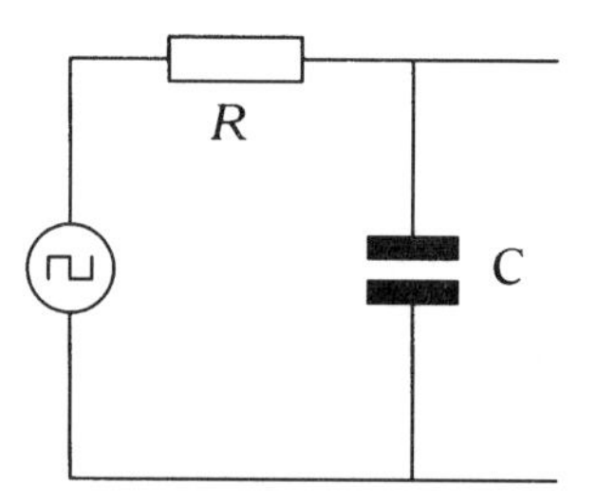

Fig. 1.31

The capacitor is initially discharged ($V_c = 0$). The step is applied and switches from 0 V to +V, an instantaneous change of voltage composed mostly of high frequencies. The capacitor has a reactance that is inversely proportional to frequency, and therefore appears as a short circuit to these high frequencies. If it is a short circuit, we cannot develop a voltage across it. The resistor therefore has the full applied voltage across it, and passes a current determined by Ohm's law. This current also flows through the capacitor, and starts charging the capacitor. As the capacitor charges, its voltage rises, until eventually it is fully charged, and no more current flows. If no more current flows into the capacitor ($I_C = 0$), then $I_R = 0$, and so $V_R = 0$. We can plot this argument as a pair of graphs showing capacitor and resistor voltage. See Fig. 1.32.

The first point to note about these two graphs is that the shape of the curve is an *exponential* (this term will be explained further later). The second point

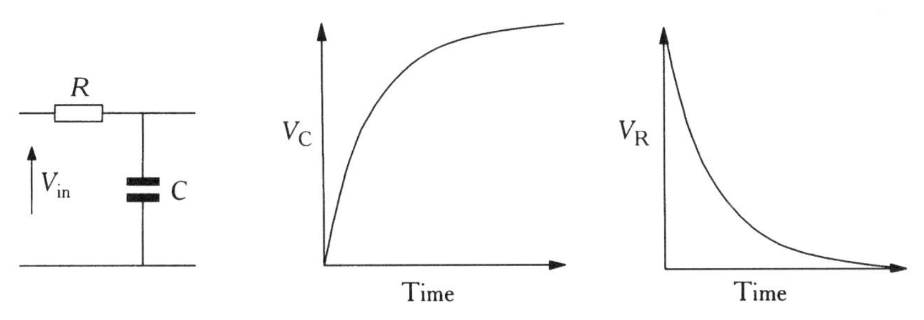

Fig. 1.32 *Exponential response of CR circuit to voltage step*

is that when we apply a step to a *CR* circuit (and even an *LR* circuit), the current or voltage curve will *always* be one of these curves. Knowing that the curve can only be one of these two possibilities, all we need to be able to do is to choose the appropriate curve.

The transient edge can be considered to be of infinitely high frequency, and the capacitor is therefore a short circuit. To develop a voltage across a short circuit would require an infinite current:

> *Infinite current is required to instantaneously change the voltage across a capacitor.*

An inductor is the *dual* or inverse of a capacitor, and so an inductor has a similar rule:

> *An instantaneous change of current through an inductor will create an infinite voltage.*

We can now draw graphs for each of the four combinations of *CR* and *LR* circuits when the same step in voltage is applied. See Fig. 1.33.

Having stated that the shape of the curves in each of the four cases is identical, we can now examine the fundamental curves in a little more detail.

Using the original *CR* circuit as an example, the capacitor will eventually charge to the input voltage, so we can draw a dashed line to represent this voltage. The voltage across the capacitor has an initial slope, and if we

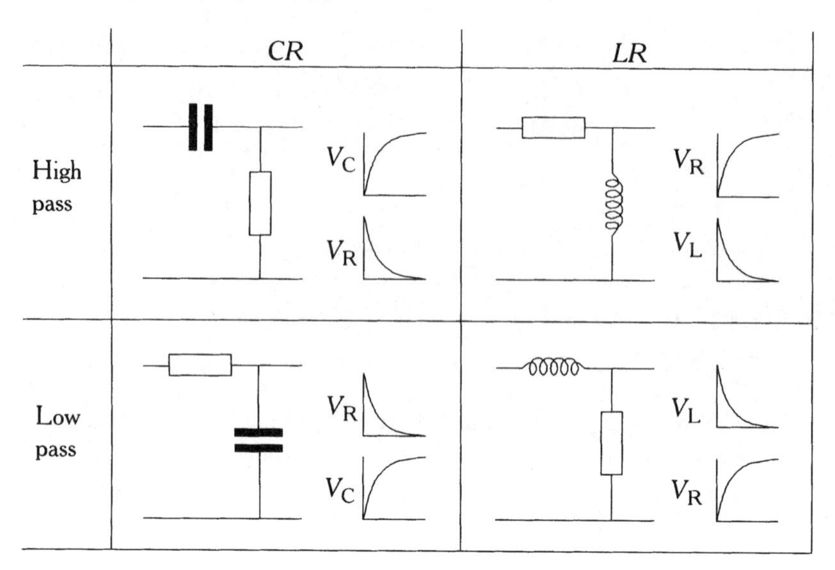

Fig. 1.33

continue this slope with another dotted line, we will find that it intersects the first line at a time that corresponds to CR, which is the *time constant* that we met earlier. The CR time constant is defined as the time taken for the capacitor voltage to reach its final value had the initial rate of charge been maintained. See Fig. 1.34.

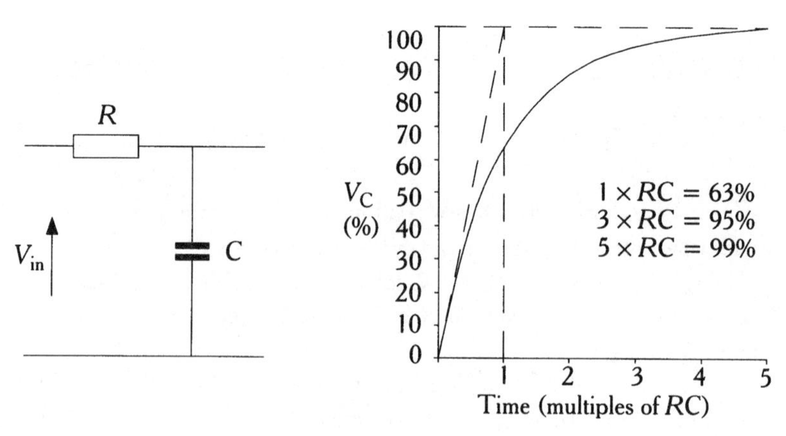

Fig. 1.34 *τ, and its significance to the exponential curve*

The equation for the falling curve is:

$$v = V \cdot e^{-t/\tau}$$

The equation for the rising curve is:

$$v = V(1 - e^{-t/\tau})$$

Where 'e' is the base of natural logarithms and is the key marked 'e^x' or 'exp' on your scientific calculator. These curves derive their name because they are based on an exponential function.

We could now find what voltage the capacitor actually achieved at various times, using the equation for the rising curve:

1τ	63%
3τ	95%
5τ	99%

Because the curves are all the same shape, these ratios apply to all four of the CR, LR combinations. The best way to use the ratios is first to decide which way the curve is heading; the ratios then determine how much of the *change* will be achieved, and in what time.

It is worth noting that after 5 τ, the circuit has very nearly reached its final position, or steady state. This is a useful point to remember when considering what will happen at switch-on to high voltage semiconductor circuits.

When we considered the response to a single step, the circuit achieved a steady state because there was sufficient time for the capacitor to charge, or for the inductor to change its magnetic field. With a square wave, this may no longer be true. As was mentioned before, the square wave is an excellent waveform for testing audio amplifiers, not least because oscillators that can generate both sine and square waves are fairly cheaply available.

If we apply a square wave to an amplifier, we are effectively testing a CR circuit made up of a series resistance and a capacitance to ground (often known as a shunt capacitance). We should therefore expect to see some rounding of the leading edges because some of the high frequencies are being attenuated.

If the amplifier is only marginally stable (because it contains an unwanted resonant circuit), the high frequencies at the leading edges of the square wave will excite the resonance, and we may see a damped train of oscillations following each transition.

We can also test low frequency response with a square wave. If the coupling capacitors between stages are small enough to noticeably change their charge within one half cycle of the square wave, then we will see tilt on the top of the square wave. Downward sag indicates LF loss, whilst upward tilt indicates LF boost. This is a very sensitive test of low frequency response, and if it is known that the circuit being measured includes a *single* high-pass filter, but with a cut-off frequency too low to be measured directly with sine waves, then a square wave may be used to infer the sine wave –3 dB point. The full derivation of the equation that produced the following table is given in the Appendix, but if we apply a square wave of frequency f:

Sag observed using a square wave of frequency 'f'	**Ratio of applied square wave frequency (f) to low frequency cut-off (f_{-3dB})**
10%	30
5%	60
1%	300

Most audio oscillators are based on the Wien bridge, and because of amplitude stabilization problems, rarely produce frequencies lower than ≈10 Hz. 10% and 5% square wave sag may be measured relatively easily on an oscilloscope, and can therefore be used to infer sine wave performance at frequencies below 10 Hz.

Another useful test is to apply a high level, high frequency *sine* wave. Provided that at all levels and frequencies the output is still a sine wave, then the amplifier is likely to be free of slewing distortion. If the output begins to look like a triangular waveform, this is because one, or more, of the stages within the amplifier is unable to fully charge and discharge a shunt capacitance sufficiently quickly. The distortion is known as slewing distortion because the waveform is unable to slew correctly from one voltage to another. The solution is usually to decrease the value of anode load, increasing the anode current of the offending stage, thereby enabling it to charge and discharge the capacitance.

Random noise

The signals that we have previously considered have been repetitive signals – we could always predict precisely what the voltage level would be at any given time. In addition to these *coherent* signals, we shall now consider *noise*.

Noise is all around us, from the sound of waves breaking on a seashore and the radio noise of stars, to the daily fluctuations of the stock markets. We are concerned with electrical noise which can generally be split into one of two categories. *White* noise, which has constant level with frequency (like white light), and $1/f$ noise whose amplitude is inversely proportional to frequency.

White noise is often known as Johnson or thermal noise, and is caused by the random thermal movement of electrons within a conductor. Because it is generated by a thermal mechanism, cooling critical devices will reduce noise, and radio telescopes cool their head amplifiers with liquid nitrogen. All resistors produce white noise, and will generate a noise voltage:

$$v_{noise} = \sqrt{4kTBR}$$

where: k = Boltzmann's constant = 1.372×10^{-23} J/K
T = absolute temperature of the conductor = °C + 273
B = bandwidth of the following measuring device
R = resistance of the conductor

From this equation, we can see that if we were to cool the conductor, to 0K or –273°C, there would be no noise because this would be absolute zero, at which temperature there is no thermal vibration of the atoms to produce noise.

The *bandwidth* of the measuring system is important too, because the noise is proportional to the square root of bandwidth. Bandwidth is the difference between the upper and lower f_{-3dB} limits of measurement. It is important to realize that in audio work, the measurement bandwidth is that of the human ear (20 Hz–20 kHz), and although one amplifier might have a wider bandwidth than another, this does not necessarily mean that it will produce more noise.

In audio, we cannot alter the value of Boltzmann's constant, or the noise bandwidth, and reducing the temperature is expensive, so our main weapon in reducing noise is to reduce resistance. We will look into this in more detail in Chapter 6.

$1/f$ noise is also known as *flicker* noise or *excess* noise, and it is a particularly insidious form of noise, because it is not predictable. It could almost be called 'imperfection' noise, because it is generally caused by imperfections such as imperfectly clean 'clean rooms' used for making semiconductors, 'dry' soldered joints, poor metal to metal contacts in connectors – the list is endless. Semiconductor manufacturers usually specify the 1/f *noise corner* where the 1/*f* noise becomes dominant over white noise for their devices, but equivalent data does not currently exist for valves.

Because noise is random, or *incoherent*, we cannot add noise voltages or currents, but must add noise *powers*, and some initially surprising results emerge. Noise can be considered statistically as a deviation from a mean value. When an opinion poll organization uses as large a sample as possible to reduce error, they are actually *averaging* the noise to find the mean value.

If we parallel '*n*' input devices in a low noise amplifier, the incoherent noise signal begins to cancel, but the wanted signal remains at constant level, resulting in an improvement in signal to noise ratio of $\sqrt{n}$ dB. This technique is feasible for semiconductors where it is possible to make 1000 matched paralleled transistors on a single chip (LM394, MAT-01, etc.), but we are lucky to find a pair of matched triodes in one envelope, let alone any more than that!

Active devices

Previously, we have investigated resistors, capacitors and inductors, but these were all *passive* components. We will now look at *active* devices, which are able to *amplify* a signal. All active devices need a power supply, because amplification is achieved by the source controlling the flow of energy from a power supply into a load via the active device.

We will conclude this chapter by looking briefly at semiconductors. It might seem odd that we should pay any attention at all to semiconductors, but a modern valve amplifier will generally contain rather more semiconductors than valves, and so we will need some knowledge of these devices in order to assist the design of the (valve) amplifier.

Conventional current flow and electron flow

When electricity was first investigated, the electron had not been discovered, and so an arbitrary direction for the flow of electricity was assumed. There was a 50/50 chance of guessing correctly, and the early researchers were unlucky. By the time that the mistake was discovered, and it was realized that electrons flowed in the opposite direction to the way that electricity had been thought to flow, it was too late to change the convention.

We are therefore saddled with a conventional current that flows in the *opposite* direction to that of the electrons. Mostly, this is of little consequence, but when we consider the internal workings of the transistor and the valve, we must bear this distinction in mind.

Silicon diodes

Semiconductor devices are made by doping regions of crystalline silicon to form areas that are known as *N-type* or *P-type*. These regions are permanently charged, and at their junction this charge forms a potential barrier that must be overcome before forward conduction can occur. Reverse polarity strengthens this potential barrier, so no conduction occurs.

The *diode* is a device that allows current to flow in one direction, but not the other. Its most basic use is therefore to *rectify* AC into DC. The arrow head on the diode denotes the direction of conventional current flow, and R_L is the load resistance. See Fig. 1.35.

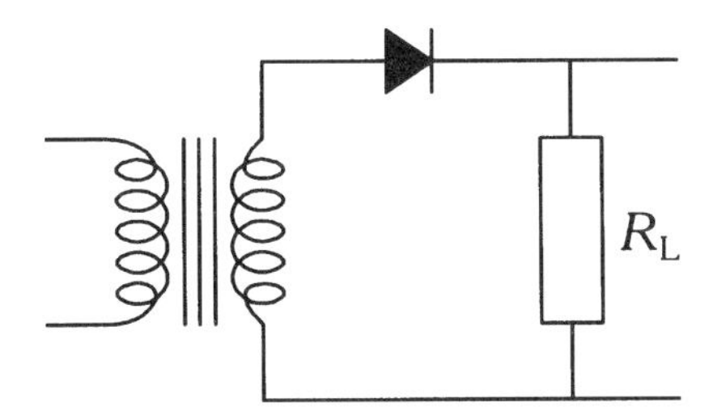

Fig. 1.35 *Use of a diode to rectify AC*

Practical silicon diodes are not perfect rectifiers, and need a certain amount of forward *bias* before they will conduct. At room temperature, this voltage is ≈ *0.6 V*. This forward bias voltage is always present, so the output voltage is always less than the input voltage by the amount of the *diode drop*. Because a voltage is always dropped across the diode, any current flow creates heat, and sufficient heat will melt the silicon. All diodes therefore have a maximum current rating.

In addition, if the reverse voltage is too high, the diode will break down and conduct, and if this reverse current is not limited, then the diode will be destroyed.

Unlike valves, the mechanism for conduction through the most common (bipolar junction) type of silicon diode is quite complex, and results in a charge being temporarily stored within the diode. When the diode is switched off by the external voltage, the charge within the diode is quickly discharged and produces a brief current pulse in the external circuit which can excite resonances. Fortunately, *Schottky* diodes do not exhibit this phenomenon, and *soft recovery* types are fabricated to minimize it.

Voltage references

The 0.6 V forward voltage drop across a diode junction is determined by the Ebers/Moll equation involving absolute temperature and current, whilst the reverse breakdown voltage is determined by the physical construction of the individual diode. This means that we can not only use a diode not as a rectifier, but also as a voltage reference.

Voltage references based on the forward voltage drop are often known as *band gap* references, whilst references based on the reverse breakdown voltage are known as *Zener* diodes. All voltage references should pass only a limited current to avoid destruction, and ideally a constant current should be passed.

Zener diodes are commonly available in power ratings up to 75 W, although the most common rating is 400 mW; voltage ratings are from 2.7 V to 270 V. Reverse biased junctions produce more noise than forward biased diodes, and so Zeners are noisier than band gap references.

Band gap references are actually a complex integrated circuit, and usually have an output voltage of 1.2 V, but internal amplifiers may increase this to 10 V or more. Because they are complex internally, band gap references tend to be more expensive than Zeners, and we may occasionally need a cheap low noise reference. Light Emitting Diodes (LEDs) operate with forward biased junctions, and are ideal cheap voltage references.

All voltage references are characterized by their *slope resistance*, which is the Thévenin resistance of the voltage reference when operated correctly. It does not imply that large currents can be drawn, merely that for small current changes in the linear region of operation, the voltage change will be correspondingly small.

The ideal Zener voltage is ≈6 V, because the sensitivity to temperature is lowest at this point, and slope resistance is also minimized. Slope resistance rises sharply below 6 V, and more gradually above 6 V, but is typically ≈10 Ω @ 5 mA for a 6.2 V Zener.

A red LED has a forward drop of ≈1.6 V @ 15 mA, and slope resistance is ≈3 Ω. Green LEDs can also be used, and their forward drop is ≈2.2 V @ 10 mA, but the slope resistance rises to ≈10 Ω.

Because band gap references usually incorporate an internal amplifier, their output resistance is much lower, and is typically ≈0.2 Ω or less.

Bipolar Junction Transistors (BJTs)

BJTs are the most common type of transistor, they are available in NPN and PNP types, and can be used to amplify a signal. The name transistor is derived from *trans*ferred res*istor*.

We can imagine an NPN transistor as a sandwich of two thick slabs of N-type material separated by an extremely thin layer of P-type material. The P-type material is the *base*, whilst one of the N-types is the *emitter* and emits electrons, which are then collected by the other N-type, which is known as the *collector.*

If we simply connect the collector to the positive terminal of a battery, and the emitter to the negative, no current will flow because the negatively charged base repels electrons. If we now apply a positive voltage to the base to neutralize this charge, the electrons will no longer be repelled, but because the base is so thin, the attraction of the strongly positively charged collector pulls most of the electrons straight through the base to the collector, and collector current flows.

The base/emitter junction is now a forward biased diode, so it should come as no surprise to learn that 0.6 V is required across the base/emitter junction to cause the transistor to conduct electrons from emitter to collector. Because the base is so thin, and the attraction of the collector is so great, very few electrons emerge from the base as base current, so the ratio of collector current to base current is high. The transistor therefore has current gain, which is sometimes known as β, but more commonly as H_{FE} for DC current gain, or h_{fe} for AC current gain. For all practical purposes, $H_{FE} = h_{fe}$, and the parameter has such wide production tolerance that the distinction is trivial.

A far more important and predictable parameter is the *transconductance*, *gm*, which is the change in collector current with base/emitter voltage:

$$gm = \frac{\Delta I_C}{\Delta V_{BE}}$$

When we look at valves, we will see that we always need to measure *gm* at the operating point using a graph. For transistors, transconductance is defined by the Ebers/Moll equation, and for small currents (less than $\approx$100 mA), *gm* can be calculated for any BJT using:

$$gm \approx 35 I_C$$

The common emitter amplifier

Now that we have a means of predicting the change in collector current caused by a change in V_{BE}, we could connect a resistor R_L in series with the collector and the supply to convert the current change into a voltage change. By Ohm's law:

$$\Delta V_{CE} = \Delta I_C \cdot R_L$$

But $\Delta I_C = gm \cdot \Delta V_{BE}$, so:

$$\Delta V_{CE} = \Delta V_{BE} \cdot gm \cdot R_L$$

$$\frac{\Delta V_{CE}}{\Delta V_{BE}} = A_V = gm \cdot R_L$$

We are now able to find the *voltage gain* A_v for this circuit, which is known as a common emitter amplifier because the emitter is common to both the input and output circuits or *ports*. Note that the output of the amplifier is *inverted* with respect to the input signal. See Fig. 1.36.

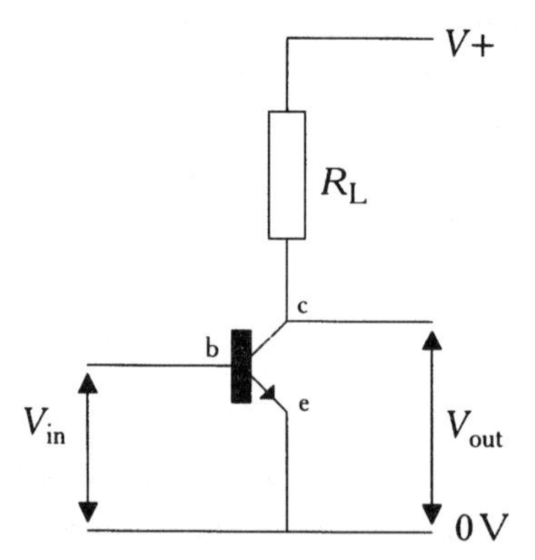

Fig. 1.36 *Common emitter transistor amplifier*

As the circuit stands, it is not very useful, because any input voltage below 0.6 V will not be amplified, and the amplifier therefore creates considerable distortion. In order to circumvent this, we bias the collector voltage to half the supply voltage, which allows the collector to swing an equal voltage both positively and negatively.

If we know the supply voltage, and the value of the collector load R_L, we can use Ohm's law to determine what value of collector current would bias the transistor correctly.

We could now use the relationship between I_B and I_C to set this optimum collector current. Unfortunately, the value of H_{FE} for a transistor is not guaranteed, and varies widely between devices. For a small-signal transistor, it could range from 50 to more than 400. The way around this is to add a resistor in the emitter, and to set a base *voltage*. See Fig. 1.37.

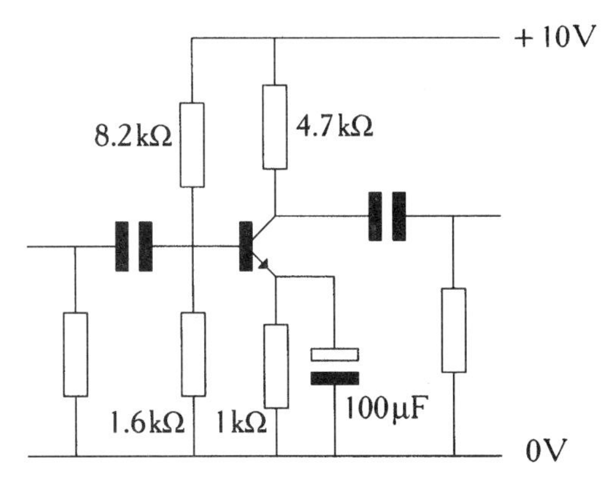

Fig. 1.37 *Stabilized common emitter amplifier*

The amplifier now has input and output *coupling* capacitors, and an emitter *decoupling* capacitor. The entire circuit is known as a stabilized common emitter amplifier, and is the basis of most linear transistor circuitry.

Considering DC conditions only: the potential divider chain passes a current that is at least 10 times the expected base current, and therefore sets a fixed voltage at the base of the transistor independent of base current. Because of diode drop, the voltage at the emitter is thus 0.6 V lower. The emitter resistor has a *fixed* voltage across it, and it must therefore pass a fixed current from the emitter. $I_E = I_C - I_B$, but since I_B is so small, $I_E \approx I_C$, and if we have a fixed emitter current, then collector current is also fixed.

The emitter is decoupled in order to prevent *negative feedback* from reducing the AC gain of the circuit – we will consider negative feedback later in this chapter.

Briefly, if the emitter resistor was not decoupled, then the change in collector current (which is the same as emitter current), would cause the voltage across the emitter resistor to change with the applied AC. The emitter voltage would change, and V_{be} would effectively be reduced, causing AC gain to fall. The decoupling capacitor is a short circuit to AC, and therefore prevents this reduction of gain. This principle will be repeated in the next chapter when we look at the cathode bypass capacitor in a valve circuit.

Input and output resistances

In a valve amplifier, we will frequently use transistors as part of a bias network, or as part of a power supply, and it is therefore important to be able to determine input and output resistances. The following AC resistances looking into the transistor do not take account of any external parallel resistance from the viewed terminal to ground.

The output resistance looking into the collector is high, typically tens of kΩ, which suggests that the transistor would make a good constant current source.

The resistance looking into the emitter is low, $r_e = 1/gm$, so $\approx 20\,\Omega$ is typical. If the base is *not* driven by a source of zero resistance ($R_b \neq 0$), then there is an additional series term, and r_e is found from:

$$r_e = \frac{1}{gm} + \frac{R_b}{h_{fe}}$$

R_b is the Thévenin resistance of all the paths to ground *and supply* seen from the base of the transistor. Note that even though we are no longer explicitly including a battery as the supply, the supply is still assumed to have zero output resistance from DC to light frequencies.

Looking into the base, the path to ground is via the base emitter junction in series with the emitter resistor. If the emitter resistor is not decoupled, then the resistance will be:

$$r_b = h_{fe}\left(\frac{1}{gm} + R_e\right)$$

If the emitter resistor is decoupled, then $R_e = 0$, and the equation reduces to:

$$r_b = \frac{h_{fe}}{gm} = h_{fe}$$

The AC input resistance due purely to the base/emitter junction is often known as h_{ie}, and is generally quite low, $<10\,k\Omega$.

If we now consider the effect of the external parallel resistances, we see that the input resistance of the amplifier is low, typically $<5\,k\Omega$. The output resistance seen at the emitter is low, typically $<100\,\Omega$ (even if the source resistance is quite high), and the output resistance at the collector is $\approx R_L$.

The emitter follower

Very occasionally, you will see this amplifier referred to as a *common collector* amplifier, although this phraseology is rare because it does not convey what the circuit is actually doing.

If we reduce the collector load to zero and take our output from the emitter, then we have an amplifier with $A_v \approx 1$. The voltage gain must be ≈ 1, because $V_E = V_B - 0.6\,V$; the emitter *follows* the base voltage, and the amplifier is *non-inverting*. Although $A_v = 1$, the current gain is much greater, and we can calculate input and output resistances using the equations presented for the common emitter amplifier. See Fig. 1.38.

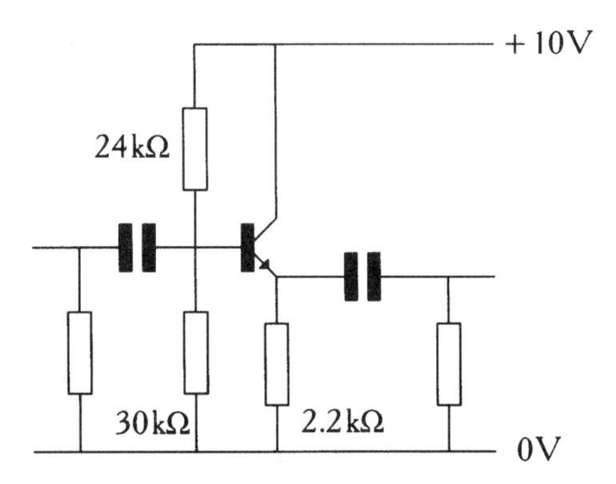

Fig. 1.38 *Emitter follower*

Because of its low output resistance and moderately high input resistance, the emitter follower is often used as a *buffer* to match high impedance circuitry to low resistance loads.

The Darlington pair

Sometimes, even an emitter follower may not have sufficient current gain, and the solution is to use a *Darlington pair*; this is effectively two transistors in cascade, with one forming the emitter load for the other. See Fig. 1.39.

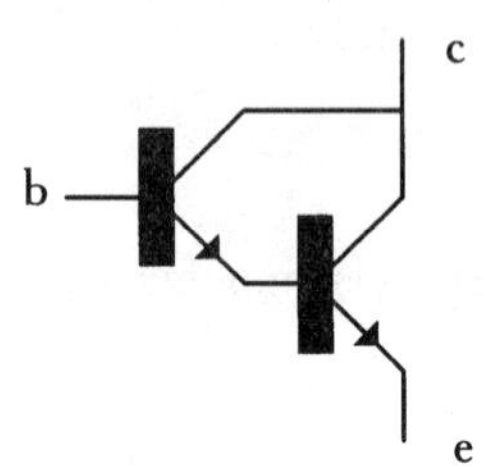

Fig. 1.39 *Darlington pair*

The two transistors form a composite transistor with $V_{BE} = 1.2\,V$, and $H_{FE(total)} = H_{FE1} \times H_{FE2}$. A Darlington pair can replace a single transistor in any configuration if it seems useful. Common uses are in the output stage of a power amplifier and in linear power supplies. Darlingtons can be bought in a single package, but it is often cheaper and better to make your own out of two discrete transistors.

General observations on BJTs

We mentioned earlier that the BJT could be considered to be a sandwich with the base separating the collector and emitter. We can now develop this model, and use it to make some useful generalizations.

As the base becomes thicker, it becomes more and more probable that an electron passing from the emitter to the collector will be captured by the base, and flow out of the base as base current. H_{FE} is therefore inversely proportional to base thickness.

High current transistors will not only pass a high collector current, but base current will also be high. In order for the base not to melt due to this current, the base must be thickened. The thicker base reduces H_{FE}, and required base current must rise yet further, needing an even thicker base. H_{FE} is therefore inversely proportional to the square of maximum collector current. High current transistors have low H_{FE}.

High voltage transistors must have a thick base in order for the base not to break down under the stress of the voltage that is being insulated from collector to emitter. High voltage transistors have low H_{FE}.

High current transistors must have a large silicon die area in order for the collector not to melt – this large area increases collector/base capacitance. The significance of capacitance in amplifying devices will be made clear when we meet Miller effect in Chapter 2, but for the moment, we can simply say that high current transistors will be *slow*.

We have barely scratched the surface of semiconductor devices and circuits. Other semiconductor circuits will be presented as, and when, they are needed.

Feedback

The feedback equation

Feedback is a process whereby we take a fraction of the output of an amplifier and sum it with the input. If, when we sum it with the input, it causes the gain of the amplifier to increase, then it is known as *positive* feedback, and this is the basis of oscillators. If it causes the gain to fall, then it is known as *negative* feedback, and this technique is widely used in audio amplifiers.

The description of feedback was deliberately rather vague because there are many ways that feedback can be applied, and they each have differing effects. Before we can look at these effects, we need a few definitions and a simple equation:

$$A = \frac{A_0}{1 + \beta A_0}$$

This is the general feedback equation that defines how the gain of an amplifier will be modified by the application of feedback. β is the *feedback fraction*, and is the proportion of the output that is fed, or looped, back to the input, and A_0 is the gain before feedback.

If βA_0 is very large *and positive* (causing negative feedback – a reduction of gain), then $\beta A_0 \approx \beta A_0 + 1$, and the gain of the amplifier becomes:

$$A = \frac{1}{\beta}$$

A_0 no longer affects A, and the *closed loop* gain of the amplifier is determined solely by the network that provides the feedback signal.

This result is very significant because it implies a great many things:

- Distortion is produced by variations in gain from one voltage level to another. If *open loop* gain is no longer part of the equation, then small variations in this gain are irrelevant, and the amplifier produces no distortion.
- If the feedback acts to maintain the correct gain *under all circumstances*, then it must change the apparent input and output resistances of the amplifier.
- If the feedback fraction β is set by pure resistors, then the equation for closed loop gain does not contain any term including frequency. Theoretically, the output amplitude is therefore independent of frequency.

In the late 1970s, when cheap gain became readily available, designers became very excited by the possibilities and implications of the feedback equation, and set out to exploit it by designing amplifiers which were thought to have very high levels of feedback. In practice, these amplifiers did *not* have high levels of feedback at all frequencies and power levels, and it was the *lack* of feedback to linearize these fundamentally flawed circuits that caused their poor sound quality.

Before we explore the expected benefits of feedback, we should therefore examine how the feedback equation could break down.

Practical limitations of the feedback equation

The feedback equation linearizes performance provided that $\beta A_0 >> 1$. If, for any reason, the open loop gain of the amplifier is less than infinite, then βA_0 will *not* be much greater than 1, and the approximation will no longer be true.

Practical amplifiers always have finite gain; moreover, this gain falls with frequency. A practical amplifier will always distort the input signal, and because the distortion reducing ability of negative feedback falls with frequency, the closed loop distortion must rise with frequency.

Crossover distortion in Class B amplifiers can be considered to be a reduction of gain as the amplifier traverses the switching point of the transistors or valves. Because of the drastically reduced open loop gain in this region, negative feedback is not very effective at reducing crossover distortion.

Although not explicitly stated in the feedback equation, the phase of the feedback signal is crucially important. If the phase should change by 180°, then the feedback will no longer be negative, but positive, and our amplifier may turn into an oscillator.

We will explore the practical limitations of feedback in Chapter 5, but we should realize that, as with any weapon wielded carelessly, it is possible to shoot oneself in the foot . . .

The effect of feedback on input and output impedances

The way in which the feedback is derived will affect the output impedance, whereas the way that it is *applied* will affect input impedance. If we make a *parallel* or *shunt* connection, then we are dealing with a *voltage*, but if we make a *series* connection, we are dealing with a *current*. Feedback confined to one stage is known as *local* feedback, whereas feedback over a number of stages is known as *global*. We may have more than one global feedback loop, with one loop enclosed by another, in which case, the loops are said to be *nested*.

As an example, we can now combine these terms to describe the global negative feedback loop of a conventional power amplifier as being parallel derived, series applied, but we could equally well describe it as being voltage derived, current applied.

Now that we have defined how the feedback is connected, we can say that for *negative* feedback, voltage feedback *reduces* impedances, and current feedback *increases* impedances. The previous example had *shunt* derived feedback to reduce the output impedance, and *series* applied feedback to increase the input impedance.

If we have *positive* feedback, all effects are reversed, and voltage feedback increases impedances whilst current feedback reduces impedances. Using a combination of positive and negative feedback it is possible to make a power amplifier with zero, or even negative, output impedance.

Unfortunately, amplifiers with negative output impedance are liable to oscillate, because they may reduce the total series damping resistance of the external (invariably resonant) load to zero. Nevertheless some early valve power amplifiers had the facility to adjust output impedance through zero to a negative value, in an effort to improve the performance of the accompanying loudspeaker. Occasionally, this idea resurfaces, but it is really only of use for the dedicated amplifiers in a loudspeaker system with an active crossover. (Active, or low level, crossovers are used *before* the power amplifiers, so that each drive unit has a dedicated power amplifier. Although this scheme may seem profligate with expensive power amplifiers, it has much to recommend it.)

Although we can describe the various modes of feedback, and their effect on impedances, we now need to quantify this effect. Impedances are changed by the ratio of the feedback factor:

$$\text{feedback factor} = (1 + \beta A_0)$$

This is also the factor by which the gain of the amplifier has been reduced, and is often expressed in dBs. An amplifier with 20 dB of global feedback has had its total gain reduced by a factor of 20 dB, and if this was a conventional power amplifier with shunt derived feedback, its output impedance would be reduced by a factor of 10.

The operational amplifier

Conventionally, when we think of computers, we think of *digital* computers, but once upon a time there were also *analogue* computers, which were hardwired to model complex differential equations, such as those required for calculating the ballistics of shells. These analogue computers were made using a basic building block that became known as the *operational amplifier.* This was a small device for its time (smaller than a brick) and used valve circuits that operated from ±300 V supplies. With suitable external components, these operational amplifiers could be made to perform the mathematical operations of inversion, summation, integration and differentiation.

Thankfully, the valve analogue computer is no longer with us, but the term 'operational amplifier', usually shortened to *op-amp*, is still with us, even if it now refers to eight-legged silicon beetles.

Op-amps attempt to define the performance of the final amplifier *purely* by feedback, and in order to do this successfully, the op-amp must have enormous gain; 120 dB gain at DC is not uncommon, although AC gain invariably falls with frequency.

In the following discussions, we will make two fundamental assumptions about op-amps:

- Gain is infinite.
- Input resistance is infinite.

These assumptions specify ideal op-amps. Real world op-amps have limitations, and will *not* achieve this ideal at all frequencies, voltage levels, etc. Provided that we remember this important caveat *at all times*, we will not run into trouble.

It is habitual in op-amp circuit diagrams to omit the (usually ±15 V) power supply lines to the op-amp to aid clarity; nevertheless, the op-amp will still need power!

The invertor, and virtual earth adder

The op-amp invertor has parallel derived, parallel applied, feedback, and since op-amp gain is infinite, both the point of derivation and application will have zero resistance; the amplifier has zero output resistance. See Fig. 1.40.

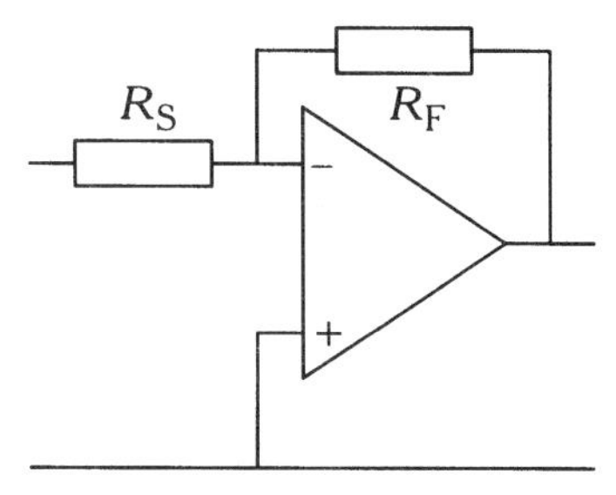

Fig. 1.40 *Inverting amplifier*

The inverting input of the op-amp also has zero resistance to earth (0 V) because of the feedback. Since the gain of the op-amp is infinite, if the non-inverting input is at earth potential, then the inverting input must also be at earth potential. In this configuration, the inverting input is therefore known as a *virtual earth*.

Although it is a virtual earth by virtue of feedback, the inverting input of the op-amp itself has infinite resistance, and *no signal current flows into the op-amp*. Input signal current from R_S can therefore only flow to ground via R_F and the zero output resistance of the op-amp. The signal currents in R_S and R_F are therefore equal, and using Ohm's law:

$$\frac{V_S}{R_S} = \frac{V_F}{R_F}$$

Since the inverting input is a virtual earth, $V_S = V_{in}$, and $V_F = -V_{out}$ (the op-amp inverts), and the voltage gain of the amplifier is:

$$A_v = \frac{V_{out}}{V_{in}} = -\frac{R_F}{R_S}$$

It is worth noting that this amplifier can achieve $A_v < 1$, and it is capable of *attenuating* the input signal. The minus sign reminds us that the amplifier is inverting.

Because the inverting node of the amplifier is a virtual ground, input resistance is equal to R_S.

When we analysed this amplifier, we considered the input signal current. There is no reason why this input current should come from only one source via one resistor, and we can sum currents at the inverting node in accordance with Kirchhoff's law. See Fig. 1.41.

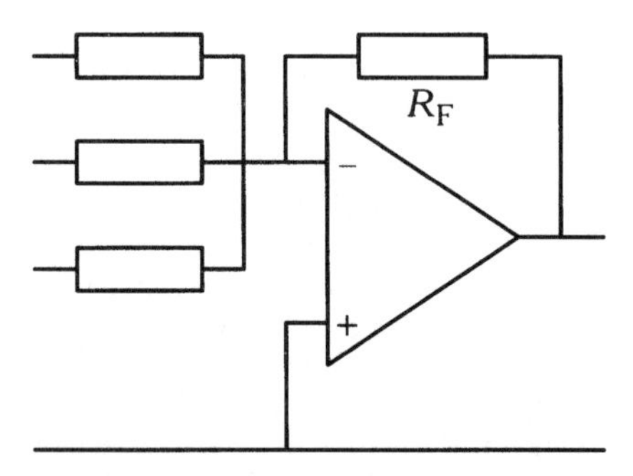

Fig. 1.41 *Virtual earth adder*

This circuit is known as the *virtual earth adder*, and is most useful in bias servo circuits where we may need to add a number of correction signals. Voltage gain for *each* input may be determined using the invertor equation. When multiple inputs are driven, it is often best to determine the output voltage by summing the signal currents using Kirchhoff's law, finding the resultant current in R_F, and using Ohm's law to determine the output voltage. This is not as tedious as it sounds.

The non-inverting amplifier and voltage follower

Frequently, we may need a non-inverting amplifier. See Fig. 1.42.

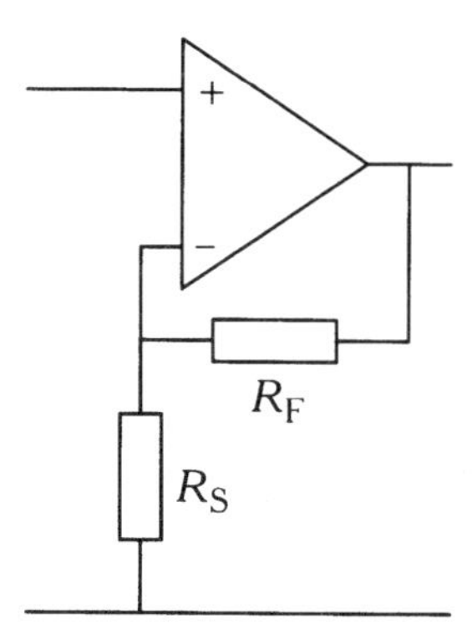

Fig. 1.42 *Non-inverting amplifier*

In this configuration, we still have R_F and R_S, but the amplifier has been turned upside down, and the far end of R_S is now connected directly to ground. The amplifier has parallel derived, series applied, feedback. Output resistance is therefore zero, whilst input resistance is infinite. R_F and R_S now form a potential divider across the output of the amplifier, and the voltage at the inverting input is:

$$V_{\text{(inverting input)}} = \frac{R_S}{R_S + R_F} \cdot V_{out}$$

Since the op-amp has infinite gain, the voltage at the inverting input is equal to that at the non-inverting input, which is V_{in}, and the gain of the amplifier is:

$$\frac{V_{out}}{V_{in}} = \frac{R_S + R_F}{R_S}$$

If we now reduce the value of R_F to zero, and discard R_S, the gain reduces to 1, and the amplifier is known as the *voltage follower*. See Fig. 1.43.

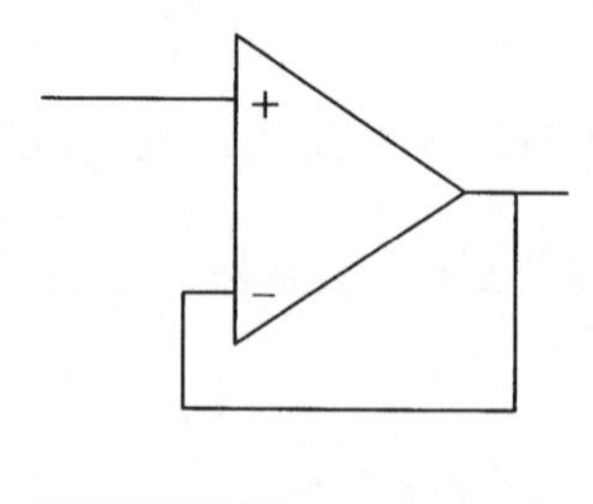

Fig. 1.43 *Voltage follower*

The voltage follower is an excellent *buffer* stage for isolating high impedance circuits from low impedance loads, and is far superior to the emitter follower that we saw earlier. If we need more current than the op-amp can provide, we can add an emitter follower, possibly even a Darlington pair, to the output, and enclose this within the feedback loop. See Fig. 1.44.

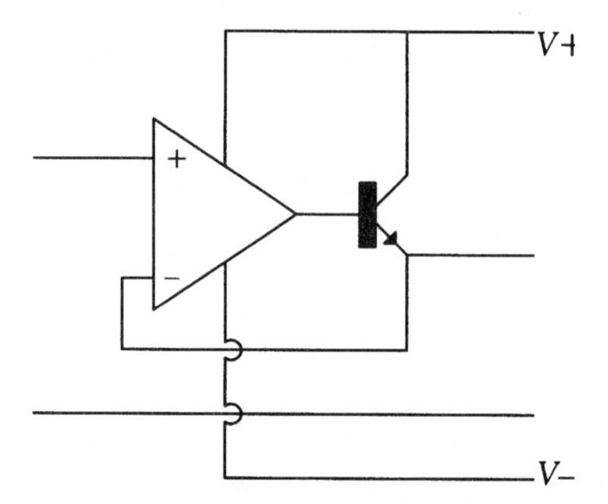

Fig. 1.44 *Addition of emitter follower to increase output current*

The integrator

This circuit is essentially an invertor with R_F paralleled by C_F, but if R_F is omitted, it becomes an *integrator* and measures *charge*. The circuit may be considered to be a low-pass filter, in which case the cut-off frequency of the filter is:

$$f_{-3\text{db}} = \frac{1}{2\pi C_F R_F}$$

We can now see that if $R_F = \infty$, the cut-off frequency = 0, and the amplifier has infinite gain at DC. Any DC offsets will gradually accumulate sufficient charge on the capacitor to cause the output of the op-amp to rise to maximum output voltage ($\pm V_{supply}$), and the op-amp is then said to be *saturated*. For this reason, practical integrators are usually enclosed by a servo control loop. The servo error signal charges C_F, but the loop aims to minimize the error, and so the voltage across C_F tends to zero.

Again, we could sum several input currents if we wished, and one possible use of this circuit would be for monitoring the voltages across the cathode resistors of several output valves, whilst the integrating function would remove any audio signal to provide a DC output voltage proportional to the *total* output valve current.

DC offsets

We briefly mentioned DC offsets when considering the integrator. Op-amps are not magical, they contain real transistors, and base or leakage current, known as *offset current*, will flow out of the inputs. The input transistors will not be perfectly matched, and there will be an *offset voltage* between the inputs. These imperfections are detailed in the manufacturer's data sheets, and should be investigated during circuit design.

Further reading

Duncan, Ben (1997) *High Performance Audio Power Amplifiers*. Newnes.
Hartley Jones, Martin (1995) *A Practical Introduction to Electronic Circuits*, 3rd ed. Cambridge University Press.
Horowitz, P. W. and Hill, W. (1989) *The Art of Electronics*, 2nd ed. Cambridge University Press.
Linsley Hood, John (1998) *The Art of Linear Electronics*, 2nd ed. Newnes.
Self, Douglas (1996) *Audio Power Amplifier Design Handbook*. Newnes.

2

Basic building blocks

In this chapter we will look mainly at the triode valve, how to choose operating conditions, and what effect these choices have on the AC performance of the stage. The analysis will use a combination of graphical and algebraic techniques, which has the advantage of being quick to use, and the results of the theory agree well with practice. This last point might seem to be an obvious requirement, but it is one that is sometimes overlooked.

The common cathode triode amplifier

The aim, in using a valve, is to provide some form of amplification of an AC signal, therefore we need to know how to configure and bias the valve so that it can amplify in a linear manner and minimize distortion. We will begin by investigating the *anode characteristics* of an ECC83/12AX7. See Fig. 2.1.

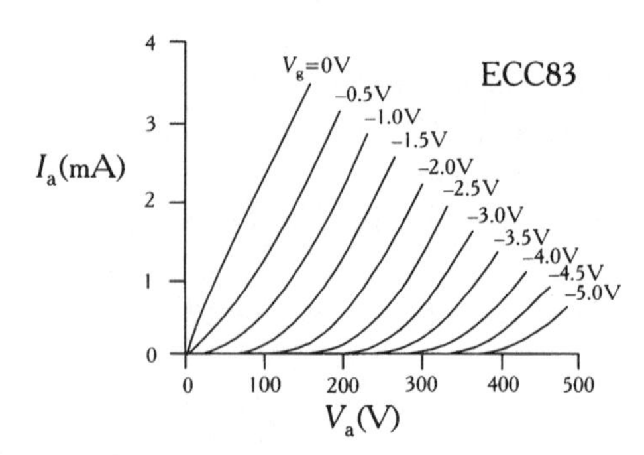

Fig. 2.1 *Triode anode characteristics*

The anode characteristics are the most useful set of curves for a valve, and plot anode current I_a against anode voltage V_a for differing values of grid to cathode voltage (V_{gk}). The first point to note is that valves operate at high voltages (typically a factor of ten greater than transistor circuits) and quite low currents. The second point is that if there is no *bias* voltage ($V_{gk} = 0$), then a large current will flow. This is known as the *space-charge limited* condition, and means that the flow of current is limited only by the number of electrons that can be released from the cathode. In contrast to the transistor, we have to turn the triode off, rather than on, in order to bias it correctly.

The basic amplifier stage has an *anode load* resistor R_L, connected between the anode and the *HT* supply (this is the historical phraseology and stands for High Tension). See Fig. 2.2.

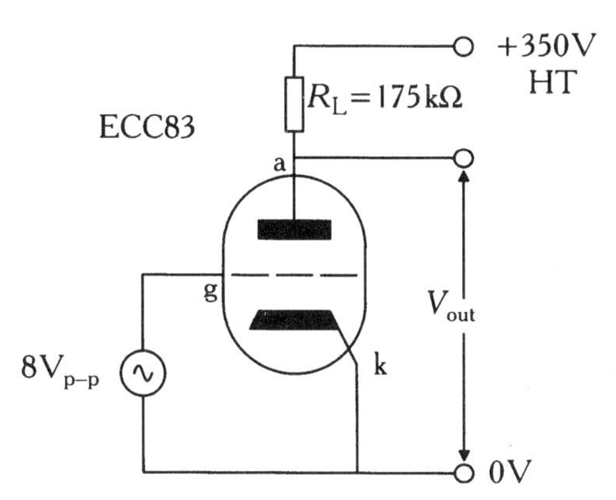

Fig. 2.2 *Common cathode amplifier*

The HT supply is assumed to have zero output resistance at all frequencies from DC to light (you may wish to consider whether this is in fact the case in a practical amplifier). By applying our input voltage between the grid and cathode, we modulate V_{gk}, and thereby control anode conditions. This is why this grid is often known as the *control grid* in multi-grid valves such as tetrodes and pentodes.

We will now use the technique of *loadlines* to link the amplifier circuit to the anode characteristics, and to extract useful information from them.

Using Ohm's law, it is apparent that if there is no current flowing through the resistor (and therefore the valve), there must be no voltage across the resistor. If there is no voltage across the resistor, then all of the HT must be across the valve, so we could mark that as a point on the graph of the anode characteristics (V_a = HT = 350 V, $I_a = I_R = 0$). Similarly, we can argue that if there is no voltage across the valve, then the HT must all be across the resistor; we can calculate the current through the resistor and therefore the valve. In this case, R_L = 175 kΩ, HT = 350 V, so the anode current I_a = 2 mA, and we can plot this point too.

Because Ohm's law is an equation that describes a straight line, if we know two points, we have completely defined that straight line. This means that we can now draw a straight line between our two plotted points, as shown. See Fig. 2.3.

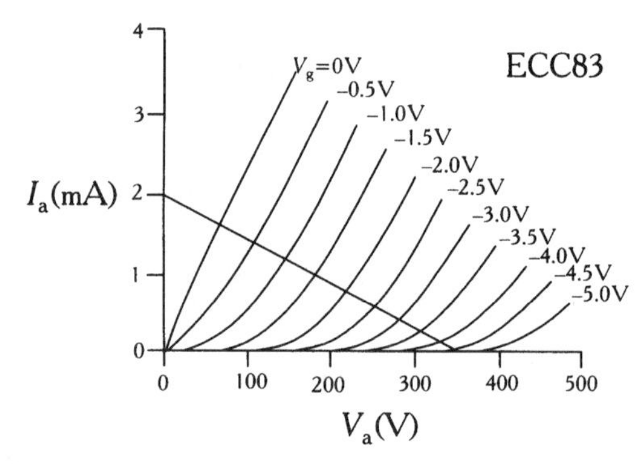

Fig. 2.3 *The loadline*

This is our *loadline*. This is perhaps the single most useful piece of analysis that can be performed on a valve stage. We have defined the anode current for any anode voltage, using an HT of 350 V and anode load of 175 kΩ. If we want to change our anode load, or HT, we must recalculate and redraw our loadline.

If we look along the loadline, we see that it is intersected at various points by the I_a/V_a curves for differing values of V_{gk}. What this means is that those differing values of V_{gk} will cause predictable changes in anode voltage, and we could therefore calculate the *gain* of the stage.

Let us suppose that we apply an 8 V_{pk-pk} sine wave to the input of the stage. If we start from 0 V, and look to see where the 0 V grid bias line intersects the loadline, this occurs at $V_a = 72$ V. We then let the sine wave swing negative to –4 V, and see that it results in $V_a = 332$ V. For an applied voltage of –4 V on the input, we produced a positive change of voltage at the anode of 260 V. The amplifier *inverts*. Since gain is defined as the ratio of output voltage to input voltage, we have just produced an amplifier with a gain of –65 (the minus sign merely reminds us that this is an inverting amplifier).

Unfortunately, it isn't very linear. If we now allow the input sine wave to continue rising past 0 V, we soon find that the anode voltage is unable to fall any further, and so the output signal no longer looks like the input signal.

We must choose a *bias* or *operating point* at which we will set the *quiescent* (no signal) conditions such that the stage can accommodate both negative and positive excursions of the signal without gross distortion.

Limitations on choice of the operating point

Not only did the previous circuit distort, but the anode DC voltage was superimposed on the output signal, so we will now add a capacitor and a resistor at the output to block the DC. See Fig. 2.4.

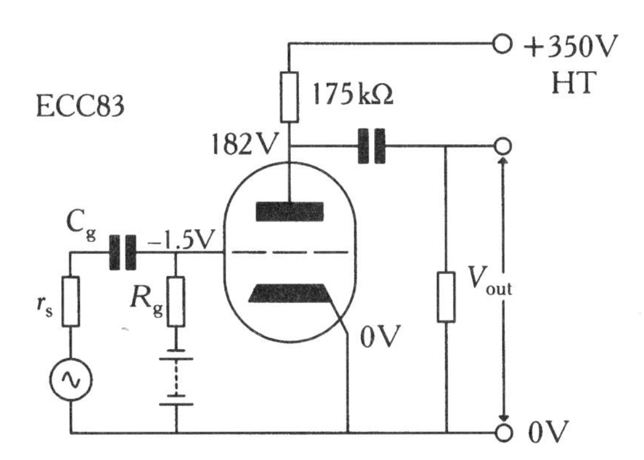

Fig. 2.4 *Grid bias using battery*

The valve may be biased by superimposing a bias voltage onto the grid via R_g which prevents the battery from short circuiting the oscillator. C_g prevents the oscillator from shorting the battery, and r_s is the output resistance of the oscillator.

Returning to the loadline, we find that as V_a rises, the grid curves become progressively bunched together, which indicates non-linearity, and is particularly severe when V_a is close to the HT voltage. This region is known as *cut-off* (because the flow of current is being cut-off). Operation near cut-off is not advisable if good linearity is required, although we will meet this mode of operation later when looking at some power stages.

Moving in the opposite direction along the loadline, we turn the valve on harder and harder, until finally there is no voltage across it. This is extreme, however, and we will encounter the problem of *positive grid current* long before that. As we make the grid less and less negative, there comes a point when the electrons leaving the cathode are no longer repelled and controlled by the grid, but are actually attracted to the grid and flow out through the grid to ground. This causes the input resistance of the valve, which could previously be regarded as infinite, to fall to a value sufficiently low that it begins to load the oscillator's (non-zero) output resistance. Because this attenuation only happens on the positive *peaks* of the input waveform, this causes distortion of the *input* signal, even though the valve is accurately amplifying what is on its grid. The onset of grid current varies with valve type (but is generally around −1 V) and is usually specified on the valve data sheet. For instance, the Mullard ECC83 used in our example specifies V_{gk} max.(I_g = +0.3 μA) as −0.9 V.

If we have a voltage across the valve, and a current flowing through it, we must be dissipating power within that valve, and there will be a limit beyond which we are in danger of melting the internal structure of the valve. This is known as the *maximum anode dissipation* and is given on the data sheet as being 1 W for the ECC83. For power valves, the curve that corresponds to this is often drawn on the anode characteristic curves, but, if we wish, we can easily add it ourselves. All we need do is to calculate the current drawn for 1 W at 0 V, 50 V, 100 V, 150 V ... We plot these results on the graph and draw a curve through the points to form a hyperbola.

The valve data sheet also specifies two more, interlinked, restrictions on the choice of bias point, maximum V_a, and maximum $V_{a(b)}$. Maximum V_a is the maximum DC voltage at which the anode may be continuously operated, whereas $V_{a(b)}$ is the maximum voltage to which the anode may be allowed to swing under signal conditions, and is effectively the maximum allowable HT voltage for that valve. Ignoring these limits will usually result in premature destruction of the valve to the accompaniment of blue flashes and bangs as residual gas in the valve is ionized and breaks down. This in itself may not cause irreversible damage, but if a path is formed between the anode and the control grid, then a large anode current will flow, and this may damage the valve permanently. You have been warned.

We can now draw all of these limitations onto the anode characteristics, and choose our operating point from within the clear area. See Fig. 2.5.

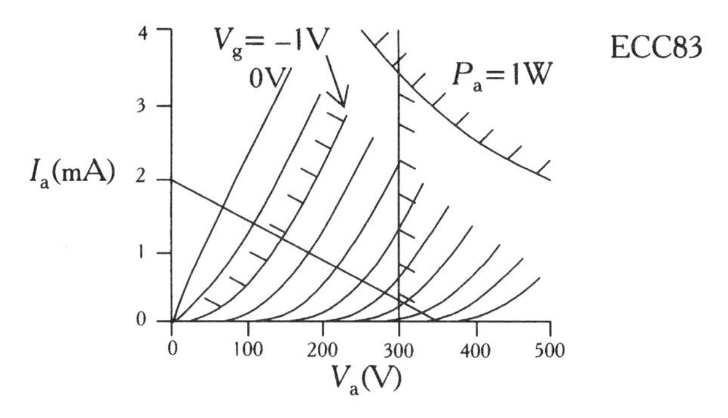

Fig. 2.5 *Determination of safe operating area*

An additional limitation is the maximum allowable cathode current $I_{k(max.)}$. One of the previous limitations usually comes into effect first, but input stages may minimize V_a, and maximize I_a in order to maximize *gm* and minimize noise, so $I_{k(max.)}$ should be checked.

Conditions at the operating point

Although the choice of operating point has now been considerably restricted, we can still optimize various aspects of performance.

In general, there are two main (and usually conflicting) factors: maximum voltage swing and linearity. If we want to bias for maximum voltage swing, then we would set the bias point at $V_a = 225$ V, to allow the anode to swing up to 300 V and down to 150 V; this would be done by setting the grid bias to −2.1 V.

However, it might be that we are interested more in linearity than in maximum swing. Triodes produce mainly even harmonic distortion, which is generated by the amplifier having more (or less) gain on the positive half cycle of the waveform than on the negative half cycle. To maximize linearity, we would look for an operating point where the distance to the first grid line either side of the operating point is, as nearly as possible, equal. In which case we might bias the anode voltage to 182 V by applying −1.5 V to the grid.

Supposing that we have chosen the linearity approach, we will now want to determine the *dynamic* or *AC* conditions of the stage to see if they satisfy our needs.

The first, and most obvious, parameter to determine is the voltage amplification (A_v), or gain, of the stage. We do this by looking an equal distance either side of the operating point to the first intersection with a grid line, noting the anode voltage. Referring to Fig. 2.5, if we move from the operating point to the right, we meet the 2 V grid line, which intersects at a voltage of 220 V, similarly, the 1 V line intersects at 148 V:

$$\text{gain } (A_v) = \frac{\text{change in anode voltage}}{\text{change in grid voltage}}$$

$$= \frac{\Delta V_a}{\Delta V_g}$$

$$= \frac{220 - 148}{1 - 2}$$

$$= -72$$

The minus sign reminds us that the amplifier is inverting, but you will often find this dropped, in the interests of clarity, since most stages invert, and the absolute polarity of any particular stage is often of little consequence.

The next important factor is the maximum undistorted voltage swing. Again, we look symmetrically either side of the operating point, but this time we look for the first limiting value. In this instance, we look to the left and see that at 148 V we are approaching positive grid current. This would not matter if our source were of zero output resistance, but this is unlikely to be the case, and so we must regard this as a limit. If we look to the right, we find that there is no practical limit until V_a = the HT.

Unfortunately, whilst this means that the valve can swing a large voltage positively, it cannot swing as far negatively. It is the *first* limit to be reached that is important. We can now see that the maximum undistorted peak to peak swing at the output is double that of the distance from the bias point to the first limit. In this example this corresponds to $72\,V_{pk-pk}$, but remembering that AC signals are specified as the RMS value of a sine wave, we should divide this figure by a factor of $2\sqrt{2}$, which results in a value of $25\,V_{RMS}$ as the maximum undistorted sine wave output, which is perhaps not so impressive.

It may be that this value of maximum output swing is insufficient, in which case we will need to go back and reselect our operating point. If we are still unable to achieve a satisfactory value, then we may need to choose a different value of R_L, HT, or both. It will now be apparent that designing valve stages requires a pencil, a clear ruler, an eraser, and a lot of photocopied anode characteristics curves.

Assuming that the stage looks promising so far, the next important parameter is the output resistance. A triode can be modelled as a voltage source coupled through a series resistance, known as the *anode resistance*, r_a. Remember that because this is an AC, or dynamic, parameter, it is given a lower case letter, and is quite distinct from R_L, the anode load. This dynamic anode resistance is then in parallel with the anode load to form the output resistance r_{out}. See Fig. 2.6.

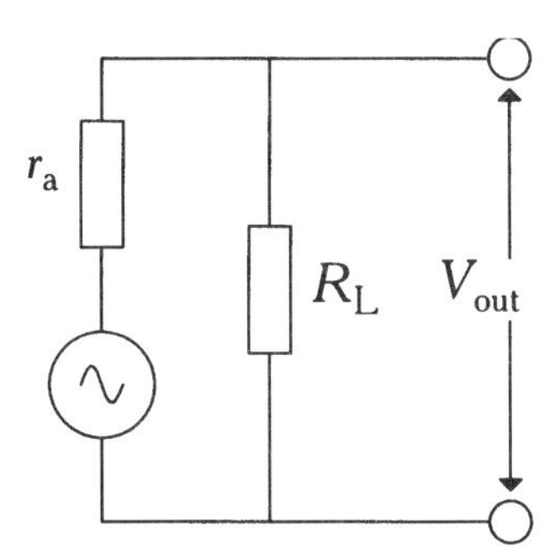

Fig. 2.6 *Thévenin equivalent of triode anode circuit*

It should be noted that the value of gain predicted for the stage by the loadline *already* includes the attenuation caused by the potential divider formed by r_a and R_L.

To find r_a, we return to the anode characteristics and draw a tangent to the curve where it touches the operating point. What we are aiming to do is to measure the gradient of the curve at that point. This is not as difficult as it sounds. A true tangent will touch, or intersect, the curve at only *one* point, and only at the correct point. A good quality transparent ruler is ideal for this purpose. Having positioned the ruler correctly, we draw a line that reaches the edges of the graduations on the data sheet, and read off the values at these points. The purpose of this is to make the resulting triangle, from which we take our figures, as large as possible in order to minimize errors. See Fig. 2.7.

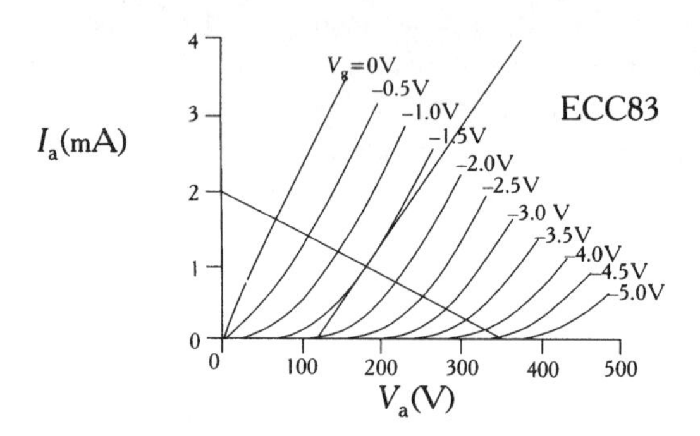

Fig. 2.7 *Determination of dynamic anode resistance r_a*

The anode resistance r_a can now be calculated from:

$$r_a = \frac{\Delta V_a}{\Delta I_a}$$

$$= \frac{V_2 - V_1}{I_2 - I_1}$$

$$= \frac{382 - 121}{4 - 0}$$

$$= 65\,\text{k}\Omega$$

You will note that the units of mA were used directly in the equation, resulting in an answer in kΩ; this is conventional practice, and saves time. R_{out} is simply r_a in parallel with R_L, which results in a value of 47 kΩ.

This is quite a high value of output resistance, and is a consequence of using a high μ (mu) valve, as they tend also to have a high value of r_a in operation.

Dynamic, or AC, parameters

So far, we have analysed the behaviour of the valve graphically, but this is not the only method. There are three parameters that can be used to completely define the characteristics of a valve, *provided that they are evaluated at the operating point*. The importance of this last point is sometimes overlooked.

These parameters are:

μ = amplification factor (no units)
gm = mutual conductance (usually mA/V)
r_a = anode resistance (kΩ, Ω)

The amplification factor is defined by:

The amplification factor (μ) of a valve is ratio of the change in anode voltage ΔV_a *to the change in grid voltage* ΔV_g, *with anode current held constant.*

$$\mu = \frac{\Delta V_a}{\Delta V_g}$$

In a more digestible form, it is the maximum possible voltage gain of the valve, and would be achieved using $R_L = \infty$. In practice, we rarely achieve a gain as high as this.

We can measure μ at the operating point by drawing a horizontal line through the operating point, which is equivalent to $R_L = \infty$, and calculating the gain as before, by noting the intersections with the grid curves. See Fig. 2.8.

$$\mu = \frac{233 - 133}{2 - 1}$$

$$= 100$$

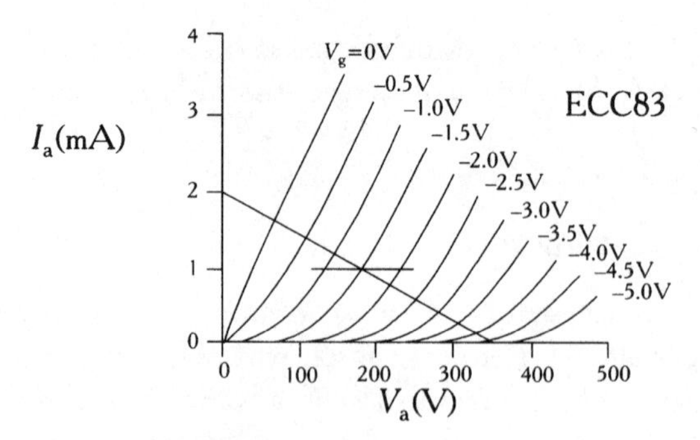

Fig. 2.8 *Determination of μ*

Note that it is usual to ignore the signs of the individual voltages measured in equations like this.

There is now a formula that we can use to determine the voltage gain A_v of the amplifier stage:

$$A_v \cong \mu \cdot \frac{R_L}{R_L + r_a}$$

$$= 100 \cdot \frac{175}{175 + 65}$$

$$= 73$$

Which is in good agreement with the value predicted by the loadline (A_v = 72). You will find that μ is one of the more stable valve parameters, and varies little with anode current (a fact that will be exploited later). However, this method is not ideal, since the accuracy of the final answer is dependent on how accurately you can draw tangents. It is, however, thoroughly recommended as a check on the general accuracy of your predictions.

$$V_{out} = -g_m V_{in} (r_a \| R_L) = -\frac{\mu}{r_a}(r_a \| R_L) V_{in}$$

$$A_v = \frac{V_{out}}{V_{in}} = -\frac{\mu}{r_a} \frac{r_a R_L}{r_a + R_L} = -\mu \frac{R_L}{r_a + R_L}$$

The second valve parameter, mutual conductance, is defined by:

The mutual conductance gm of a valve is the ratio of the change in anode current ΔI_a to the change in grid voltage ΔV_g, with anode voltage held constant.

$$gm = \frac{\Delta I_a}{\Delta V_g}$$

Although we did not define it earlier, we should now define r_a:

The anode resistance r_a of a valve is the ratio of the change in anode voltage ΔV_a to the change in anode current ΔI_a, with grid voltage held constant.

$$r_a = \frac{\Delta V_a}{\Delta I_a}$$

There is a very useful equation which links these three parameters together:

$$gm = \frac{\mu}{r_a}$$

Obviously, we can rearrange this equation as necessary to find the third parameter if we know the other two, but it is specified this way round, because although we can always predict μ and r_a reasonably accurately from the anode characteristics, we cannot directly predict an accurate value for *gm* (yet valve testers very rarely measure any parameter other than *gm*). See Fig. 2.9.

In theory, to find *gm*, we simply draw a vertical line through the operating point (hold V_a constant), and measure the change in anode current. However, it is immediately apparent that the change from 1.5 V to 1 V is considerably greater than the change from 1.5 V to 2 V, and

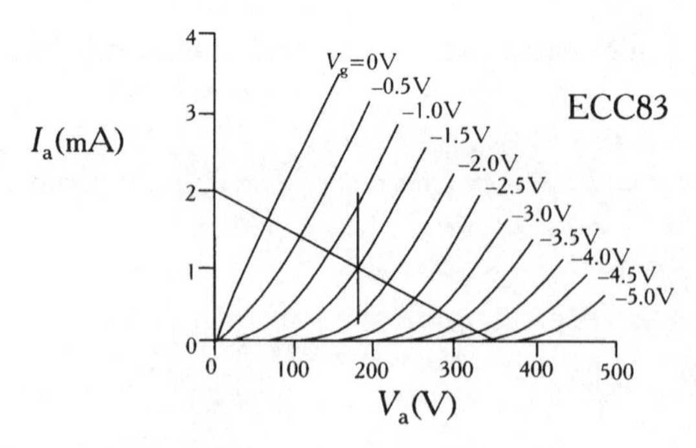

Fig. 2.9 *Determination of gm*

taking the *average* value from 1 V to 2 V does not give an accurate figure of *gm* at the operating point.

$$gm = \frac{1.72 - 0.3}{1 - 2}$$

$$= 1.42\,\text{mA/V}$$

Using the equation, however, with accurate values of μ and r_a (they must be, because they agreed well with the loadline prediction of gain), we find:

$$gm = \frac{100}{65}$$

$$= 1.54\,\text{mA/V}$$

Which means that the previous value was almost 8% low. We will return to use *gm* later.

Cathode bias

Now that we have chosen our operating point and evaluated the dynamic characteristics of our amplifier stage, we need to look at practical ways of implementing the stage. Whilst we could bias the stage using a battery, it is inconvenient to have to periodically disassemble the amplifier to change the battery. Having said that, with the advent of easily available lithium batteries with a shelf life of 10 years, replacement of the battery could perhaps be less frequent than replacement of the valve.

Another way of providing *grid bias* would be to have a subsidiary negative power supply, and use potential dividers to determine the bias to individual valves. This is frequently done on power stages, but could cause noise and stability problems with small-signal stages.

An alternative method is to insert a *cathode bias* resistor between the cathode and ground, and connect the grid to ground via a *grid leak* resistor. Conveniently, the grid is now at 0 V, and we no longer need an input *coupling* capacitor. See Fig. 2.10.

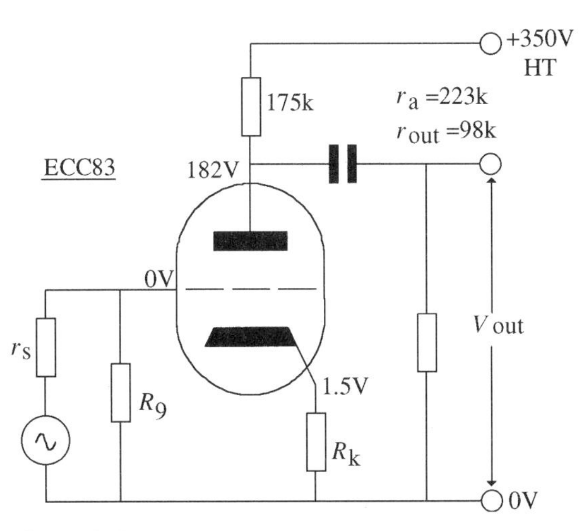

Fig. 2.10 *Cathode bias*

To understand the operation of this stage, we will assume a perfect valve that does not pass grid current even if $V_{gk} = 0$.

Initially there is no current flowing through the valve. If this is the case, there will be no voltage drop across the cathode bias resistor, and the cathode will be at 0 V. The grid is tied to 0 V, so V_{gk} must be 0 V. This will cause the valve to conduct heavily, but as it does so, the anode current (which in a triode is equal to the cathode current), flows through the cathode bias resistor, causing a voltage drop across it. This voltage drop causes the cathode voltage to rise, V_{gk} falls, and an equilibrium anode current is reached.

We know our operating point, therefore we know anode, and hence cathode, current. We know what value of V_{gk} we need. If the grid is at 0 V, then the cathode must be at $+V_{gk}$. If we know the voltage across, and the current through, an unknown resistor, then it is a simple matter of applying Ohm's law to find the value of that resistor. In our example we chose to place our operating point at 182 V. We could read off the anode current directly, but it is more accurate in this instance to calculate the current using Ohm's law. (This is because we can read off the value of V_a with greater accuracy.)

$$I_a = \frac{HT - V_a}{R_L}$$

$$= \frac{350 - 182}{175}$$

$$= 0.96\,\text{mA}$$

We know that the cathode voltage is 1.5 V, so the cathode bias resistor will be:

$$R_k = \frac{1.5}{0.96}$$

$$= 1.56\,\text{k}\Omega$$

Again, note that the equation used mA directly, resulting in a resistance in kΩ.

The effect on AC conditions of an unbypassed cathode bias resistor

Whilst the cathode bias resistor *stabilized* and set the DC conditions of the stage, it did so by means of negative feedback, so we should expect it to affect the AC conditions such as gain and output resistance. We can use the universal feedback equation to determine how much effect it will have.

$$A_{fbk} = \frac{A_0}{1 + \beta \cdot A_0}$$

The feedback fraction β in this case is the ratio R_k/R_L.

$$A_{fbk} = \frac{72}{1 + \frac{1.56}{175} \cdot 72}$$

$$= 44$$

The gain has been considerably reduced. The feedback is series derived, and series applied, so it will raise both input and output resistances. Since the input resistance of a valve is virtually infinite anyway, this won't be affected, but the anode resistance r_a will be raised.

Although the feedback equation is very handy for quickly determining the new gain, it is not quite so easily used for finding the new r_a.

Looking down through the anode, the only path to ground is the cathode, via the anode resistance r_a. Since, in this direction, resistances are *multiplied* by $(\mu + 1)$, we see an effective anode resistance of:

$$r'_a = r_a + (\mu + 1)R_k$$

$$= 65 + (100 + 1) \times 1.56 = 223\,\text{k}\Omega$$

The value of r_a rises from 65 kΩ to 223 kΩ. In parallel with R_L, this gives a new output resistance of 98 kΩ, as opposed to 47 kΩ. Incidentally, there is no reason why we should not calculate the new r_a *first* and use that new value in the standard gain formula to determine the new gain:

$$A = \mu \cdot \frac{\cancel{r_a}\ R_L}{R_L + r'_a}$$

$$A_v = 100 \cdot \frac{175}{175 + 223} = 44$$

It is most important to appreciate that the feedback affected only the valve's internal r_a. The anode load resistor R_L was external to the feedback, and therefore not affected.

Having evaluated the new values of gain and output resistance, we may decide that they are no longer satisfactory. We could either choose a new value of R_L, and try a new operating point, or we might even choose a new valve. However, there is another avenue open to us.

The cathode decoupling capacitor

The addition of the cathode bias resistor caused negative feedback, and reduced gain. This may not always be desirable, so we will now consider how to prevent this feedback.

Because the output signal is derived from changing I_a through R_L, and I_a also flows through R_k, we must also develop a signal voltage across R_k. The signal voltage across R_k is in phase with the input signal, but because the valve responds to changes in V_{gk}, which is the *difference* between V_g and V_k, we have effectively reduced the available driving voltage to the valve.

In order to restore full gain, we must suppress the feedback voltage produced at the cathode with a *decoupling* or *bypass* capacitor. The capacitor should be of low enough reactance that it is a short circuit at all AC frequencies of interest. In conjunction with the output resistance at the cathode, this will form a local low-pass filter. See Fig. 2.11.

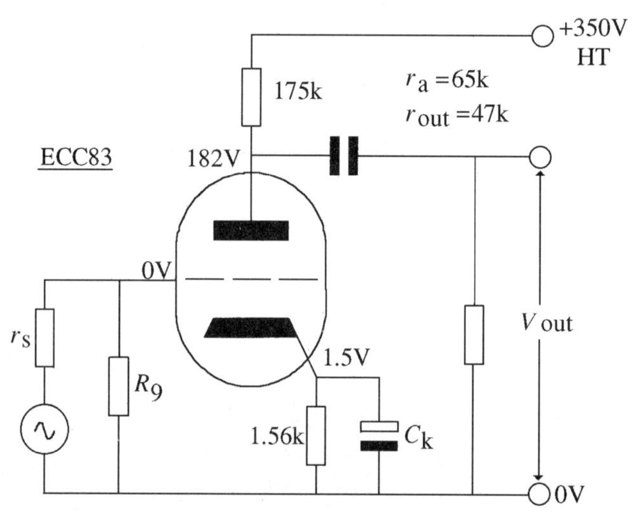

Fig. 2.11 *Cathode decoupling*

We now need to know what resistance the capacitor 'sees' from its positive terminal to ground. Clearly it sees the resistor R_k, but it also sees the cathode resistance of the valve. The resistance looking into the cathode is:

$$r_k = \frac{R_L + r_a}{\mu + 1}$$

We see the HT supply (AC ground) through the series resistance of r_a and R_L, but this is *divided by* the factor (μ + 1) due to the voltage gain of the stage. If we now put some numbers into the equation, we have:

$$r_k = \frac{175 + 65}{100 + 1}$$

$$= 2.38\,k\Omega$$

This in parallel with the 1.56 kΩ cathode bias resistor gives a total resistance (r'_k) of 946 Ω.

In audio, we usually consider frequencies down to 20 Hz (although a 32 foot organ stop will produce 16 Hz). Compact disc is certainly capable of reproducing these frequencies, as are very large transmission-line loudspeakers. In addition, there will be a number of stages to the amplifier, each with filters, so the effect is cumulative. The filter will be made using electrolytic capacitors, which are not known for their initial tolerance or stability of value, so the filter frequency should be much lower than any other filter frequency in the amplifier, in case it changes. It has also been argued that a good low frequency response is required not merely to maintain correct amplitude response, but to ensure that the effects on phase and transient response (which can extend to ten times the filter cut-off frequency) are kept to a minimum. Bearing all these factors in mind, it is not unreasonable to design for a cut-off frequency of 1 Hz.

$$C_k = \frac{1}{2\pi f r'_k}$$

$$= 170\,\mu F$$

The nearest value is 220 μF, and this is what we would use. You will note that quite a large value of capacitor is needed; valves with a lower r_k are not uncommon, and would require a correspondingly larger value of capacitor.

Choice of value of grid leak resistor

Although we have shown the grid leak resistor in place previously, we have not assigned it a value. Historically, it has generally been 1 MΩ for small-signal stages, but somewhat lower for power stages.

It is in our interests to make the grid leak as large as possible, for two reasons.

The grid leak forms a potential divider in conjunction with the output resistance of the preceding stage, and therefore causes a loss of gain. This loss is generally quite small, but it accumulates, so that at the output of a four-stage amplifier, the gain could be significantly less than predicted if this loss is not taken into account.

The second reason is that a large value of grid leak will allow the interstage coupling capacitor to be as small as possible for a given low frequency cut-off.

Once again, if we consult the valve data sheet, we find that there is a limit on the maximum value of grid leak resistor. Usually, two values are given, one for cathode bias, and one for grid bias, the value for grid bias is invariably lower (2.2 MΩ versus 22 MΩ for the ECC83). The reason that the grid bias value is so much lower is that there is *no* stabilization of operating conditions in this mode. We set grid voltage and the anode current is solely determined by the characteristics of that particular valve.

The clue to the interaction between the valve and the resistor is in the name 'grid leak'. In practice there is always a very small leakage current flowing from the grid to ground, because there will always be some contamination of the grid with the oxide coating used to form the cathode emissive surface. This leakage current produces a potential across the grid leak resistor, and the grid becomes positive. V_{gk} is therefore reduced and if the value of grid leak resistor is sufficiently high, this change in V_{gk} will become significant, and anode current will rise. The increase in anode current will raise the internal temperature of the valve, the heater will emit more electrons, and the process becomes self-sustaining until the valve is destroyed.

This process becomes more likely at higher anode currents, but fortunately valve manufacturers' limits for grid leak resistor are fairly conservative and cover all possible uses. With caution, therefore, it may sometimes be permissible to raise the value of the grid leak resistor beyond that in the data sheet. This approach should only be tried with a very good reason, and due regard for the possible consequences. The author has not used it, but has observed it in some published designs.

Choice of value of output coupling capacitor

This is actually something of a misnomer, since it is actually protecting the input of the next stage from the anode voltage of the first stage, but because input stages of valve amplifiers are invariably DC coupled, the capacitor becomes associated with the first stage.

The first and most obvious point to observe is that the capacitor should be able to withstand the anode voltage applied to it. What is not so obvious is that it should also be able to withstand the maximum *likely* HT voltage. Modern amplifiers are usually built using silicon rectifiers for the HT. This means that at the instant of switch-on, the cathodes of the valves may be cold, and there will be no anode current. The HT will therefore be unloaded and will rise to its maximum possible value. This voltage will then be applied directly across the coupling capacitors. If they fail catastrophically, then when the valves begin to warm up, they will find a large positive bias on their grids, and conduct heavily. They may be destroyed.

Using high voltage capacitors may be expensive, but it is cheaper than having to replace an expensive valve (or loudspeakers).

The only other way around this is to ensure that the HT is *never* present before the heaters are warm. Usually this means leaving the heaters on permanently; which is practical, and beneficial, for pre-amplifiers, but we would not wish to leave power amplifier heaters on permanently. A delay is needed, and valve rectifiers solve this problem at a stroke.

The other choice is the value of capacitance of the capacitor. Since we will only be using poly-something capacitors, which are stable in value, we do not mind if they define the low frequency cut-off of the amplifier. However, all the other arguments used for the cathode decoupling capacitor still apply, and so a choice of 2 Hz for cut-off frequency is not unreasonable. Incidentally, the traditional value of 1 MΩ grid leak and 0.1 μF forms a filter whose –3 dB point is 1.6 Hz.

Some modern designs use much larger values than this, and we will consider the rationale for this later.

Miller capacitance

So far, we have looked at the external, wanted, components of our amplifier stage. We will now turn to an unwanted component, *Miller capacitance*.

There will always be some capacitance between the anode and the control grid. In a tetrode or pentode it is still there, but greatly reduced. This capacitance is reflected into the grid circuit and forms a low-pass filter in conjunction with the output resistance of the preceding stage. See Fig. 2.12.

We now have two identical stages of the type that we have just designed, connected in *cascade* to form a two-stage amplifier.

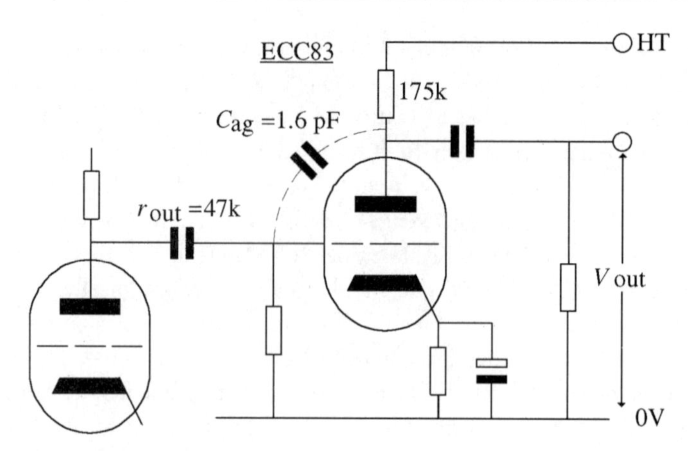

Fig. 2.12 *Miller capacitance*

Miller capacitance acts like this. When the second valve amplifies the signal, it has to charge and discharge the anode to grid capacitance C_{ag}. That charging current cannot flow into the grid (because the grid is high resistance), so it must be sunk or sourced by the preceding stage. Now, suppose that this capacitance requires a current i to charge it to 1 V. We apply a 1 V step to the input of the amplifier, and the anode moves negatively by 1 V × the gain of the amplifier, in this case by –72 V. The *total* voltage change across the capacitor is therefore $(A + 1)$ V = 73 V.

The total current required to be sunk or sourced by the preceding stage is now $(A + 1) \times i$, or $73i$. We could now reflect this capacitance into the grid by saying that exactly the same current would flow from the source if there was a capacitance between the grid and ground that was $(A + 1)C_{ag}$, hence the Miller equation:

$$C_{Miller} = (A + 1) \cdot C_{ag}$$

It is clear that a quite small value of anode to grid capacitance can have an alarming effect on the high frequency response of an amplifier combination. In our particular case, we find that we have a Miller capacitance of 115 pF (C_{ag} = 1.6 pF for ECC83). In combination with the output resistance of the previous stage, this gives a high frequency (HF) –3 dB point of 29 kHz. Stray capacitance will reduce this frequency even further.

There are various ways that we can combat this problem:

- Reduce the output resistance of the preceding stage.
- Reduce C_{ag} by screening the grid from the anode (tetrode or pentode).
- Reduce the gain to the offending anode (cascode or cathode follower).

This problem of HF response is so important that we will investigate all of these methods of improving the performance of the common cathode triode stage.

Reducing the output resistance of the preceding stage is possible, but only to a limited extent. Choosing an E88CC/6922 and operating it correctly will reduce the output resistance to a typical value of 10 kΩ. If we also change the second stage to E88CC, the Miller capacitance is then lower, typically 50 pF (due to the gain falling to 30), giving a –3 dB point of approximately 300 kHz. However, we have reduced the combined gain of the two-stage amplifier from 5184 (72^2) to 900 (30^2).

The tetrode

The tetrode was invented to overcome the HF response problem of the triode, and works by placing a second grid, g_2, initially connected to 0 V, between the anode and grid which *screens* the grid from the anode, thus lowering C_{ag}. Unfortunately, the screening is so effective that electrons are no longer attracted to the anode. To restore anode DC current, the *screen grid*, g_2 is

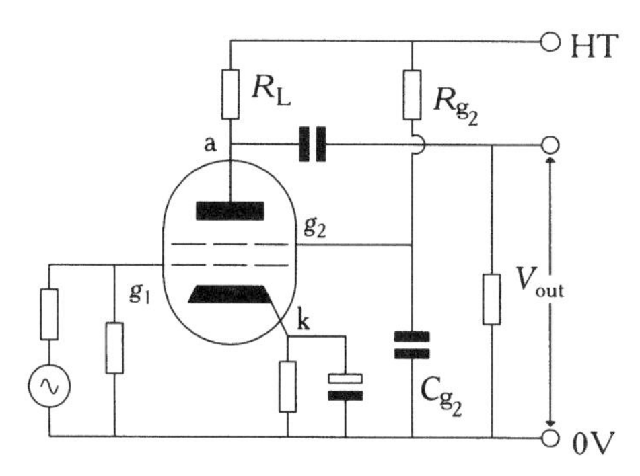

Fig. 2.13 *The tetrode*

connected via a resistor to the HT such that $V_{g2} \geq V_a$. Meanwhile, a capacitor from g_2 to ground maintains g_2 at AC ground. See Fig. 2.13.

It will come as no surprise to learn that this tinkering with the internal structure changes the anode characteristics of the valve.

The reason for adding g_2 was to screen the changing voltage at the anode from the control grid, but this was so effective that it was necessary to place g_2 at a positive potential simply to attract electrons. It is therefore the potential at g_2 in combination with V_{gk} that determines I_a, not V_a. We should therefore expect the anode characteristics of a perfect tetrode to be constant current, and consist of straight horizontal lines, implying infinite anode resistance.

In practice, the anode *cannot* be perfectly screened by g_2 and an increased V_a assists in attracting electrons, thus I_a rises with V_a, so $r_a < \infty$. See Fig. 2.14.

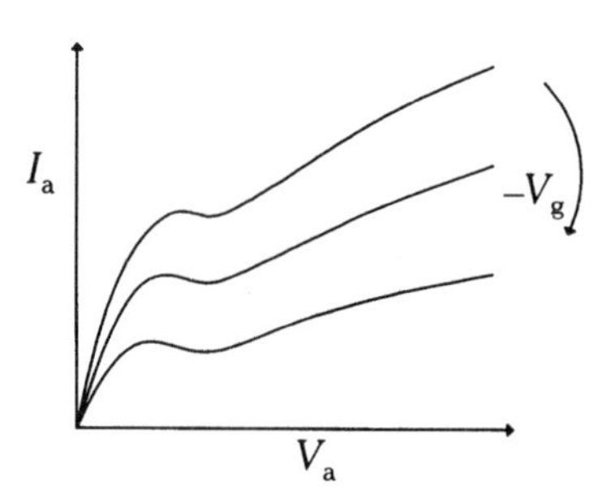

Fig. 2.14 *Anode characteristics of the tetrode*

The kink in the curves is caused by *secondary emission*. At low anode voltages, electrons are emitted by the cathode in the normal way, and collected by the anode. At higher anode voltages an electron will hit the anode with such velocity that instead of merely being absorbed by the anode, it may dislodge *two* low velocity electrons, which are then easily attracted to the higher potential of the screen grid. The anode has effectively emitted one electron, and anode current has fallen. As anode voltage rises still further, electrons may still be dislodged from the anode, but they are returned to the anode because the screen grid is not at such a high relative potential, and anode current rises once more.

The kink in the anode curves caused distortion of the output signal, so the pure tetrode was soon superseded.

The beam tetrode and the pentode

These two valves sought to keep the advantages of the tetrode (low C_{ag}) but without the disadvantage of the kinked anode characteristic. The pentode works by placing a very coarse grid, the *suppressor grid*, g_3, connected to the cathode, between the screen grid and the anode, to screen g_2 from the anode. The result of this is that the high velocity electrons emitted from the cathode pass straight through the suppressor grid, but the low velocity secondary electrons emitted from the anode are screened from g_2, and are attracted back to the anode. Secondary emission of the anode does not then cause electrons to be attracted to g_2, and the kink in the anode characteristics is avoided.

The operation of the beam tetrode is different to the pentode in order to avoid patent infringement. Instead of the electrons leaving the cathode from all points of the compass and flowing to the anode, the electrons are directed into two narrow beams of high electron density by the *beam anodes*, which are connected to the cathode. Each beam is further focused and divided into thin streams because of the alignment of the g_1 and g_2 windings, which increases electron density still further. Electrons attempting to leave the anode by secondary emission are now repelled by the incoming flood of electrons and are quickly returned to the anode. Because the dynamics of this space are very similar to the space-charge stabilized emission from the cathode, it is known as a *virtual cathode*.

Some electrons may succeed in leaving the anode and travelling a limited distance, and to avoid them reaching g_2, the anode to g_2 distance is rather greater than in the pentode (which is why the anode of the beam tetrode KT66 is of a larger diameter than that of the pentode EL34, despite their very similar ratings).

The necessary alignment of the g_1 and g_2 windings in a beam tetrode focuses the streams of electrons such that they mostly pass between the wires of g_2, thus reducing g_2 current compared to the pentode, and improving efficiency in a power valve, although there is no reason why a pentode should not adopt the same strategy. In practice, when we come to use a beam tetrode, or a pentode, we see little difference in their electrical characteristics and so we can treat them both as pentodes. (Thorn–AEI describe the PCL82 as a triode/beam tetrode, yet Mullard describe it as a triode/pentode.)

The beam tetrode offers some interesting possibilities. For instance, if we had *two* anodes, and individually connected beam anodes, we could modulate the voltage on the beam anodes to control the ratio of current split between the two anodes; this effect was exploited in some valve tuners in their mixer stages.

The significance of the pentode curves

If we now investigate the anode characteristics of the EF86 small-signal pentode for V_{g2} = 100 V, we can see that the anode curves are nearly horizontal. See Fig. 2.15.

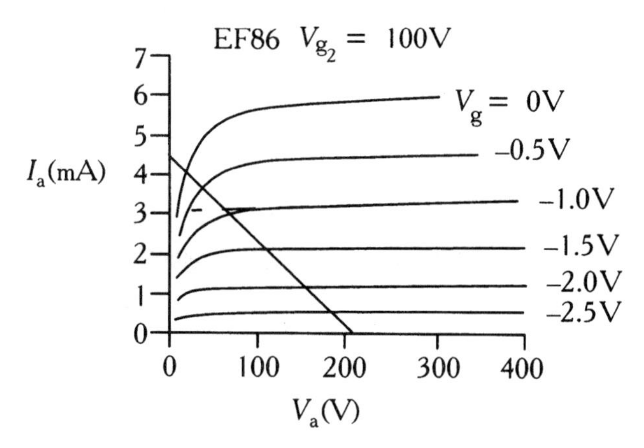

Fig. 2.15 *Anode characteristics and determination of gain of the pentode.*

We can now make some useful observations from these curves.

First, pentode characteristics are very similar to bipolar transistor characteristics, and indicate an anode resistance that is sufficiently high that for most practical purposes, it may be taken to be infinite. The output resistance of the pentode stage is therefore $\approx R_L$.

Second, the anode is able to swing much closer to 0 V than the triode, and so we can obtain a greater peak to peak output voltage. This has significant implications for efficiency, and makes the pentode a good choice for high voltage stages.

Third, the shape of the V_a, I_a curves for the pentode (and transistor) is exponential so that:

$$I_a \propto (1 - e^{-kVa})$$

This relationship results in the pentode producing predominantly odd harmonic distortion.

By contrast, the shape of the triode anode curve is a power law:

$$I_a \propto V_a^{3/2}$$

This equation can be approximated using a binomial series, and although it contains both odd (x^3, x^5, . . .) and even (x^2, x^4, . . .) terms, indicating odd and even harmonics, the terms die away very rapidly. We can therefore expect the triode to produce predominantly second harmonic distortion.

The type of distortion produced is significant because the ear is far more tolerant of even harmonic distortion than odd, partly because the ear itself produces even harmonic distortion, but also because the higher odd harmonics are no longer musically related to the fundamental and sound discordant. The measured distortion of a pentode amplifier will therefore need to be much lower than the measured distortion of a triode amplifier, because the subjective effect is so much greater, and we will generally use plenty of negative feedback.

Using the EF86 small-signal pentode

We can now consider how we would use the EF86. R_L is chosen in the normal way, in conjunction with loadlines and the 210 V HT; in this example R_L = 47 kΩ and the operating point is at 108 V. See Fig. 2.15.

When we come to calculate the gain, we find that the anode characteristic begins to curve as we reach its intersection with the loadline. It is perfectly valid to treat the anode curve as a straight line, and to project this line onto our loadline in order to find the *small-signal* gain, thus giving a gain in this example of 90. See Fig. 2.16.

R_{g2} is chosen either by a detailed perusal of the full data sheets, or by observing that, in general, the anode current is a fixed ratio of the g_2 current. For the EF86 this ratio is ≈4:1. Therefore, if the anode voltage and the g_2 voltage are to be the same, the g_2 resistor should be equal to $4R_L$, and 180 kΩ is therefore appropriate. The latter method is much quicker, but for power valves we must resort to the data sheet.

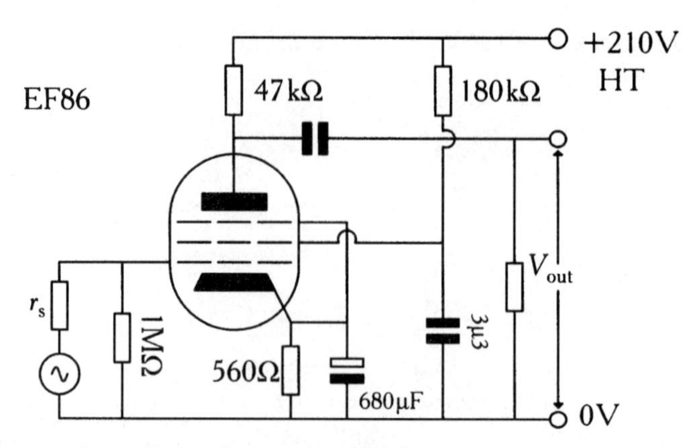

Fig. 2.16 *A small-signal pentode amplifier*

Although termed a grid, g_2 behaves as an anode in that it receives electrons, and it will therefore have an 'anode' resistance. We will need to know this resistance in order to calculate the value of capacitor required to hold g_2 at AC ground potential. Unfortunately, the data sheets for pentodes do not always give μ_{g1-g2}, gm_{g2}, or r_{g2}, but these can be deduced from triode connected valve data (g_2 connected to anode):

$$\mu_{g1-g2} \approx \mu_{\text{triode}}$$

Remembering that gm describes the controlling effect of V_{gk} on I_k, once the electrons have left the control grid/cathode region, their numbers are fixed, and the density of the g_2 mesh simply determines how the cathode current is split between anode and g_2. Thus:

$$gm_{g2} = \frac{I_{g2}}{I_a} \cdot gm_{a\ (\text{pentode})}$$

$$r_{g2} = \frac{I_a + I_{g2}}{I_{g2}} \cdot r_{a\ (\text{triode})}$$

Using the *triode* curves for the EF86, at $V_a = 108\,\text{V}$, $V_g = 1.5\,\text{V}$, $r_a \approx 14\,\text{k}\Omega$, so $r_{g2} \approx 70\,\text{k}\Omega$. This $70\,\text{k}\Omega$ is in parallel with R_{g2} ($180\,\text{k}\Omega$), giving a final resistance of $\approx 50\,\text{k}\Omega$, and so for a 1 Hz cut-off, $C_{g2} = 3.3\,\mu\text{F}$.

For a pentode, $I_k \neq I_a$, and we must sum I_a (2.17 mA) and I_{g2} (0.54 mA) to find I_k (2.71 mA), before we can calculate R_k. $V_{gk} = 1.5$ V, so the cathode bias resistor must be 560 Ω.

Evaluating *gm* from the anode characteristics, by holding anode voltage constant, and measuring the change in anode current for grid voltage, produces a value of about 1.95 mA/V. For the pentode, the cathode resistance $r_k = 1/gm$, and allowing for the 560 Ω R_k, we would need a 680 μF decoupling capacitor for a 1 Hz cut-off.

We can also use this value of *gm* in an alternative method of calculating the gain, which is given by the following equation:

$$A_v = gm \cdot R_L$$
$$= 1.95 \times 47$$
$$= 92$$

The loadline gave a gain of 90, so the agreement is good. Note that this equation does not work for triodes because it assumes infinite r_a.

C_{a-g} for the EF86 is given as <50 mpF, which is a rather quaint way of writing 50 fF (femtofarads, 10^{-15} F). You might wish to consider how Mullard measured a value of capacitance this small in 1955. Clue: You wouldn't measure it directly.

$$C_{Miller} = (90 + 1) \times 50 \times 10^{-15}$$
$$= 4.6 \text{ pF}$$

This is a dramatically reduced value compared to the triode, but because it is so small, we will now have to consider stray capacitances that were previously insignificant.

Since the control grid g_1 is near to the cathode, it must have significant capacitance to the cathode, which, since we have bypassed it with a capacitor, is at ground potential. In the data sheet, a value for C_{in} is given, which is the capacitance from the grid to all other electrodes *except* the anode, and is therefore the value of stray capacitance within the valve. For the EF86, C_{in} is 3.8 pF, which gives a total input capacitance (due to the valve) of 8.4 pF. Realistically, we ought to add a few pF for wiring capacitance, so a value of 11.5 pF would be a reasonable total figure.

The ECC83 triode gave a value of 115 pF, so in this respect, the pentode is ten times better. In summary, the pentode has greater gain, greater output voltage swing, and dramatically reduced input capacitance compared to a triode. So why don't we use them all the time?

We mentioned earlier the form of distortion that a pentode generates, but the real killer for small-signal pentodes is noise.

The EF86 is described as a 'low-noise pentode', and in a very strict sense this is true, because it *is* low noise by pentode standards. By triode standards, it really isn't very good, because of *partition noise*.

This is the *additional* noise, compared to the triode, that is generated by the electron stream splitting to pass either to the anode, or to g_2. This additional noise is related to the ratio of anode to screen grid current and to the mutual conductance of the *screen* grid; typically this makes a given pentode 6 dB to 14 dB noisier than that pentode connected as a triode. (Connected as a triode, the EF86 is actually quite a good triode.) Even worse, partition noise has a $1/f$ frequency distribution, which means that its amplitude rises as frequency falls, which has been found to be particularly irritating to the ear.

The cascode

What we would like is a valve, or a *compound device*, that gives the advantages of the pentode with none of its disadvantages – this compound device is known as the *cascode*. See Fig. 2.17.

The cascode bears considerable similarity to the pentode in that there is an arrangement of components (R_1, R_2, C_1) that looks very much like a screen grid bias supply, and indeed this is what it is. The device has a very high r_a, approximately equal to the r_a of the lower valve, multiplied by $(\mu + 1)$ of the upper valve.

Operation is as follows: The upper valve has an anode load R_L, as usual, but instead of modulating V_{gk} by varying the grid voltage, and holding the cathode constant, we vary the cathode voltage, and hold the grid constant. The upper grid is biased to whatever voltage we feel is necessary for linear operation of the upper valve, and is held at AC ground by the capacitor. This is significant, because it means that the cathode is screened from the anode by the grid, and so Miller capacitance is not a problem. Because we are changing the *cathode* rather than the grid voltage, this part of the stage is non-inverting.

Although the upper valve has a grid in the way of the electron stream, it draws no current, and partition noise does not occur.

The lower valve operates as a normal common cathode stage, except that it has as its anode load the cathode of the upper valve. Because the dynamic resistance looking into the cathode is low, the gain of the lower

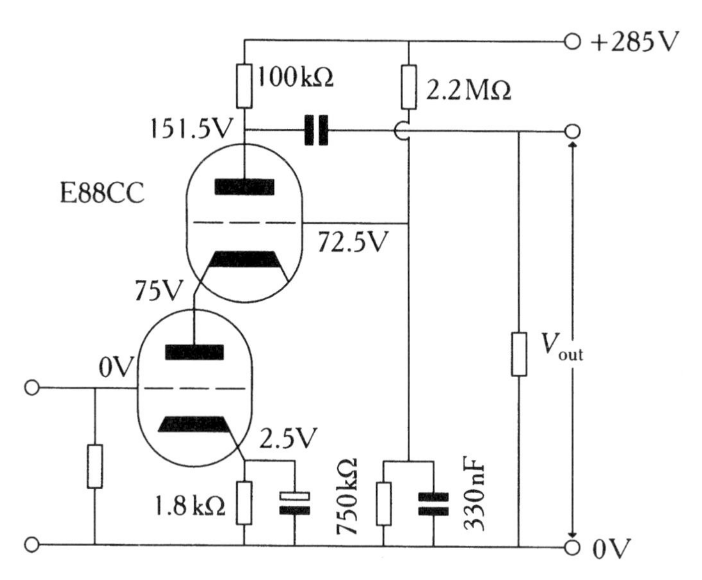

Fig. 2.17 *The cascode*

valve to its anode is low, and so its Miller capacitance is also low. Another way to view the cascode is to consider that both the cascode and the pentode seek to screen the changing voltage across R_L from the sensitive input circuit, and thereby reduce C_{in}. The pentode does this by adding an internal screen between the input grid and anode, and directly reduces C_{ag}, whereas the cascode grounds the grid of the upper valve (which then acts as a screen) and drives the upper cathode from the lower valve.

Because the lower valve has a low value of load resistance, it could generate considerable distortion if it were to be allowed swing very many volts. Fortunately, most of the gain is provided by the upper stage, and so distortion of the lower stage should not be a significant problem.

An important point to note with cascodes is that there is only one valve that is really suitable for use as a cascode; this is the ECC88/6DJ8 or E88CC/6922 (special quality version), which was *designed* to be used as a cascode. Try other valves by all means, but do not expect the performance to be as good.

We will now see how to design a cascode. It is usual to operate the lower anode at about 75 V, so if we have a 285 V HT, this will leave 210 V across the upper valve. See Fig. 2.17.

We can choose an anode load for the upper valve, and draw a loadline in the usual way. In this case R_L = 100 kΩ, and V_g = −2.5 V, causing V_a = +76.5 V, which gives a particularly linear operating point. The anode current is therefore 1.34 mA. See Fig. 2.18.

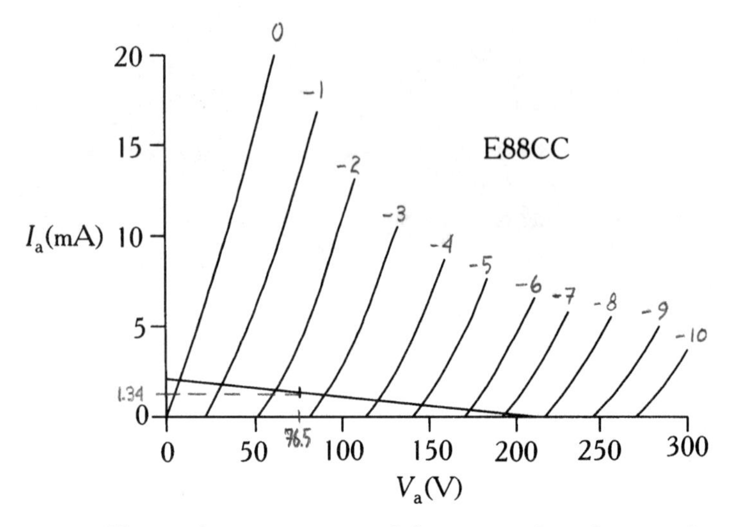

Fig. 2.18 *Choice of operating point of the upper valve of a cascode*

If the anode of the lower valve is to be operated at 75 V, and the upper valve has a V_{gk} of −2.5 V, then the grid of the upper valve must be at 72.5 V. Since the grid of the upper valve does not draw any current, its voltage is set by the potential divider, and completely determines the conditions of the upper stage, which is working in grid or *fixed* bias mode. We still have to be careful not to exceed the maximum permissible grid leak resistance of the upper valve, which for an E88CC/6922 is 1 MΩ, but the Thévenin resistance of the potential divider is 560 kΩ, so we are well within limits. (We assume that the DC resistance of the power supply is zero in making this calculation.) We only need a 0.33 μF capacitor to make the grid a short circuit to ground as far as AC is concerned ($f_{-3\,dB}$ = 1 Hz), compared to 3.3 μF for the EF86 g_2 capacitor.

Attempting to investigate the lower stage using anode characteristic curves is not very helpful. Instead, we will use the *mutual characteristics* of anode current against *grid* voltage. See Fig. 2.19.

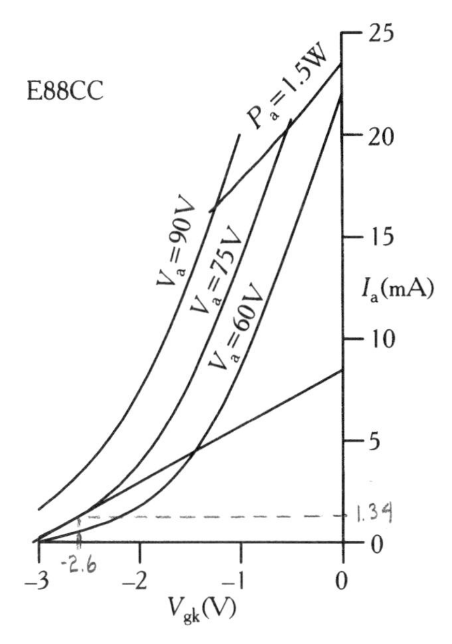

Fig. 2.19 *Triode* mutual *characteristics* (transconductance curves)

We know that V_a of the lower valve is 75 V, so we can look along the curve for $V_a = 75\,V$ until we come to the point where $I_a = 1.34\,mA$ (upper and lower anode currents are equal), this is the operating point of the lower valve, and gives a V_{gk} of about 2.6 V. Plotting the point $V_a = 75\,V$, $I_a = 1.34\,mA$ on the anode characteristics gives $V_{gk} = 2.4\,V$, so the agreement is not too bad. From this we could take an average value of 2.5 V, and calculate the value of R_k at 1.8 kΩ.

Because the cascode is made up of one stage that is non-inverting, and one that inverts, the output is inverted with respect to the input. The gain of a cascode, where V_1 is the lower valve and V_2 is the upper valve, with equal anode currents is:

$$A_v = \frac{1}{\dfrac{1}{gm_1 \cdot R_L} + \dfrac{r_{a2} + R_L}{R_L} \cdot \dfrac{1}{\mu_1(\mu_2 + 1)}}$$

This is frequently approximated to $gm_1 \cdot R_L$. From the equation, we see that we need to find gm for the lower valve. This is easily done using the mutual characteristics, by measuring the gradient at the operating point.

$$gm = \frac{\Delta I_a}{\Delta V_g}$$

$$= \frac{8.35}{3.08}$$

$$= 2.7 \text{ mA/V}$$

We need r_a for the upper valve, but we are not sitting conveniently on a grid line, so we must interpolate (guess). We can either do this by taking an average of the values either side of the operating point (if they are symmetrically about the operating point), or we could use a French curve to draw a new grid line where we need it (quite a good method). In this instance, we will take an average value. See Fig. 2.20.

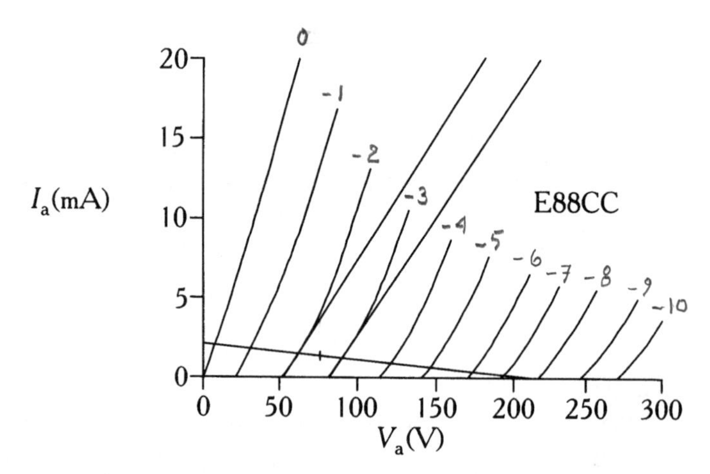

Fig. 2.20 *Averaging two values of r_a to find intermediate value*

$$r_{a\,(V_g\,=\,-2\,V)} = 5.5\,\text{k}\Omega$$

$$r_{a\,(V_g\,=\,-3\,V)} = 6.45\,\text{k}\Omega$$

Therefore we will say that at $V_g = 2.5\,\text{V}$, $r_a \approx 6\,\text{k}\Omega$.
At the operating point of both valves, $\mu = 32.5$.

Putting all these values into the equation yields a gain of 214. Using $gm_1 \cdot R_L$ would have given a gain of 270, which is 2 dB high; nevertheless, the approximation is useful, because it tells you whether it is worth pursuing the design any further.

We can now use this value of total gain to calculate the gain of the lower stage (if we wish). This is a useful exercise, because it allows us to find the voltage swing on the lower anode. From this we can check linearity (which might be problematic), and Miller capacitance. We can read the gain of the upper stage from the loadline, which gives us a gain of 30, so the gain of the lower stage must be 7.1. $C_{a-g} = 1.4\,pF$ for E88CC, so the Miller capacitance is:

$$C_{Miller} = (7.1 + 1) \times 1.4\,pF$$

$$= 11.3\,pF$$

As with the pentode, we should add the strays, 3.3 pF for the internal (valve) strays and 3 pF for external strays, to give a total value of 18 pF. This is not quite as good as the pentode that we saw earlier, but if the pentode had used a 100 kΩ anode load, its gain and Miller capacitance would have doubled and the answer would then have been comparable.

The values of cathode bias resistor and decoupling capacitor are calculated in the normal way for a triode.

We need not use equal values of anode current in the upper and lower valves. This is useful, because using a higher value of anode current for the lower stage will increase gain (by increasing gm_1) and improve linearity. See Fig. 2.21.

Cascodes, in common with other topologies that involve operating cathodes at voltages significantly above ground, have problems because of heater/cathode leakage currents and the maximum allowable cathode to heater voltage V_{k-h} (see Chapter 3). It is not uncommon for the cathode of a valve to be unbypassed and therefore have signal voltages on it. If, as in the cascode, the gain to the cathode of the upper valve is low, and we are using the device because of its good noise performance, then it is likely that the signal voltage on that cathode is very small, perhaps only a few millivolts.

Leakage currents via the heater/cathode insulation become worse as V_{hk} rises, and the combination of $V_{hk} = 75\,V$, and a small-signal voltage, means that the effects can be significant. The author once made a circuit using valves that were rated at $V_{hk(max.)} = 150\,V$, operated the valves at $V_{hk} = 120\,V$ and suffered low frequency noise, which was only cured by sitting the relevant heaters on a 150 V DC supply.

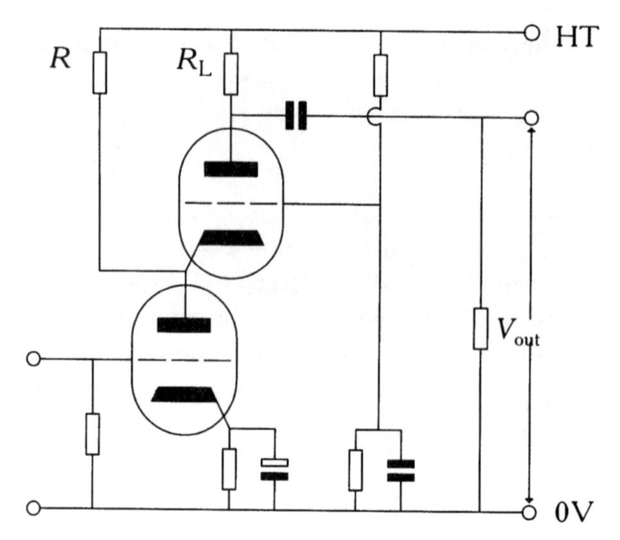

Fig. 2.21 *Increasing I_a of the lower valve in a cascode*

There is an understandable reluctance to do this, because it means that we need two or more heater supplies, one connected to ground as normal, and another connected to an elevated voltage. We will return to this practical problem later.

The cathode follower

The circuits that we have considered up until now have been concerned exclusively with providing voltage gain. Sometimes we will need a *buffer* stage that provides high input and low output resistance. The cathode follower has a voltage gain of slightly less than 1, a low output resistance, typically $\leq 1\ \text{k}\Omega$, a high input resistance ($\approx 500\ \text{M}\Omega$ in valve microphones), and is non-inverting. We will consider the fixed bias version of the cathode follower first. See Fig. 2.22.

We have changed the position of the load resistor, and the output is now taken from the cathode, but the circuit can still be analysed in the same way as before, using loadlines. See Fig. 2.23.

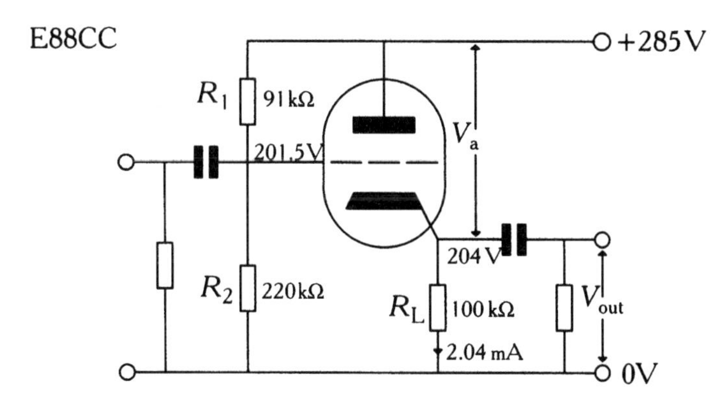

Fig. 2.22 *Fixed bias cathode follower*

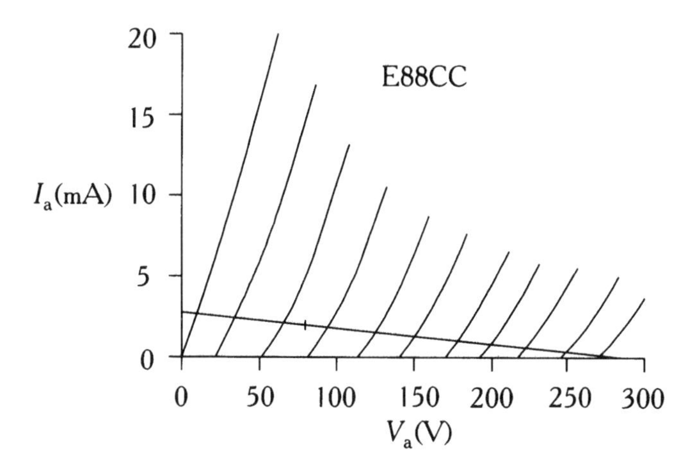

Fig. 2.23 *Operating point of fixed bias cathode follower*

$R_L = 100\,\text{k}\Omega$, and so we draw the appropriate loadline, $V_g = -2.5\,\text{V}$, with $V_a = -81\,\text{V}$, because of the excellent linearity in this region. Remembering that V_a is actually the *anode to cathode* voltage, the cathode is now at $285\,\text{V} - 81\,\text{V} = 204\,\text{V}$, and because $V_{gk} = -2.5\,\text{V}$, the grid must be at 201.5 V to bias the valve to this condition. This voltage is set by the potential divider R_1, R_2.

We can now calculate the gain of the stage. The cathode follower is simply a special case of the common cathode amplifier with 100% negative feedback (parallel derived, series applied). So we can use our normal technique of measuring the gain from the loadline (A_v = 28.5), and then apply the feedback equation:

$$A_{fbk} = \frac{A_0}{1 + \beta \cdot A_0}$$

Since we have 100% feedback, $\beta = 1$, and the gain of our example becomes 28.5/29.5 = 0.97.

We saw earlier that the resistance at the cathode was:

$$r_k = \frac{R_L + r_a}{\mu + 1}$$

But for a cathode follower, R_L from the anode to the HT = 0, so this equation can be approximated to $1/gm$. From the anode characteristics, $gm \approx 5$ mA/V, this gives an output resistance of ≈200 Ω. This is not a particularly accurate answer, since the method of determining *gm* was crude, but this does not matter, since it is usual to operate a cathode follower with a ≈1 kΩ resistor in series with its output to ensure stability – this then swamps the slight inaccuracy. Nevertheless, 1.2 kΩ is a low output resistance for a valve stage.

As shown, the stage does not have a high input resistance, although this configuration is useful for making Sallen & Key active filters (see Appendix). We need to rearrange our bias to achieve a high input impedance. See Fig. 2.24.

We now have cathode or *self* bias provided by the 1.3 kΩ resistor, whose value is calculated in the normal way. You will note that by adding this resistor, we have slightly increased the value of R_L, and indeed this was also the case in the common cathode amplifier, but this ≈1% increase has a negligible effect on circuit conditions.

At first sight, this configuration is very little better than the fixed bias configuration, as the input resistance appears to be only 1.1 MΩ. However, the 1 MΩ grid leak resistor has been *bootstrapped*, which is to say that the entire input signal does *not* appear across it.

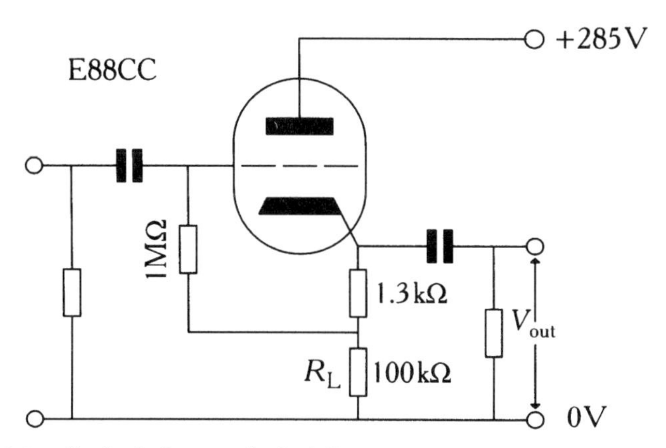

Fig. 2.24 *Cathode bias cathode follower*

It works like this. We have just calculated the gain A_v to the cathode as being 0.97. We can calculate the attenuation of the potential divider formed by the cathode bias resistor and R_L as being 0.987, therefore the proportion of input signal voltage at the lower end of the grid leak resistor is $0.96\,V_{in}$. Now, since the output of a cathode follower is *non-inverting*, this means that there is only $0.04\,V_{in}$ *across* the grid leak resistor. The *signal* current through this resistor will therefore be only 4% of what it would have been, had the grid leak resistor been connected directly to ground. It presents an input resistance equivalent to 1 MΩ/0.04 = 25 MΩ. Formalizing this argument:

$$r_{input} = \frac{R_g}{1 - A \cdot \dfrac{R_L}{R_L + R_k}}$$

Note that A is the gain of the cathode follower, not the original loadline gain. A similar argument can be used to determine the input capacitance of the cathode follower:

$$C_{input} \approx C_{ag} + (1 - A) \cdot C_{g-k}$$

Note that this is an approximate value because there will be significant strays. Using our example with the E88CC:

$$C_{input} \approx 1.4\,pF + (1 - 0.96) \times 3.3\,pF$$
$$\approx 1.5\,pF$$

We should add a few pF for wiring strays, as we did before, which brings the likely input capacitance of the cathode follower to 4.5 pF, which is rather less than half the value of the cascode or pentode.

It has been suggested that the linearity of the cathode follower is questionable. It is hard to see how this accusation can be true, particularly if the operating point has been chosen carefully, as in the previous example, since the stage operates under 100% negative feedback. This means that any non-linearity will be reduced in proportion to the feedback factor $(1 + \beta A_0)$, which in our example gives a reduction of 30:1.

Nevertheless, it *is* possible to do even better. We mentioned earlier that μ was one of the more stable valve parameters, whereas r_a varies considerably with anode current. This is significant, because it is the variation of r_a that causes distortion, and we can see why this is if we look at the equation for the gain of a common cathode amplifier:

$$A_v = \mu \cdot \frac{R_L}{R_L + r_a}$$

If we can make R_L very large, ideally infinite, then r_a will become insignificant in comparison and will no longer cause distortion. Provided that we have chosen a sensible operating point where μ does not vary greatly, we will then have a very low distortion amplifier. Unfortunately, if we merely make R_L very large, we find that there is such a voltage drop across it that we need an HT of over 2 kV! See Fig. 2.25.

We need a way around the problem of excessively large values of R_L, and to do this, we need to examine some definitions.

Sources and sinks: definitions

A current or voltage source is a supply of energy (such as a battery) capable of supplying energy into a load whose other terminal is connected to ground, whereas a sink may *control* the characteristics of an external source of energy, but provides none of its own. Audio electronics often needs real world

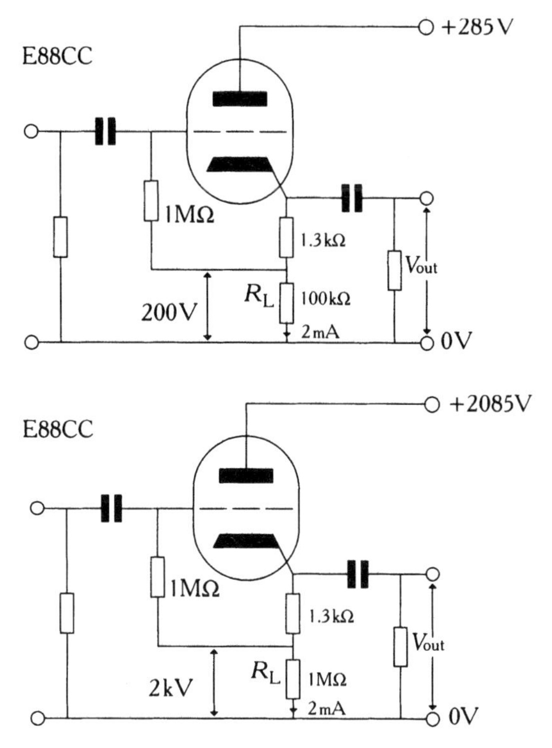

Fig. 2.25 *Effect of increasing R_L in a cathode follower*

approximations to these hypothetical devices in order to improve the AC performance of the surrounding circuit, so the following definitions are couched in terms of their common effects on AC performance, even though practical implementations using active devices may be equally effective at DC.

A perfect constant voltage source/sink is a short circuit (zero resistance) to AC, and Ohm's law therefore ensures that no matter what AC current passes through it, no AC voltage will be seen across it. Although active devices such as voltage regulators are frequently better, suitably sized capacitors are often inescapably used as AC approximations to constant voltage sources/sinks. Common audio examples include the reservoir capacitor in a capacitor input power supply (source), or the cathode bypass capacitor (sink).

Conversely, a perfect constant current source/sink is an open circuit (infinite resistance) to AC, and even an infinite AC voltage across it is incapable of causing an AC current to pass through it. Active constant current sources/sinks are becoming more common, but inductors are also used as AC approximations to constant current sources/sinks, the main audio examples being the choke in a choke input power supply (source), or the primary inductance of an output transformer (sink).

Although it has been suggested that capacitors and inductors may be used as approximations to perfect sources or sinks, the implication is that topologies traditionally using these approximations may have their approximations replaced by active devices which can be more nearly perfect. It is probably true to say that the primary difference between valve amplifiers designed in the 'Golden Age', and modern designs is that the modern designs are likely to replace passive components with active devices to more closely approximate perfect sources and sinks.

The common cathode amplifier as a constant current sink

We saw earlier that if we used a common cathode amplifier with R_k unbypassed, r_a rose due to negative feedback. We can exploit this effect deliberately to create a constant current sink. See Fig. 2.26.

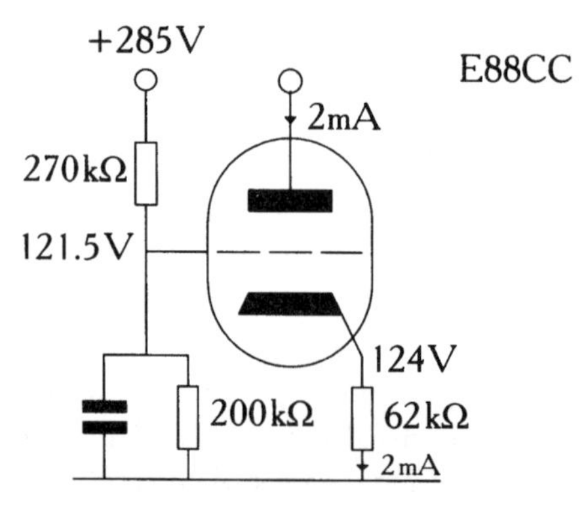

Fig. 2.26 *Constant current sink*

Let us suppose that we need to sink a current of 2 mA, using an E88CC valve and that in that condition, the voltage across the sink will be 204 V. We can draw a loadline to show this condition. See Fig. 2.27.

Plotting V_a = 204 V, I_a = 0 is easy, but we don't yet know where the other end of the loadline will be. However, we *do* know a point on the

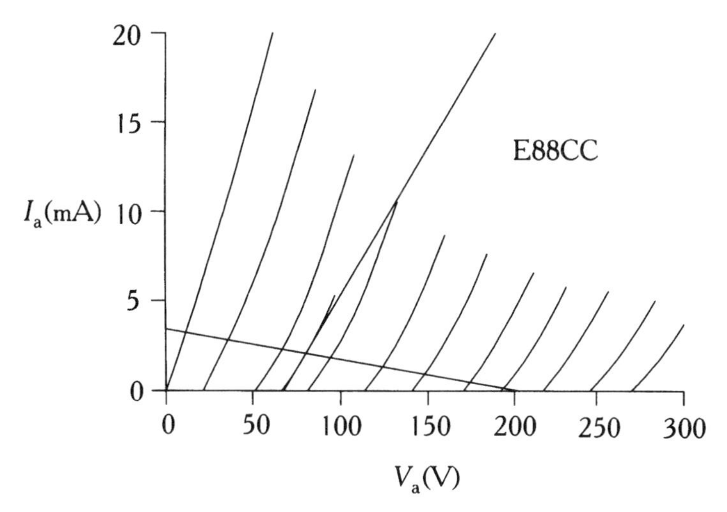

Fig. 2.27 *Operating conditions of constant current sink*

line; we know that $I_a = 2\,\text{mA}$ at the operating point, although we do not know the voltage. It is up to us to *choose* a voltage, and $V_a = 81\,\text{V}$ is a good choice for linearity. Linearity is still important in a constant current sink because in the complete circuit, we will probably modulate the anode voltage with an audio signal. If linearity is poor, this indicates non-constant r_a, which is part of the term that governs the output resistance of the sink. If the output resistance varies with applied voltage whilst it is being used as an active load for another valve, it will cause distortion in that valve. If we now draw our loadline, we can find the current through R_L when $V_a = 0$. From this we can calculate the value of R_L, which is then $60\,\text{k}\Omega$. The nearest value to this is $62\,\text{k}\Omega$, and this is what we would use.

Since $I_a = 2\,\text{mA}$, we know that the cathode of the valve will be at 124 V. $V_{gk} = -2.5\,\text{V}$, so the grid needs to be at 121.5 V. This voltage is set in the normal way using the potential divider/capacitor combination.

The resistance, looking into the anode of this device, is:

$$r_{sink} = r_a + (\mu + 1) \cdot R_k$$

For our design, this gives a value of slightly more than $2\,\text{M}\Omega$. This resistance in parallel with C_{out} and C_{ag} will cause sink gain to fall, so the sink impedance will fall as frequency rises. (C_{out} is the capacitance from anode to all other electrodes *except* the grid.) Nevertheless, achieving this result with a pure resistance would require a 4 kV HT supply.

Pentodes are even better as constant current sinks because of their high μ, and are particularly useful if the allowable voltage drop across the sink is quite low.

If we needed a constant current sink of 10 mA, but were only allowed a 100 V drop across it, an E88CC could only achieve an output resistance of $\approx$100 kΩ, which is still a ten-fold improvement over a 10 kΩ resistor, but a pentode can do rather better. The pentode section of an ECF80 can manage an $r_a \approx 1$ MΩ with the cathode bypassed, but removing the bypass capacitor gains a further ten-fold improvement. Since the value of μ for a pentode is difficult to measure, we have to guess it using $\mu = gm \cdot r_a$. If $r_a \approx 1$ MΩ, and gm for the ECF80 pentode $\approx$ 5 mA/V, then μ is very approximately 5,000. If we leave even a 2 kΩ cathode resistor unbypassed, this will increase the output resistance to >10 MΩ. This is a stunningly good constant current sink, but it should be remembered that pentodes tend to be noisy, so this would not be a good choice in the first stage of a sensitive pre-amplifier. See Fig. 2.28.

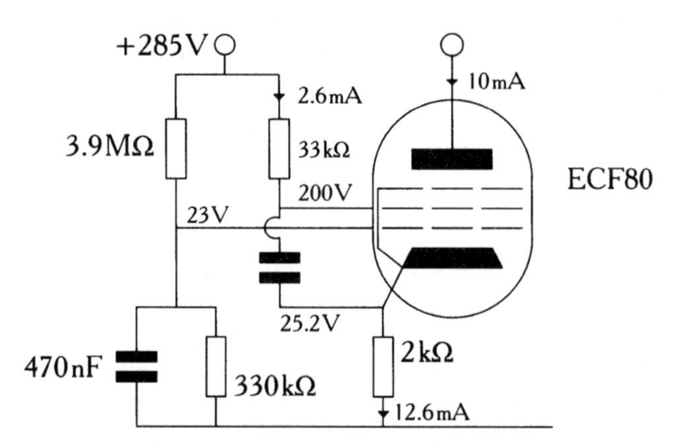

Fig. 2.28 *Pentode as constant current sink*

When using a pentode as a constant current sink, it is vital to remember that the cathode resistor passes not only the desired constant current, but *also* the g_2 current. Note also that the g_2 decoupling capacitor must be taken to the cathode, and not to ground. This is because we want cathode feedback to increase r_a, but we do not want the voltage between g_2 and the cathode to vary, as this would cause positive feedback that would reduce r_a.

The cathode follower with active load

You will probably have realized that the requirements for the triode constant current sink were set by the cathode follower designed earlier, so we can now combine them to form a cathode follower with *active load*. See Fig. 2.29.

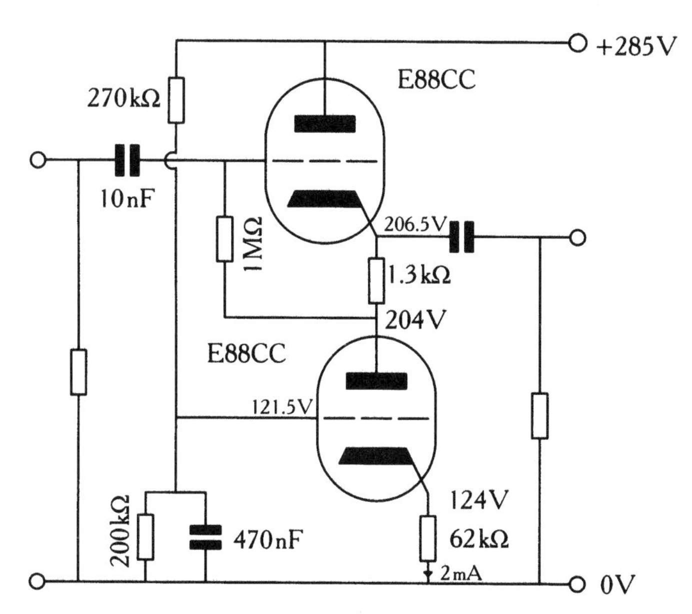

Fig. 2.29 *Cathode follower with active load*

Because the value of R_L for the upper stage is now so large, we can say that the gain becomes:

$$A_v = \frac{\mu}{\mu + 1}$$

The gain is therefore 0.97, which is only a little higher than before, but the distortion is reduced. It is possible to make calculations about what the distortion will be, but these are of very doubtful value, since *real* valves do not behave in the nice mathematical fashion that these equations require to generate sensible answers.

To sum up, a carefully designed cathode follower with a resistive load will produce very low distortion – replacing this with an active load is gilding the lily.

The White cathode follower

Named after its inventor, the White cathode follower is the basis of all output transformer-less power amplifiers because of its extremely low output resistance. The circuit comes in two forms, one self-contained, and the other requiring an external phase splitter.

Analysis of the self-contained White cathode follower

The lower valve is fed with a signal from the upper valve, which in turn, it feeds back into the cathode/grid circuit of the upper valve. At the input to the lower valve, the circuit may be considered to be a cascode amplifier. See Fig. 2.30.

Provided that μ is reasonably large, and the cathode is bypassed:

$$A_v \approx gm \cdot R_L$$

This gain will be devoted to reducing the output resistance at the cathode of the upper valve:

$$r_k = \frac{R_L + r_a}{\mu + 1}$$

Combining these:

$$r_k' \approx \frac{R_L + r_a}{(\mu + 1)gm \cdot R_L}$$

μ is usually rather greater than 1, even for power triodes, and if we substitute $\mu = gm \cdot r_a$:

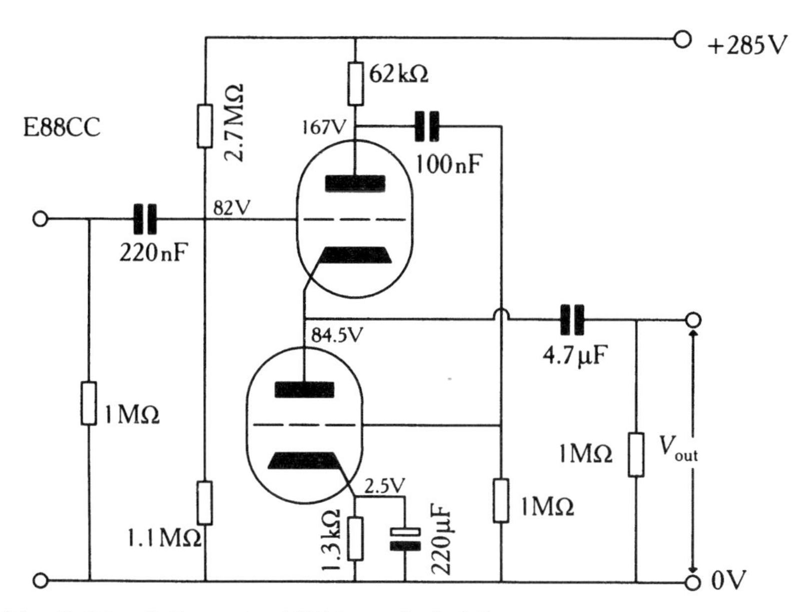

Fig. 2.30 *Self-contained White cathode follower*

$$r'_k \approx \frac{R_L + r_a}{gm^2 \cdot R_L \cdot r_a}$$

We can now recognize the R_L, r_a terms as the parallel combination of R_L and r_a. This is significant because it indicates that there is a point beyond which increasing R_L has no effect, and that final output resistance is limited by r_a:

$$r'_k \approx \frac{1}{gm^2 \cdot R'}$$

It should be noted that two rather dubious approximations were made to derive this result, both of which relied on a high μ.

The example in Fig. 2.31 was optimized for low output resistance, and $R_L \approx 10r_a$, beyond which limit no practical improvement is possible; with $gm \approx 5$ mA/V, this gave a theoretical output resistance of 6 Ω. Since $\mu \approx 32$ for the E88CC, the approximations made previously are valid, so this output resistance may be regarded as valid, and the stage would make an excellent output cable driver for a pre-amplifier.

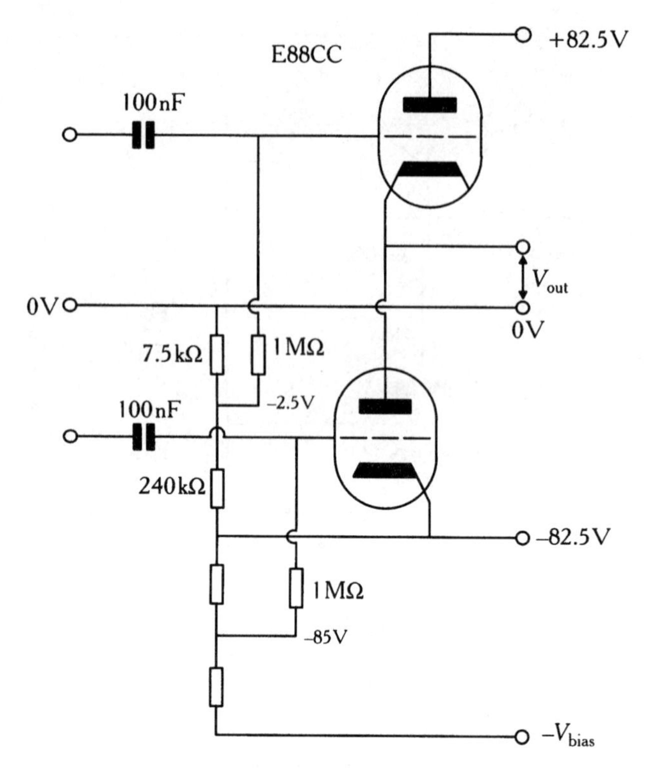

Fig. 2.31 *Push–pull input White cathode follower*

Note that because the feedback that causes the low output resistance is AC coupled, output resistance rises at low frequencies not to $1/gm$, but to:

$$Z_{out} = \frac{R_L + r_a}{\mu + 1} \parallel r_a$$

$$= \frac{r_a(R_L + r_a)}{r_a(\mu + 1) + R_L}$$

In this instance, Z_{out} rises to 1.5 kΩ, rather than 200 Ω, which is what a normal cathode follower would achieve. The practical implication is that the stage will not short circuit induced output cable noise (such as mains hum) as effectively as a stage with a true 6 Ω output resistance from DC to light.

The White cathode follower as an output stage

The primary use of the White cathode follower is as an output stage for output transformer-less amplifiers. A series resistor in either of the HT rails is a serious waste of power, so we must use the version preceded by a phase splitter. See Fig. 2.31.

The gain of the lower valve is:

$$A_{v(lower)} = \frac{\mu \cdot R_L}{R_L + r_a}$$

The upper valve no longer has a resistor in its anode circuit, so $r_k = 1/gm$, and this is the anode load of the lower valve. Substituting:

$$A_{v(lower)} = \frac{\mu \cdot \frac{1}{gm}}{\frac{1}{gm} + r_a}$$

Multiplying by *gm* and simplifying:

$$A_{v(lower)} = \frac{\mu}{\mu + 1}$$

A cathode follower with $R_k = \infty$ would have the same gain, and because the lower valve strives to produce exactly the same signal as a standard cathode follower if it saw $R_k = \infty$, there is no voltage difference between the two valves, and the upper valve *does* see $R_k = \infty$. However, the input to the lower valve must be inverted, requiring an external phase splitter.

The lower valve no longer reduces the output resistance of the upper valve, since with a gain of 1 it cannot apply feedback to the upper valve, which is why output transformer-less amplifiers need very large amounts of global feedback to bring their output resistance down to a sensible value for driving loudspeakers.

The μ-follower

This is a design which has attracted considerable interest since its rediscovery a few years ago. (There is nothing new under the sun.) Essentially, it is a common cathode amplifier with an active load. Unlike the cathode follower, where it is arguable whether this is really necessary, the common cathode amplifier can definitely benefit from this sort of treatment. See Fig. 2.32.

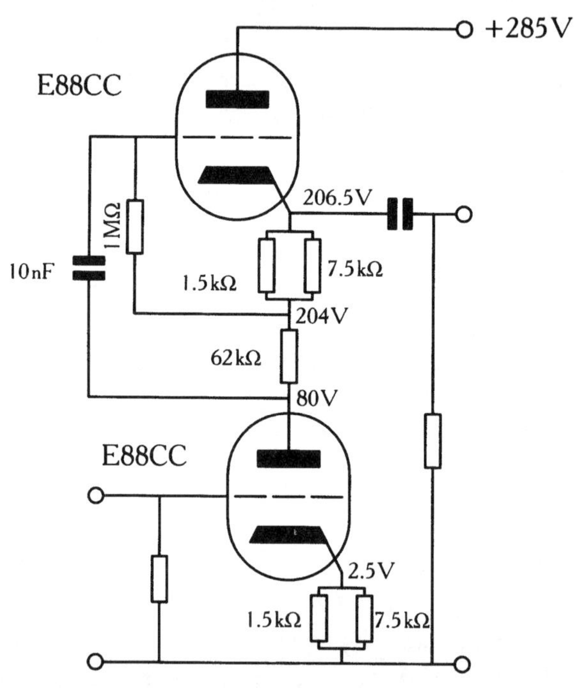

Fig. 2.32 *μ-follower*

The top valve is a self-biased cathode follower which has its input capacitively coupled from the anode of the common cathode lower stage. Since the cathode follower has an $A_v \approx 1$, and is non-inverting, the signal at its cathode will be nearly equal to that at the anode of the lower valve. If this is the case, then there will be little, or no, signal voltage across the upper resistors. Little, or no, signal current flows, implying a high resistance active load, or constant current source. The lower valve has a voltage gain≈ μ, and it produces low distortion (r_a is no longer a factor). As a bonus, we have two output terminals, either the direct output from the lower anode, or the low resistance output from the cathode follower. It should be noted, however, that the high resistance active load actually only operates at AC, since the coupling capacitor forms a high-pass filter in conjunction with the (admittedly high) cathode follower input resistance.

If the upper valve is a constant source (even if only at AC), then we can plot the loadline for the lower valve as a horizontal line. See Fig. 2.33.

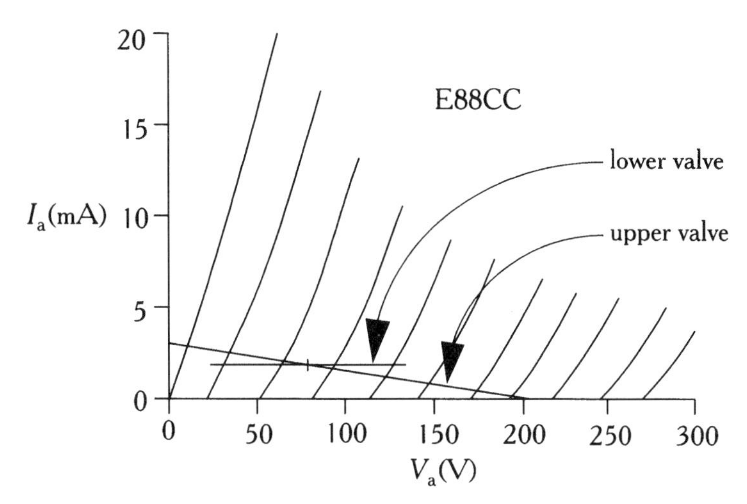

Fig. 2.33 *Operating conditions of lower valve in μ-follower*

This is an example of an AC *loadline*, where the slope of the loadline does not relate to DC conditions, although it must pass through the DC operating point. We can move this line to any operating point that we like. I have chosen an anode current of 2 mA, and I shall bias the anode voltage to 80 V; this gives μ = 32.5, and so we would expect a gain of ≈32.

We now have to determine the operating point for the upper stage. We will supply the compound stage from an HT of 285 V. (When we investigate power supplies, we will see why 285 V is such a convenient HT voltage.) This leaves 205 V for the upper stage. Since the anode currents are equal, I_a for the upper valve must also be 2 mA. If we now choose an anode voltage for the upper valve (I have chosen 80 V) we can plot the loadline. At $V_a = 0$, we have a current of 3.25 mA, which corresponds to a 63 kΩ total cathode load for the upper valve. V_{gk} for the upper valve is 2.5 V, for $I_a = 2$ mA, we will need a 1.25 kΩ cathode bias resistor. We have now established the DC conditions of the stage.

If we know the gain of the cathode follower, we can determine the value of the active load that it presents, and we will also be able to find its input resistance, which will enable us to choose an appropriate value for the coupling capacitor.

From the loadline, the gain before feedback is 29, so the gain of the cathode follower is 29/30, which is 0.97. The lower valve will see an anode load of:

$$r_{\text{load}} = \frac{R_L + R_k}{1 - A}$$

This gives a value of ≈2 MΩ, so our earlier assumptions about the gain and linearity of the lower stage were justified. We can use our earlier formula to determine the input resistance at the grid of the cathode follower:

$$r_{\text{input}} = \frac{R_g}{1 - A \cdot \dfrac{R_L}{R_L + R_k}}$$

This gives an input resistance of ≈19 MΩ. If we need a 1 Hz cut-off, then 10 nF will be perfectly adequate.

The cathode bias resistor for the lower valve was calculated in the normal way. Although the high value of load resistance for the lower valve causes β to be so small that the reduction of gain due to negative feedback is insignificant, we should still include a decoupling capacitor, otherwise the lower anode resistance will rise. Increased anode resistance is undesirable because it reduces the anode's ability to drive stray capacitance and shunt sources of noise such as induced hum or resistor noise.

An extremely useful advantage of the μ-follower is its increased immunity to noise on the HT supply, known as *Power Supply Rejection Ratio*

(PSRR). At the output of any common cathode amplifier, PSRR can be found:

$$\text{PSRR} = \frac{R_L + r_a}{r_a}$$

This is quite simply because r_a forms a potential divider with R_L. For maximum rejection of HT noise and ripple, R_L should be as high as possible compared to r_a. A pentode has $r_a > R_L$ and therefore has *no* rejection of HT noise.

Cathode feedback considerably increases r_a, but does not reduce total gain by a proportionate amount, and therefore destroys HT rejection. In our (bypassed) example, r_a for the lower valve = 6 kΩ, and the active load ≈2 MΩ, resulting in 50 dB rejection of HT noise, but removing the bypass would raise r_a for the lower valve to 47 kΩ, and reduce HT rejection to 33 dB, despite leaving gain relatively unchanged.

Strictly, we should include the loss of the cathode follower in any calculation of gain to the low resistance output ($A_{total} = \mu \times A_{cathode\ follower}$), giving a gain of 31.5 in this instance.

The importance of the AC loadline

Up until now we have tacitly assumed that the input resistance of the following stage had little or no effect on the performance of the preceding stage. This would not be true if we used the anode output of the μ-follower, because the value of the following grid leak (typically ≈1 MΩ) is not merely comparable with the value of R_L, it is actually *less* than R_L, and therefore lowers the effective value of R_L from ≈2 MΩ to 670 kΩ. This will have a negligible effect on gain, but will treble the distortion – use of the anode output is not recommended.

Whilst $R_g \geq 10R_L$, it is legitimate to ignore its effect on the preceding stage, but once it becomes smaller than this, we should consider drawing an AC loadline to investigate whether it will cause a problem. Stages with active loads *must* take into account the input resistance of the following stage.

An accurate AC loadline is easily drawn. First, we find the AC load, which is usually just the anode load and the following grid leak in parallel. We know that the AC loadline must pass through the operating point, so all we need is a second point. The simplest way to do this is to move a convenient number of squares horizontally (change the voltage by 100 V or so), and calculate the increase, or decrease, in current through the AC load to give our second point. The line through these points is then the AC loadline, and inspection of this line will give the gain, and linearity, of the stage *including* the effects of the following load resistor.

The differential pair

All of the circuits that we have so far studied have been *single-ended*, which is to say that they have one input and one output. (The μ-follower was single-ended, because although it had two outputs, they were of the same phase.)

By contrast, the *differential pair* has two inputs, and amplifies the *difference* between them, to provide two outputs, one inverted with respect to the other; this makes the differential, or *long-tailed*, pair a very useful device.

A differential pair can be made using the basic common cathode triode amplifier, or using cascodes. (The μ-follower is not suitable because differential pairs attempt to exploit the normally large ratio between R_L and R_k.) For simplicity, we will analyse the differential pair using the basic common cathode triode amplifier. See Fig. 2.34.

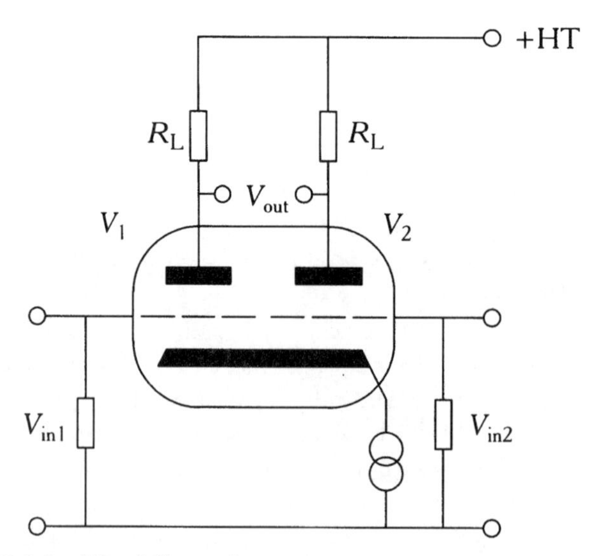

Fig. 2.34 *The differential pair*

The circuit consists of two identical triodes, often in the same envelope, with their cathodes tied together, passing anode current to ground via a constant current sink, and each driving equal value anode load resistors.

Suppose that we apply an input signal such that the voltage on the anode of V_1 rises by 1 V. The current through V_1 must therefore have fallen, but since both valves are sitting on a constant current sink, this can only occur if the current through V_2 has *risen* by an *equal* amount. Since the anode load resistors are equal, it follows that the voltage on the anode of V_2 must have fallen by 1 V.

The outputs of the two anodes are equal in voltage, but one is *inverted* with respect to the other.

Turning to the inputs. If we short circuit g_{V2} to ground, and apply a sine wave to g_{V1}, then the cathode will 'follow' that signal, because, ignoring the anode loads, the circuit is a cathode follower. This means that V_2 is driven by its *cathode*, an amplified sine wave appears on its anode, and therefore an equal and opposite signal appears on the anode of V_1. The argument works in the same manner for a signal applied to g_{V2}.

Gain

When driven by a signal connected between the two grids, the gain of the differential pair is identical to that of a standard common cathode stage, but the output voltage is found *between* the anodes of the stage. Therefore, if we look between one anode and ground, we only see half of the output voltage, and the gain appears to be halved.

If we use the differential pair as a phase splitter, and apply the same input voltage as before between one grid and ground, instead of each grid seeing half the input voltage, one sees the entire input voltage and the other none. Because the voltage difference between the two grids is the same, the gain remains the same.

Output resistance

Provided that the output of the differential pair is not unbalanced *in any way*, r_{out} at each terminal is identical to that of a simple common cathode amplifier ($r_a \parallel R_L$).

However, if only one output is loaded, the output resistance rises considerably. Working backwards from the path to ground (HT supply) via the first R_L, we see:

$$r_k = \frac{R_L + r_a}{\mu + 1}$$

But we now also see a path R_k to ground (0 V), which is in parallel with r_k:

$$r'_k = \frac{R_k \cdot \dfrac{(R_L + r_a)}{\mu + 1}}{R_k + \dfrac{(R_L + r_a)}{\mu + 1}}$$

Multiplying through by $(\mu + 1)$:

$$r'_k = \frac{R_k(R_L + r_a)}{R_k(\mu + 1) + R_L + r_a}$$

Looking down through the second anode we see r_a in series with this multiplied by a factor of $(\mu + 1)$:

$$r'_a = r_a + \frac{R_k(\mu + 1)(R_L + r_a)}{R_k(\mu + 1) + R_L + r_a}$$

If we divide by $R_k(\mu + 1)$, we obtain:

$$r'_a = r_a + \frac{R_L + r_a}{1 + \dfrac{R_L + r_a}{R_k(\mu + 1)}}$$

As R_k tends to ∞, the right-hand term on the bottom line reduces to zero, resulting in a maximum value of r'_a:

$$r'_a \approx R_L + 2r_a$$

This high value of r_a will become significant when we investigate the PSRR of the differential pair.

If $R_L >> r_a$, then the output resistance (only one terminal loaded) is:

$$r_{out} \approx \frac{R_L(R_L + 2r_a)}{2(R_L + r_a)} \approx \frac{R_L}{2}$$

Common-mode rejection ratio (CMRR)

If we now apply +1 V to *both* grids, the cathode voltage simply rises by 1 V, the cathode current remains constant, and the anode voltages do not change, because we have not modulated V_{gk}. The amplifier only responds to differences between the inputs, or *differential* signals. Applying the same signal to both grids is known as a *common-mode* signal.

This property of rejecting common-mode signals is significant, since it means that the circuit can reject hum on power supplies, or common-mode hum on the input signal, so we will investigate it further.

The signal at each output can be expressed in terms of currents using Ohm's law:

$$V_{out(1)} = i_1 \cdot R_{L(1)}$$

$$V_{out(2)} = i_2 \cdot R_{L(2)}$$

Each output will be an exact inverted replica of the other *if* $i_1 = i_2$, provided that the two load resistors are equal. There are two main ways in which this nirvana may be eroded:

- If signal current is lost through an additional path to ground. A signal current i_1 flowing down V_1 out of its cathode must split, with some current being lost down R_k, and the remainder flowing up into the cathode of V_2, to become i_2. However, if $R_k = \infty$, no current can flow down R_k, and $i_1 = i_2$. If $\mu_1 = \mu_2$, and $R_{L(1)} = R_{L(2)}$, then:

$$\text{CMRR} \approx \frac{\mu R_k}{R_L + r_a}$$

  This indicates that we should use high μ valves, and maximize the *ratio* of R_k to R_L. As an example, the second stage of the balanced RIAA pre-amplifier (Chapter 6) has an EF184 constant current source ($R_k \approx 1\,\text{M}\Omega$), E88CC differential pair ($\mu = 32$), and $R_L = 47\,\text{k}\Omega$, so CMRR ≈ 57 dB.
- Results from the above equation will be degraded if $\mu_1 \neq \mu_2$, or if $R_{L(1)} \neq R_{L(2)}$, or a combination of the two. With the easy availability of low cost, accurate digital multimeters, mismatching of load resistors is avoidable, but matching the valves is harder. If $\mu_1 \neq \mu_2$ then:

$$\text{CMRR} \propto \frac{\mu_1 \cdot \mu_2}{\mu_1 - \mu_2}$$

which indicates that high μ valves are still desirable, but that matching is important.

Because the simple equation for CMRR ignores mismatched valves, mismatched load resistors, and stray capacitances, any predictions of CMRR > 60 dB should be treated with caution. Neverthless, it is handy for checking that your tail resistance R_k is sufficiently high to ensure that predicted CMRR > 40 dB, since 40 dB is an easily achievable CMRR in practice.

Power supply rejection ratio (PSRR)

Since hum and noise on the power supply line is a common mode signal, it will be attenuated by the CMRR. We might also expect the potential divider formed by r_a and R_L to give significant additional attenuation. However, at either terminal, the only path to ground is via the other anode and R_L up to the HT supply, which is exactly the scenario we investigated when determining r_{out} with only one output loaded, therefore:

$$r'_a \approx R_L + 2r_a$$

And the attenuation of power supply noise (solely due to potential divider action) is thus:

$$\text{Attenuation} = \frac{R_L + 2r_a}{2(R_L + r_a)} \leq 6\,\text{dB}$$

If $R_L >> r_a$, we achieve the maximum attenuation of 6 dB! Our previous example ($R_L = 47\,\text{k}\Omega$, $r_a = 4.95\,\text{k}\Omega$) attenuates by 5.2 dB, and together with the 57 dB due to CMRR, PSRR = 62 dB.

It is now well worth comparing the PSRR of the common cathode stage, μ-follower, and differential pair (same DC conditions for the amplifying valve).

Stage	PSRR
Common cathode (R_L = 47 kΩ)	20 dB
μ-follower (dynamic $R_L \approx$ 742 kΩ)	44 dB
Differential pair ($r_{sink} \approx$ 1 MΩ)	62 dB

The differential pair is the best, and will remain the best, since an improved constant current source for the μ-follower could be adapted to become an improved constant current sink for the differential pair.

PSRR enables us to design power supplies correctly because it gives an indication of the allowable hum on the HT supply. This is best demonstrated by example.

We would like 100 Hz power supply hum to be 100 dB quieter than the maximum expected audio signal in our example stage of the balanced pre-amplifier. At this point, the signal has not received RIAA 3180/318 μs correction, so the level at 100 Hz is 13 dB down compared to 1 kHz levels. However, peak levels from LP are +12 dB compared to the 5 cm/s^{-1} line-up level, so the maximum audio signal at 100 Hz is 1 dB lower than the 1 kHz calculated signal level at the anode (2.2 V_{RMS}) = 2 V. We want 100 dB signal to hum, but 62 dB will be provided by PSRR, so we only need the hum on the power supply to be 38 dB quieter than 2 V, and we could tolerate 25 mV of hum on the power supply – which is easily achievable.

Transistor constant current sinks

The differential pair has demonstrated the need for constant current sinks, but the pentode constant current sink is profligate with HT voltage (although it is a very good sink), and a differential pair with grids at ground potential would require a subsidiary negative supply for the sink of –100 V. This is clearly undesirable, and a solution is needed.

Unlike the original valve designers, we are in the fortunate position of being able to use transistors, and even op-amps if we consider them to be necessary. This is a perfect example of where a transistor or two can be very helpful.

The simplest form of a transistor constant current sink is very similar to our triode version. The red LED sets a constant potential of ≈1.6 V on the base of the transistor. V_{be} is 0.6 V, so the emitter resistor has 1 V held across it. If we needed to sink 5 mA, we would therefore use a 200 Ω sense resistor.

Note that an expensive 2 W resistor is required to bias the LED. See Fig. 2.35a.

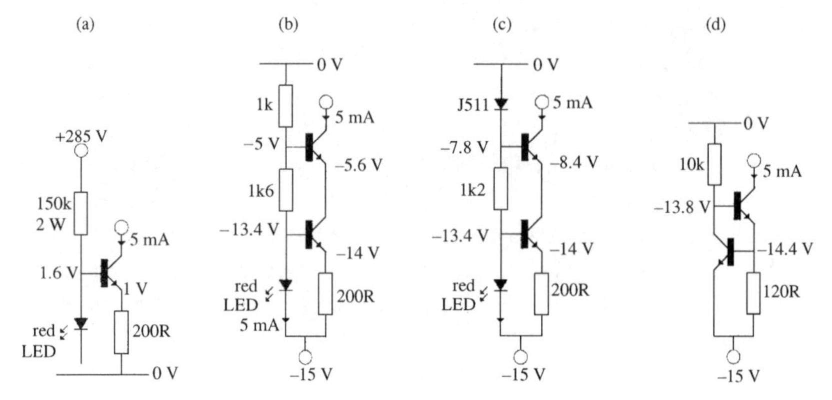

Fig. 2.35 *Semiconductor constant current sources*

The simple circuit can easily be improved upon, and since silicon is cheap, it seems worthwhile to do so.

A transistor cascode is broadly similar to a pentode, but a practical circuit requires a negative supply, although this may not be a problem in a power amplifier, because there is often a (negative) bias supply for the output valves that we can use. This circuit has a higher output resistance than the previous circuit, but its main advantage is that its output port may be taken down to 0 V without linearity problems. High frequency stability of this circuit is excellent. See Fig. 2.35b.

As shown, the cascode current source is relatively sensitive to hum and noise on the negative supply because of current changes through the voltage reference. This sensitivity can be greatly reduced by modifying the circuit to include a current regulator diode in the chain that feeds the voltage reference. See Fig. 2.35c.

The 'ring of two' circuit works by holding 0.6 V across the 120 Ω sense resistor. If that voltage rises, due to increased current through the resistor, T_1 turns on harder, which causes the base voltage of T_2 to fall. T_2 begins to turn off and so the current through the 120 Ω resistor, and therefore the sink current, is held constant. Because this circuit uses feedback applied over two transistors, there is a possibility of oscillation at high frequencies due to stray capacitances. See Fig. 2.35d.

All of these circuits may be made with whatever general purpose transistors that you have to hand, although you may wish to replace the sense resistor with a variable resistor to allow the current, and therefore anode voltage, of your differential pair to be set accurately.

Output impedance at high frequencies is determined by the capacitance seen at the collector of the transistor, which will partly be determined by strays, but also by the transistor itself. In general, high voltage/high power transistors will have a larger silicon die area, and consequently greater capacitance. To minimize this capacitance, we should use only as large a transistor as is actually required, BC549 (NPN) and BC558B (PNP) are ideal 30 V, 100 mA, 500 mW, 200 MHz transistors. 2N3904 (NPN) and 2N3906 (PNP) are very similar, but are 200 mA rated, and h_{fe} is likely to be lower. If we must use a power device, the MJE340 is a popular choice, but video driver transistors are also a possibility.

Another possibility for a constant current sink is to use a purpose designed programmable constant current sink such as the LM334Z. This device does not need any subsidiary supplies, and operates correctly with only 1.2 V across it. The typical application for this device is in the tail of a differential pair, but stability can be a problem if the differential pair is momentarily driven into overload, since the supply voltage for the internal operational amplifier may then fall so low that it no longer operates linearly, and it oscillates. You have been warned.

Further reading

Agger, L. T. (1955) *Principles of Electronics*, 2nd ed. Macmillan.

Duncan, Tom (1994) *Advanced Physics*, 4th ed. John Murray.

Eastman, A. V. (1949) *Fundamentals of Vacuum Tubes*, 3rd ed. McGraw-Hill.

Kimmel, Alan (1993) The Mu Stage. *Glass Audio P12*, Vol. 5, No. 2.

Langford-Smith (1953) *Radio Designers Handbook*, 4th ed. (Reprinted 1999).

Radio Society of Great Britain. (1976) *Radio Communications Handbook*, 5th ed.

Ryder, J. D. (1964) *Electronic Fundamentals and Applications*, 3rd ed. Prentice-Hall.

Terman, F. E. (1955) *Electronic and Radio Engineering*, 4th ed. McGraw-Hill.

The use of electronic valves (1962) British Standard Code of Practice CP1005.

Wallman and Valley (1948) *Vacuum Tube Amplifiers*. McGraw-Hill.

3

Component technology

In the last chapter, we began to design simple circuits, which will later be combined to form complete systems. In doing this, we calculated values for components. We now need to know how to specify voltage or thermal ratings of components.

Correct specification of individual components is extremely important. An underspecified component may fail prematurely, and cause further damage as it does so, whereas an overspecified component may waste money that could have made an improvement elsewhere. To be able to specify components correctly requires knowledge of the stresses that will be applied to the component (electrical, thermal and mechanical), and knowledge of the imperfections of that breed of component. (No components are perfect, although some are more equal than others.)

Much has been said about the 'sound' of components, particularly capacitors. This has caused such polarization of the Engineers vs Audiophiles debate that rational speech has only rarely been observed. This is curious, since there are well-known physical imperfections in components, and it seems reasonable to suppose that they could have an influence on sound quality. On the other hand, whilst components are not magical, there *are* purveyors of snake oil.

This chapter will help you avoid the more obvious pitfalls, but it is *not* a substitute for detailed manufacturers' data sheets and the application of intelligence.

Resistors

Preferred values

So far, we have calculated resistor values, and then picked the nearest preferred value. These preferred values are known by their *E series* (E6, E12, E24 and E96), whose values are given in the Appendix. Each series denotes the number of different values in a decade.

For example, E6 contains the values 1, 1.5, 2.2, 3.3, 4.7 and 6.8, making a total of six values per decade. If we now consider that we will probably need values from 1 Ω to 10 MΩ, then this is seven decades, and we will need 43 different values (10 MΩ is the start of a new decade). For a complete set of E24 (the most commonly used range) we would need 169 different values.

The E series is loosely related to the tolerance of the component; 20% tolerance components are E6. The reason is that the upper limit of tolerance of one value just meets the lower limit of tolerance of the next highest value. This means that there are no gaps in the range.

The argument begins to fall down when we look at E24, since 1.3 + 5% ≠ 1.5 – 5%. Also, 1% resistors are commonly available only in the E24 series, rather than E96, which leaves large gaps between values.

Heat

Resistors convert electrical energy into thermal energy. The amount of energy converted per second is the power, and this will determine the temperature rise. A signal resistor is unlikely to be a problem, but an anode load resistor could be dissipating significant power. We can easily calculate the power dissipated as V^2/R, and select an appropriate component. This is not actually quite as easy as it sounds, and there is plenty of scope for getting it wrong.

Resistor manufacturers typically specify power ratings at a component temperature 70°C (158°F). If your equipment is being operated in a domestic temperature of 20°C (68°F), then the *internal* temperature must be higher than this, since the equipment is consuming energy and is not 100% efficient. An average internal temperature of 40°C (104°F) would be quite likely, whilst areas of localized heating (hotspots) could be considerably higher. If you are fortunate enough to live somewhere warmer, then an external temperature of 35° (95°F) might not be unusual, and the internal temperature would rise accordingly.

We can only lose heat from a higher temperature to a lower one, and we can make a useful electrical analogy:

Temperature difference, ΔT (°C) is equivalent to potential difference.
Power dissipation q (W) is equivalent to current.
Thermal resistance R_θ (°C/W) is equivalent to electrical resistance.

From this we can derive a thermal 'Ohm's law':

$$\Delta T = q \cdot R_\theta$$

This tells us that a given thermal resistance will create a greater temperature rise above ambient as more power is dissipated. Resistor specifications give a value for the thermal resistance R_θ, but this value assumes that the flow of air to cool the resistor by convection is *not restricted*.

In practice, we often mount the resistor on a printed circuit board (PCB), which considerably restricts the flow if the board is mounted horizontally. Even mounted vertically, there may still be large components, such as capacitors, that block the air flow.

Combining the arguments of restricted air flow and high ambient temperature, it is not generally advisable to operate resistors at more than one-third of their 70°C rating unless you are able to do a detailed thermal analysis.

Even with this proviso, a resistor being used at one-third of its rating will become significantly warmer than its surroundings; if it changes in temperature, then we should expect its electrical parameters to change too. And they do.

Electrical resistance will change with temperature, determined by the temperature coefficient of the resistor, generally given as a parts per million change of value per °C. This may sound small, but a 30°C rise in temperature can cause a significant change in value. Therefore, if we have gone to the expense of using 0.1% resistors in a critical part of a circuit, we should not allow *any* significant power to be dissipated in them if we want their value to remain substantially the same. Maximum dissipation of one-eighth of full rated power would not be unreasonable. In addition, we should ensure that the resistor is not heated by other components.

Resistors are available in two broad types: metal film resistors and wirewound. Despite their recent minor cult status, carbon film resistors are an anachronism and will not be considered, as their tolerance and noise specifications are so very poor.

Metal film resistors

The control of the quality of the materials and processes in the manufacture of metal film resistors determines their performance, so it is worth detailing their construction.

The process starts with the individual insulating ceramic rods onto which the resistive film is to be deposited. These rods must have a smooth surface, as excessive surface roughness causes variations and discontinuities in the resistive layer which will produce electrical noise. Although the ceramic material is chemically inert, it may have picked up surface contamination, such as grease or packaging materials, so this is burnt off by passing the rods through an oven at a temperature of >1000°C.

Whilst the rods are still hot, they are transferred to a drum in batches of up to 50 000 at a time. The drum is located within a high vacuum sputtering system (effectively a big valve). The rods are then tumbled in the drum whilst a film of nickel–chromium alloy is deposited on them by secondary emission from a solid 'target' (anode) of the alloy. The duration in the drum determines the thickness of the film, and this is the first process that determines the resistance of the final resistor.

The thickness of the applied film affects the noise of the final resistor, with thinner films being noisier than thicker films. If the nickel–chromium alloy contains impurities, it will become more granular, and this also causes noise. If the adhesion of the film to the ceramic rod is poor, the film will lift, causing noise, instability and open circuit failure.

A simple nickel–chromium film cannot achieve a 5 ppm temperature coefficient unaided, but proprietary techniques can improve this by adjusting the chemistry of the film, if necessary.

End caps are fitted next, to allow connection to the resistive film. These end caps are an interference fit onto the rods, and their precise fitting is critical. If they are too tight, then as they are pushed on, they will damage the film, but if they are too loose, then they will not make a good contact. Either of these defects will cause noise in the finished resistor. Because the end cap is of a dissimilar metal to the resistive film, the interface between the two is a thermocouple, which can generate noise (as thermal EMF), so care must therefore be taken in selection of the cap material.

Commonly, ferrous end caps are used, but some manufacturers, such as MEC Holsworthy, use non-magnetic end caps on their Holco range, and it is thought that this may be a contributory factor to their good sound.

Sputtering is not a particularly precise process, and the resistance of the films typically has a spread of ±10–20%. Now that the end caps have been fitted, it is possible to measure this resistance and grade the rods into batches. The purpose of this is to ensure consistency of the amount of helixing (see below) and hence the performance of the product.

Although the rods now have a resistive element, it is quite low resistance, and this must be increased. This is done by cutting a helix through the film from one end cap to the other in order to lengthen the resistive path, whilst at the same time making it narrower. If there are more turns to the helix, this makes a longer, narrower path, and the resistance of the final resistor will be proportionately higher; resistor manufacturers refer to this parameter as *gain*, and we will return to this later.

Traditionally, the helix was cut using a diamond edged circular saw whose depth of cut was critical. If the cut was too shallow, then the resistive film would be incompletely removed, leaving traces of material bridging adjacent turns of the helix. If the cut was too deep, the saw would be damaged on the ceramic rod, and subsequent resistors would be cut poorly. Either defect caused noise in the finished resistor.

The modern technique is to use a YAG laser to cut the helix, which produces a narrower, more precise cut, but even this process is not without its pitfalls. If the energy directed by the laser is insufficient, the resistive layer will be incompletely burnt away, causing bridging. If the energy received from the laser is too great, then the resistive film at the edges of the cut will be disrupted and have an uneven edge. Both defects cause noise.

As the gain of the resistor rises, the track narrows and edge imperfections become more and more significant. This is reflected by the published noise performance, which shows that the excess noise generated by film resistors rises for values $>100\,k\Omega$. This effect is particularly noticeable for resistors of low power rating, because their smaller physical size demands a higher gain for a given value.

Film resistors also have a maximum *voltage* rating which is independent of their power rating. This is determined by the maximum allowable potential across the gap from one turn of the helix to the next. As the applied voltage rises, it becomes more and more likely that *tracking* (intermittent voltage dependent conduction) will occur across the gap due to imperfect removal of the film in the gap. Taken to its extreme, a sufficiently high applied DC will cause arcing between turns of the helix and will permanently damage a film resistor. When using film resistors as anode loads, it is not sufficient simply to ensure that the power dissipation is satisfactory, voltage rating must also be checked. Typically, higher power components have higher voltage ratings and lower excess noise.

At much lower voltages this mechanism is partly responsible for the inclusion of an excess noise specification for the resistor, which is typically given in terms of μV of noise per volt of applied DC. For minimum noise with film resistors, the applied DC across them should be minimized. A typical value for this excess noise is 0.1 μV/V, which is −140 dB; however, this can rise to 1 μV/V or more for higher resistance values, which would then be −100 dB.

By this means, applying a *signal* voltage across a film resistor generates a signal level dependent noise or *modulation noise*, since amplifiers contain many resistors, modulation noise could conceivably rise above the thermal noise floor in a low noise amplifier, but be masked in a noisier amplifier.

Laser cutting of the film resistor produces a precisely toleranced resistor, which then has tinned copper wire leads welded to its end caps before being coated with an insulating protective epoxy film onto which the value is marked.

It will be seen that almost every process can cause noise if carried out incorrectly, and for this reason, resistor manufacturers routinely measure noise or third harmonic distortion as a means of quality assurance. Unfortunately for audio designers, they generally use a 1 kHz bandwidth filter centred on 1 kHz rather than a 20 Hz–20 kHz audio band filter. Nevertheless, the figure is a useful guide to product ranges from one manufacturer.

The resistor need not have endcaps and leads fitted; it may be a surface mount resistor, where the ends of the resistor are plated with a silver palladium alloy. When soldering surface mount resistors, it is essential to use a silver loaded solder to prevent the silver leaching out from the plating and reducing solderability.

Metal film resistors are commonly available in values from 1 Ω to 10 MΩ, although values up to 50 GΩ are now available off the shelf.

Wirewound resistors

Wirewound resistors are generally designed for dissipating significant amounts of power, with 50 W components being readily available, but ratings up to 1 kW are possible. Resistance values cover as many decades as metal film resistors, but the maximum value available is typically 100 kΩ.

Again, the process begins with a ceramic rod as a former for the resistive element, but this time, resistance wire or tape is wound onto the rod and welded to the end caps, to which leads are then welded. Smaller components (<20 W) are then coated with a ceramic glaze to prevent movement of the windings and also to seal the component. Larger components may have screw terminal end caps and be fitted into an aluminium extrusion which serves as a means of conducting the heat from the resistive element to an external heatsink. However, high value resistors require many closely spaced turns of fine resistance wire, and the possibility of tracking between adjacent turns defines a voltage rating which can easily override the power rating.

Aging wirewound resistors

Scroggie[1] points out that resistance wire is wound under tension to ensure a consistent wind, that this sets up strains within the wire which relieve with time, thus causing the resistor's value to change. He further suggests that the process could be accelerated by heating the resistors in an oven heated to 135°C for 24 hours. The author tested the theory by measuring his entire stock of aluminium clad wirewound resistors, leaving them in the kitchen oven on its minimum setting for a day, allowing them to cool slowly in the oven, and measuring them again. Within the limits of measurement (3½ digit DVM), resistors over four years old showed no change, but the newest resistors changed by up to ½% in value. It therefore seems sensible to age wirewound resistors intended for anode loads in differential pairs *before* matching.

Noise and inductance of wirewound resistors

Because the resistive element in film resistors is a thin track, they develop excess noise proportional to the DC voltage drop across them (typically 0.2 μV/V). By contrast, surface imperfections of the resistance wire in a wirewound resistor form a very small proportion of its cross-sectional area, and excess noise is virtually non-existent, making them ideal as anode loads in low noise pre-amplifiers.

Wirewound resistors are wound as a coil, and even though $\mu_r \approx 1$ for the ceramic core (making it comparable with an air core), all coils have inductive reactance which may be significant compared to the resistance.

The resistance of a conductor is:

$$R = \frac{\rho L}{A}$$

where: ρ = resistivity of conductor
L = length of conductor
A = cross-sectional area of conductor

But the wire is of circular cross-section, and the area of a circle is:

$$A = \frac{\pi d^2}{4}$$

Substituting:

$$R = \frac{4\rho L}{\pi d^2}$$

To make resistors cheaply, the resistance wire is wound onto standard sized cores. To ensure efficient heat transfer to the surroundings, and to reduce the possibility of hotspots, the core will normally be completely covered with one layer of wire from end to end with an infinitesimal gap between turns. The number of turns of wire that will completely fill a core of length C is:

$$n = \frac{C}{d}$$

The length of this wire is:

$$L = \pi n D = \frac{\pi C D}{d}$$

Substituting into the resistance equation, π cancels, and the resistance achievable by a single layer wirewound resistor is:

$$R = \frac{4\rho C D}{d^3}$$

Simplifying:

$$R \propto \frac{1}{d^3}$$

Inductance is proportional to n^2, and since n is proportional to $1/d$:

$$L \propto \frac{1}{d^2}$$

As observed earlier, it is the *ratio* of L to R that is important, not absolute values, so:

$$\frac{L}{R} \propto \frac{\frac{1}{d^2}}{\frac{1}{d^3}} \propto d$$

This shows us that L/R rises as we use thicker wire, so we can expect low value wirewound resistors to possess significant inductance. This theory was tested on a component analyser, which produced models for a selection of wirewound resistors. See Fig. 3.1.

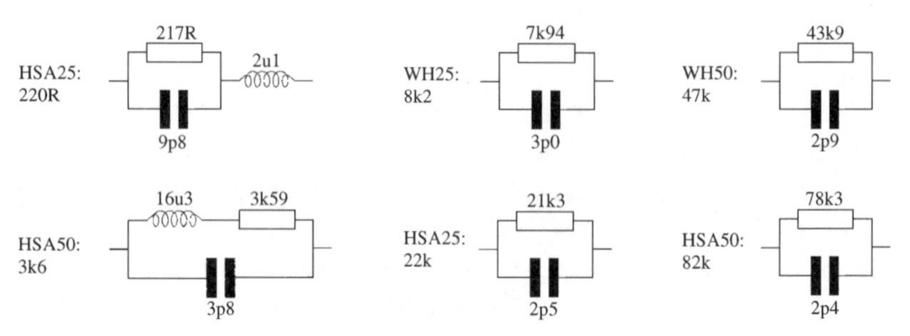

Fig. 3.1 *Equivalent circuits of some practical resistors*

It should be noted that the resistors were all aluminium clad, and the shorted turn of the casing coupled to the coil probably reduced inductance, but it is noticeable that only the lower values of resistance required the addition of a series inductor to model them accurately.

In addition to deriving models, each resistor was swept from 100 Hz to 100 kHz whilst measuring phase deviation from a perfect resistor; only the 220 Ω showed a deviation (0.2°).

All the resistor models required a small shunt capacitor, and once the resistor values were typical of anode load resistances, this shunt capacitor settled to a value of 3 pF ± 1 pF, a value commensurate with the strays that one would expect to find in a practical circuit.

Summarizing, the inductance of aluminium clad wirewound resistors is entirely negligible provided that their value is greater than 10 kΩ, but, as predicted, inductance becomes significant as the resistance falls. This outcome is very fortunate, since in a typical valve stage, the anode load resistor, $R_L >$ 10 kΩ, and dissipates significant power, requiring an aluminium clad component, whereas the cathode bias resistor, $R_k <$ 1 kΩ, but dissipates little power, and can therefore be a non-inductively wound component normally intended for component bridges.

General considerations on choosing resistors

Tolerance

- Is the *absolute* value important? If the resistor is part of a network that determines a filter or equalization network, then we will need a close tolerance (perhaps even 0.1%) component to minimize frequency response errors.

- Matching: is the component part of a pair? Anode loads in differential pairs should be matched, and so should corresponding components in filter networks for each stereo channel.

Heat

Will the resistor be heated by other components; causing its value to change? Will this matter?

Voltage rating

- Is the voltage rating of the component sufficient, even under conditions of maximum signal? (Grid leak resistors for low μ power valves, such as 845, may need to consider this factor.)
- Will the DC voltage drop across the resistor develop an unacceptable level of excess noise? If so, a wirewound or bulk foil type should be considered.

Power rating

Is the power rating of the component adequate under all conditions? Will the (changing) *audio* signal heat the resistor sufficiently that the value will change and cause an error? If a power component is required, what provision has been made to lose the heat that this component will generate? Will it heat other, sensitive, components?

Capacitors

Capacitors store charge. This charge is stored in the electric field between two plates having a potential difference between them. If there is no potential difference between the plates, then there is no stored charge, and the capacitor is said to be discharged.

Capacitors for electronic circuits are made of two fundamental components. The *dielectric*, which is the insulating material separating the plates, and a pair of conducting plates. In its simplest form, a capacitor could be a pair of parallel plates separated by a vacuum.

The parallel plate capacitor

Unsurprisingly, the capacitance of a parallel plate capacitor is proportional to the area (A) of the plates, and inversely proportional to the distance (d) between the plates. We should expect this, since if we move the plates an infinite distance apart, they can no longer 'see' one another, and a plate on its own is not much of a capacitor. If the charge is stored between the plates, then it is reasonable to suppose that the interposition of any material (k) between the plates will affect the capacitance. We can formalize these arguments by combining them into a proportionality:

$$C \propto \frac{Ak}{d}$$

To calculate real values in electronic units, we must add some fudge factors, to generate the equation below:

$$C = \frac{A \cdot \varepsilon_0 \cdot \varepsilon_r}{d}$$

This equation looks a lot more impressive, but ε_0 is simply a fudge factor to make the real world fit into our system of units, and is known as 'the permittivity of free space'; it has a measured value of $\approx 8.854 \times 10^{-12}$ C^2/Nm. Whilst ε_r (also known as 'k') is the *relative* permittivity of whatever material we insert as a dielectric, compared to the value for a vacuum, and is always >1.

A quick calculation using this equation shows that a parallel plate capacitor in a vacuum (although air is almost identical) with plates 1 m^2, separated by 10 cm, would have a capacitance of 88.5 pF. If we are going to make practical valve amplifiers, we are going to have to do something about the size of this capacitor.

Reducing the gap between the plates and adding plates

One obvious method of increasing capacitance is to reduce the gap between the plates, typical commercial capacitors use gaps of 5 μm or less.

Another method is to add more plates, in the form of a stack with alternate plates connected together. This doubles capacitance over what might at first be expected because we are now using *both* sides of each plate (except for the two outermost plates). This form of construction is used for silvered mica capacitors and for stacked film/foil capacitors. See Fig. 3.2.

Fig. 3.2 *Cross-section of general form of parallel plate capacitor*

Cutting squares of dielectric and plates and assembling them to form a capacitor is an expensive business, so many capacitors are constructed by winding two long strips of plates and dielectric together to form a cylinder, and then connecting a wire to each plate.

The dielectric

Maintaining a precise air gap of 5 μm between a set of plates would be virtually impossible, and so an insulating spacer is needed. This insulating dielectric will have $\varepsilon_r > 1$, which will further reduce the physical size of the capacitor for a given value of capacitance.

Unfortunately, we gain this increase of capacitance at the expense of other parameters, and so we should investigate these. The dielectric has three important properties; relative permittivity ε_r, *dielectric strength*, and *dielectric loss*.

Relative permittivity, ε_r, has been mentioned earlier, and is effectively the factor by which the capacitance of a capacitor is increased by the insertion of the new dielectric.

Dielectric strength refers to the maximum field strength, measured in volts per metre, that can be applied to a given insulator before it breaks down and conducts. It is this limit that sets voltage ratings for capacitors.

Dielectric loss refers to how closely the dielectric approaches a perfect insulator at voltages *below* breakdown. One way of specifying this loss is to measure the leakage current, in μA, that flows when the maximum rated voltage is applied across the capacitor – this method is typically used for aluminium electrolytic and tantalum capacitors. Film capacitors are typically rather less lossy, and so the *insulation resistance* or *leakage resistance* of the capacitor may be specified. Dielectric loss may be different for AC to DC, and so a more useful measurement is to measure *tan* δ, which is the ratio of the total resistive component of the capacitor to the reactive component at a specified frequency or frequencies. Note that tan δ makes no distinction between the parallel leakage resistance of the dielectric and any series resistance, such as lead or plate resistance.

Lead and plate resistance are collected together as one term, *effective series resistance* (*ESR*). In components such as high capacitance electrolytic capacitors for power supplies, or cathode bypasses, the ESR is highly significant, since it may be an appreciable fraction of the total impedance of the capacitor. In power supplies, significant currents may flow in the reservoir capacitors, which will cause self-heating of the internal structure. For this reason, a parameter is quoted that is very closely linked to ESR, and this is *maximum ripple current*.

The leads will have series inductance, and unless precautions are taken, the plates will also have inductance. We can now draw a simple equivalent circuit for a real capacitor. See Fig. 3.3.

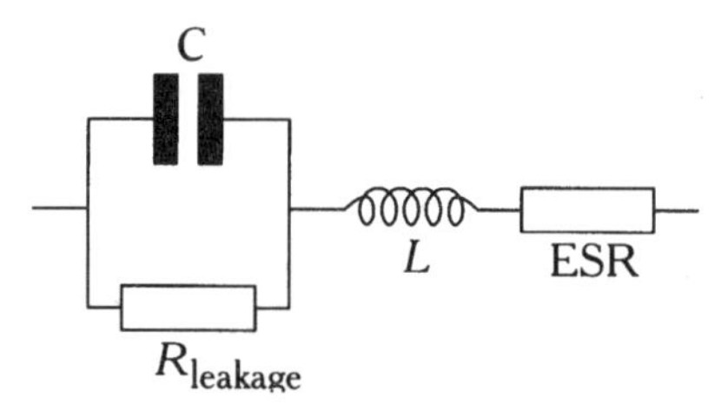

Fig. 3.3 *Basic equivalent circuit of practical capacitor*

It is immediately apparent that we are dealing with a resonant circuit, and for electrolytic capacitors, this self-resonant frequency is often specified in the manufacturers' data sheets, and we will return to this later.

Different types of capacitors

With the various ways of making the plates or the dielectric, there are many combinations of capacitor construction available. See Fig. 3.4.

This tree of capacitors shows the various possibilities available. The first branching is between polarized and non-polarized capacitors; a capacitor that is polarized will be damaged by having DC applied in the reverse direction. Non-polarized capacitors branch into their plate construction; self-supporting

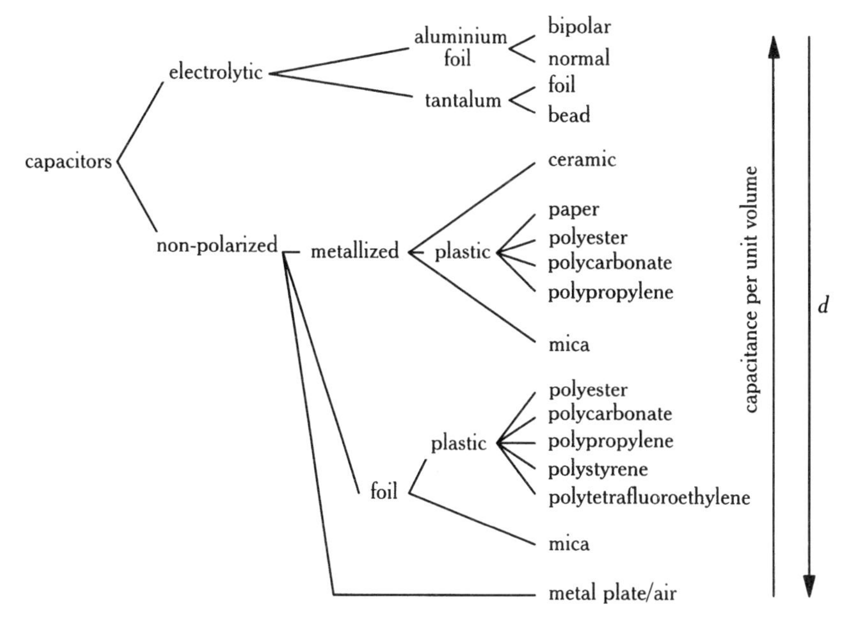

Fig. 3.4 *Comparison of different types of capacitor*

plates, foil, or a surface coating of metal sputtered directly onto the dielectric. The final branchings deal with the dielectric; it will be noticed that although some dielectrics are represented in both categories, others are not, due to their manufacturing impossibility.

Broadly speaking, the more nearly perfect capacitors are at the bottom of the tree, whilst high capacitance per unit volume capacitors are at the top of the tree. This can be further generalized by observing that high quality capacitors tend to be physically large for their value of capacitance.

Air dielectric, metal plate capacitors

Because of the difficulty of supporting plates that are very closely separated, air dielectric capacitors have low values of capacitance and are not usually larger than 500 pF. These capacitors are invariably constructed as trimmer or variable capacitors, with sets of intermeshing semicircular rigid plates, and are primarily used in radio frequency (RF) circuits, although they are occasionally useful in audio.

Plastic film, foil plate capacitors

This is the most important class of capacitors for use in valve amplifiers, as we will use these for coupling stages together and for precise filters. These capacitors are very nearly perfect, and so their imperfections are usually described by the value of tan δ, or '*d*'. There appears to be a strong correlation between the subjective sound quality of capacitors and their value of '*d*', with low '*d*' capacitors being subjectively superior.

Polytetrafluoroethylene (PTFE), sometimes known by the trade name Teflon, has the lowest value of '*d*', followed by polystyrene, polypropylene, polycarbonate and polyester. Cost rises exponentially from polyester (cheap) to PTFE (outrageously expensive).

The significance of '*d*' in engineering terms is not simply that the capacitor has a leakage resistance across it, but that the capacitor is actually a ladder network of capacitors separated by resistors that extends indefinitely. See Fig. 3.5.

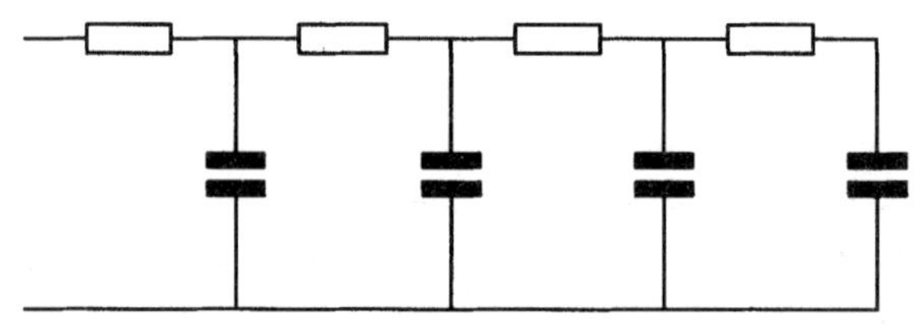

Fig. 3.5 *Equivalent circuit of practical capacitor to model dielectric absorption*

If we charge a capacitor whilst monitoring its terminal voltage with a voltmeter of infinite resistance, then discharge it by briefly short circuiting it, we would expect the capacitor voltage to remain at 0 V. However, we actually see the voltage rise from 0 V the instant that the short circuit is removed. This is because we discharged the capacitor that is 'near' to the terminals, but the other capacitors were isolated by series resistors, and were not discharged. Removing the short circuit allowed the undischarged capacitors to recharge the 'near' capacitor, and the voltage at the capacitor terminals rose. This effect is known as *dielectric absorption*, and it is more pronounced as the value of '*d*' rises.

If we apply a pulse to a capacitor, this is equivalent to instantaneously charging and discharging the capacitor, and any voltage left on the capacitor at the end of the pulse is distortion. Music is made up of a series of transients, or pulses, and it may be that dielectric absorption is one cause of capacitor 'sound'.

Some plastics are *polar*, this does not mean that the capacitor is sensitive to the polarity of applied DC, but that at a molecular level within the dielectric, there are permanently charged electric dipoles, similar to the magnetic dipoles in a magnet. Under the influence of an external electric field, these dipoles will attempt to align themselves to that electric field. If we apply an AC field, energy is absorbed as we successively align these dipoles first one way and then the other, with the result that we incur a loss that rises with frequency. As frequency rises still further, we are no longer able to align the dipoles, and the loss begins to fall.

The dipoles may be imagined as individual blocks of identical mass, pivoted at their centre and resting on a rubber surface, that are rotated by pulling on rubber bands tied to their ends. As we pull on an individual rubber band, nothing happens until we overcome stiction (static friction), but once a block moves, it flips quickly to the new orientation because dynamic friction is always lower than stiction. Initially, as we pull on all of the rubber bands, only a few blocks rotate, because the stiction between individual blocks and the rubber surface is not constant. As we try to flip the dipoles faster and faster, we use more energy in overcoming stiction, but there will come a point when the applied frequency is so high that we simply vary the tension of the rubber bands and progressively fewer blocks move because they are decoupled. As fewer blocks move, we overcome less stiction, less energy is absorbed, and so the losses begin to fall.

This mechanism is so similar to magnetic hysteresis that it is sometimes known as electrostatic hysteresis, and the above model is equally good for explaining magnetic losses.

Because of this frequency dependent loss, which begins to show at mid audio frequencies and peaks at low radio frequencies, capacitors using polar plastics as their dielectric are not ideal for audio use. By comparison, non-polar dielectrics have losses that are independent of frequency almost up to UHF. Polyester and polycarbonate are both polar.

Film/foil capacitors are constructed by laying together four alternate layers of dielectric and foil which are then wound into a cylinder. Guiding these four layers whilst winding the capacitor is not a trivial task, and is partly responsible for the higher price of these capacitors. The foils are wound so that they are slightly displaced laterally relative to one another, so that one end of the cylinder has a spiral of foil that is one plate, whereas the spiral at the other end is the other plate. Each end is then sprayed with zinc to connect all points of the spiral together (which reduces inductance) and leads are welded or soldered to this layer.

Polystyrene capacitors often have one end of the capacitor delineated by a red or yellow band. This does not mean that they are polarity sensitive, but that the marked end is connected to the outer foil. This is significant because one end of the capacitor may be connected to a less sensitive part of circuitry than the other. For instance, if a small polystyrene capacitor was used as part of an active crossover network, and connected as a series coupling capacitor (high-pass filter), then the banded end should be connected to the source to reduce induced hum. Alternatively, if one end were connected to ground, then the banded end should go to ground to reduce stray capacitance to active signals (strays to ground rarely cause problems, but Miller effect can cause other strays to be significant).

Metallized plastic film capacitors

Because of the difficulty of winding the layers of dielectric *and* foil, most film capacitors are made by sputtering one side of the film with a layer of aluminium up to 12 μm thick to form the plate. Not only does this make the capacitor easier to construct, but a far higher capacitance per unit volume is obtained because the plate is so much thinner.

Because the plate is thinner, ESR rises. Since ESR for a plastic capacitor is only significant at very high frequencies, when it becomes comparable with the reactance of the capacitor, this is not too much of a problem. However, *foil* capacitors are often described as being more suitable for high frequency pulse applications for this reason.

If there is granularity due to impurities in the film of a metal film resistor, this generates excess noise, and film resistors are always noisier than wirewound resistors. Since the plates in a metallized film capacitor are also produced by sputtering, it is not unreasonable to suppose that they will suffer from the same quality control problems – with the difference that capacitors are not routinely tested for modulation noise. Subjectively, foil capacitors seem to sound better than metallized film, and it is possible that this is the reason.

Metallized paper capacitors

Metallized paper was the traditional dielectric for capacitors in classic valve amplifiers, and depending on the impregnant in the paper, the performance ranged from poor to quite good. Unfortunately, if the seals of the capacitor are less than perfect, humidity will get in and the capacitor will become electrically leaky. The author once bought a 30-year-old Leak Stereo 20 which had paper coupling capacitors, every one of which had gone leaky.

Oil or epoxy resin impregnating the paper will improve matters considerably, such that the resulting capacitor is almost as good as a polypropylene capacitor. Because paper capacitors are inherently self-healing, they are widely used in the power generating industry. In the event of an over-voltage spike, the paper breaks down at its weakest point and the metallization at that point is vaporized, thus preventing a short circuit and catastrophic failure.

Silvered mica capacitors

This was the traditional small value capacitor used for RF circuitry or for audio filters where excellent stability of value was important. Mica is a crystalline rock that can be easily cleaved into fine sheets, which are then coated with silver, and a stacked construction gives low inductance.

Since mica is a natural material, it is subject to all the accompanying vagaries of inconsistency. Top quality silvered mica capacitors have slightly lower losses than polystyrene, and theoretically should sound better, but are extremely expensive.

Because of cost and variability, silvered mica capacitors have now been almost totally superseded by polystyrene, which is beginning to be superseded by polypropylene.

Ceramic capacitors

These have no place in analogue audio circuitry!

Up until now, the dielectrics that we have seen have had $\varepsilon_r < 10$, but 'high-k' ceramic capacitors can achieve ε_r (or k) $\approx 200\,000$! Commonly, ceramic capacitors are made of barium or strontium titanate, which are both piezo-electric materials. This means that they will generate a voltage when mechanically stressed (these materials are used in the ceramic cartridges found in inferior 'music centres').

Ceramic capacitors excel as high frequency bypasses in digital circuitry where their poor stability of value and low 'd' is not a problem.

Electrolytic capacitors

These capacitors are polarized. Reverse biased, they form quite a good quality short circuit, and will damage the driving circuitry, whilst the capacitor expires to the accompaniment of heat, smoke and noxious fumes. Aluminium electrolytics may even explode and shower surrounding circuitry with wet electrolyte and aluminium foil, causing further damage.

Some people have religious convictions against using electrolytic capacitors, but with all their faults, they are still useful components, and our choice of design is severely restricted if we refuse to use them. Many of the faults ascribed to electrolytics relate to inappropriate usage.

Electrolytic capacitors take high capacitance per unit volume to the limit, and they do this by attacking all parts of the parallel plate capacitor equation. The gap between the plates is minimized, surface area is maximized, and $\varepsilon_r \approx 8$ for aluminium oxide, as opposed to ≈ 3 for the plastic films. The principle of operation is broadly similar for all types, so only the aluminium type will be described in detail.

Aluminium electrolytic capacitors

The aluminium foil of one plate is anodized to form an insulating layer of aluminium oxide on the surface (≈ 1.5 nm/V of applied polarizing voltage), and it is this micro-thin layer that is the dielectric of the capacitor. Since anodizing is an electrochemical process, and the aluminium oxide is an insulator, it follows that there must be a maximum thickness of aluminium oxide that can be produced before the insulation of the layer prevents deeper anodizing. This means that there is an absolute maximum voltage limit for electrolytic capacitors. Reliable capacitors can be made up to 450 VDC working; any electrolytic capacitor claiming a higher limit than this should be looked upon with grave suspicion, particularly if it is old.

Although by anodizing the aluminium foil, we have both a plate and the dielectric, we still need the other plate. We could use another piece of aluminium foil pressed tightly to the first, but any gap between the two foils would negate the advantage of the micro-thin dielectric. The second plate is therefore made of thin, soggy paper or simply a gel, which because it is wet, makes perfect contact with the anodized surface of the first plate, and this is the electrolyte from which the component derives its name. The electrolyte is not a particularly good conductor of electricity, and so a second aluminium foil is laid on top of the paper to allow a low resistance plate to be made.

We now have two aluminium foils separated by electrolyte that can be rolled into a cylinder to make our capacitor. If, before anodizing the aluminium foil, we had etched its surface, this would roughen the surface, and would greatly increase the surface area on a microscopic level. Since the electrolyte plate is in perfect contact with this surface, we have now dramatically increased the area of the plates of the capacitor, and the capacitance rises accordingly.

Unfortunately, the electrolytic capacitor does have its disadvantages. Electrolyte resistance is significant, so deep etching of the foil increases the resistance from the bulk of the foil to the extremities that form the plate, and we can expect the highest capacitance per unit volume components to have a higher ESR.

ESR rises because the narrow paths to the anode have limited current carrying ability before they heat significantly and cause the electrolyte to evaporate. Compact capacitors therefore not only have high ESR, but also low ripple current rating.

However, newer types, such as Sanyo's 'OS-CON' range of capacitors, use an organic semiconductor electrolyte to significantly reduce ESR. Reduced electrolyte ESR allows deep etching, resulting in higher capacitance per unit volume, and therefore reduced inductance. OS-CONs therefore have significantly better high frequency characteristics, and would be ideal as cathode bypass capacitors if it were not for their extremely high price.

Electrolytic capacitors have poor tolerance. Historically tolerance was quoted as +100% −50%. Although modern types are much better, we should *never* use an electrolytic in a position where its value could not be safely doubled or halved without upsetting operation of the circuit.

All of the previous wound capacitors could have leads connected to their plates by spraying molten zinc onto their ends to connect all parts of each plate together, which reduced inductance. This is not possible with an electrolytic capacitor, and connections to the plates are made with individual tabs of foil at various points along the spiral.

Because it is not possible to connect to all parts of the plates, this adds series inductance. The inductance is low, therefore its inductive reactance is also low ($X_L = 2\pi fL$), but for a large capacitor, X_c is also low, and the inductance of an electrolytic capacitor is significant. Manufacturers' data sheets generally describe the problem by specifying the self-resonant frequency for each value of capacitor. In general, the higher capacitance types have a lower resonant frequency, which may be as low as tens of kHz.

Electrolytic capacitors are lossy. When they are first manufactured, a *polarizing* voltage is applied, and this causes a current to flow through the capacitor, forming the aluminium oxide layer on the plate. Once this oxide layer has been *formed*, very little current flows. However, this micro-thin layer gradually deteriorates with time and will need to be *re-formed*. Provided that the capacitor always has DC applied across it, the capacitor balances itself by always passing the minimum necessary current to maintain the oxide layer for the applied DC voltage.

If the equipment is switched off, then when power is re-applied, a higher than normal leakage current will flow until the oxide layer has been re-formed. The longer the elapsed time without bias volts applied, the greater duration and amplitude this initial leakage current will be, and there is a danger that this current will cause serious heating of the electrolyte. As the electrolyte is heated, it will evaporate more readily, and the resulting gas may build up sufficient pressure to cause the can to explode.

Because of this, it is wise to use a Variac to gently apply power to equipment containing electrolytic capacitors that has lain idle for some time.

Modern capacitors have safety pressure seals to vent the gas, this may be a rubber bung in the base of the component, or the aluminium can may be deliberately weakened at the top with a series of indentations which allows controlled rupturing for the gas to escape. Either of these occurrences signifies the demise of the component, but they do prevent damage to other components, with the bonus of a simple visual inspection of the health of components.

More gentle heating will tend to evaporate the electrolyte through the seals of the capacitor (no seals can be perfect), and as the quantity of electrolyte falls, we make less and less contact with the nooks and crannies of the etched plate, and capacitance falls.

Because of the problem of evaporation of the electrolyte, electrolytic capacitors are very sensitive to temperature, and a doubling of capacitor life results from a 10°C temperature reduction.

Applied voltage also has an effect on electrolytic capacitor life. If there is no bias voltage, then the oxide layer will not be re-formed and will gradually deteriorate, causing the capacitor to become leaky. This is a well-known fault in sound mixers using symmetrical + and – supplies with operational amplifiers coupled by electrolytic capacitors, which then have little or no polarizing voltage.

There is a class of aluminium electrolytics available for use at AC known as *bipolar*. These are often found in loudspeaker crossovers because of their cheapness compared to plastic film capacitors of comparable capacitance. Their construction is effectively two electrolytic capacitors back to back. See Fig. 3.6.

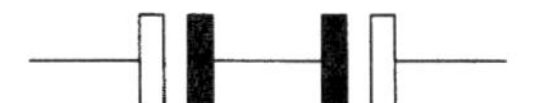

Fig. 3.6 *The bipolar electrolytic capacitor*

There is no constant polarizing voltage, and each individual capacitor has to be twice the value of the final capacitor. Defects are thereby multiplied by a factor of four over the normal unipolar electrolytic capacitor, and performance is poor.

Provided that there is a polarizing voltage, operating electrolytic capacitors below their maximum rated voltage increases their life significantly:

$$\text{Life}_{(\%\text{ of rated})} = \left(\frac{V_{\text{max.}}}{V_{\text{applied}}}\right)^5 \times 100\%$$

Using this relationship, we see that operating an electrolytic capacitor at 87% of its rated voltage will double its life. It is wise not to read too much into this formula, since we could easily use it to predict a lifetime measured in centuries by lowering the operating voltage sufficiently. A good engineering rule of thumb is that, if possible, electrolytic capacitors should be operated at two-thirds of their maximum voltage rating; this would give a theoretical eight-fold increase in life expectancy, which is probably at the limit for which this formula is valid.

Many valve amplifiers have electrolytic capacitors with more than one component concentrically wound in a single can. The outer capacitor is marked with a red spot, and in an amplifier using cascaded RC smoothing, this capacitor should be connected to the most positive potential. The logic for this is that the highest potential will have the greatest ripple, and there is no field within a conductor, so this ripple is not coupled to the other circuitry. Wiring the capacitors in the wrong order will result in excessive hum.

If the aluminium foil is replaced with tantalum foil, two advantages result from the increased chemical inertness of the resulting oxide layer. First, ESR can be reduced, by using a lower resistance electrolyte that would have corroded aluminium foil. Second, because the oxide layer is more inert, leakage current is reduced. However, tantalum foil capacitors are expensive and not readily available, whereas aluminium electrolytic capacitors are improving all the time.

Tantalum bead capacitors are only available in small values and low voltages but their reduced inductance compared to aluminium electrolytic capacitors makes them ideal for bypassing voltage regulators or logic chips. Unfortunately, they are only available in quite small values, and are often not large enough to be used as cathode bypasses. They are expensive.

General considerations in choosing capacitors

Voltage rating

- Will the voltage across the capacitor change polarity, or is it simply a varying DC? If the voltage across the capacitor is AC, then electrolytic capacitors are eliminated.
- Will the capacitor withstand the applied DC voltage plus the maximum expected signal voltage (V_{pk}, not V_{RMS})?
- Could the capacitor withstand the maximum possible HT voltage? If not, what arrangements have been made to ensure that its rated voltage is never exceeded?

Tolerance

- Is the *absolute* value important? If it is part of a filter or equalization network, then we will need a close tolerance component, such as polystyrene or silvered mica.
- Matching. Is the capacitor part of a pair, such as coupling capacitors in a push–pull amplifier, or the corresponding component in the other stereo channel; if it is, then they should be matched if possible.

Heat

Will the capacitor get warm? Will the consequent change in value matter? In general, capacitors should not be operated at more than 50°C (because insulation resistance falls with increasing temperature), but even this temperature could reduce the life of an electrolytic unacceptably.

Capacitance

Each type of capacitor is only available in a limited range of values, and if we need 330 μF, then only an electrolytic can provide this value.

Leakage, and '*d*'

- Is leakage important? A cathode bypass or HT smoothing capacitor may be allowed to pass a small leakage current. A grid coupling capacitor may not be allowed to be leaky under any circumstances.
- Is this component important for final sound quality? Capacitors in the obvious signal path are important, but the signal *current* has to return through the HT supply, so HT smoothing and bypass capacitors are equally important. Smoothing capacitors for bias circuitry may be less important if there is no audio signal on them.

Microphony

All capacitors are microphonic to a greater or lesser extent. The reason for this is very simple. Suppose that we have stored a fixed charge on a capacitor:

$$Q = CV$$

The capacitance of the capacitor is:

$$C = \frac{kA}{d}$$

Combining these equations, and solving for V:

$$V = \frac{Qd}{kA}$$

Since Q, k and A are constants, if we vary the spacing between the plates, the voltage across the capacitor *must* change. It is exactly this principle that is used to make the capacitor microphones used in studios, and the ubiquitous electret microphone found in cassette recorders.

The principle is reversible, and varying the applied voltage across a capacitor will alter the attractive forces between the plates, and if they are free to move, this will cause vibration. This is the basis of the electrostatic loudspeaker.

It might be thought that the plates of a plastic film capacitor are sufficiently tightly wound that no movement is possible, but the author once built a stabilized HT power supply where the output bypass capacitor whistled loudly at about 2 kHz. The circuit was thereby diagnosed as unstable even before the oscilloscope had warmed up!

The problem of capacitor microphony can be tackled in three ways, listed in descending order of desirability:

- Avoid using capacitors. To a limited extent, this is feasible.
- Isolate capacitors from vibration. Capacitors carrying low level signals will be proportionately more sensitive to microphony than those carrying high level signals. Pre-amplifiers are therefore most susceptible, and it is well worth isolating them from vibration. This is easily done at the design stage, but is much harder once built.
- Capacitors are physical objects, and so we should expect them to have mechanical or acoustical resonances. If we excite these resonances, we should expect them to be audible in the same way that striking a tuning fork produces an audible note. If we mechanically damp the capacitor by gluing it to another surface, then these resonances will be reduced. Provided that the capacitor can survive the heat, the soft glue used by hot-melt glue guns is ideal.

There is no reason why we should not use all three methods in combination if it seems that microphony is likely to be a problem. A good test for microphony is to tap each component with a plastic insulating pen (to avoid shock risk) with the equipment turned on whilst listening to the loudspeaker. The results may surprise you!

Bypassing

All capacitors become inductive as frequency rises, but if we bypass them with parallel capacitors of a smaller value, the composite component can more nearly approximate a perfect capacitor. A practical rule of thumb is that the ratio of main capacitor to bypass capacitor should be ≈100:1.

It is possible that more than one bypass may be required. For instance, a 10 000 μF electrolytic would be bypassed by a 100 μF electrolytic, which would, in turn, be bypassed by a 1 μF plastic film capacitor. It might even be beneficial to bypass the 1 μF with a 10 nF, but there is little point in carrying the principle any further because the inductance of the leads used to connect the capacitor together will become very significant.

Wires have inductance, and it should be realized that there is no point in having a beautifully bypassed capacitor if it is physically distanced from its point in the circuitry. Thus, power supply capacitors should be bypassed *at the load*, not at the main electrolytic. We often cannot connect all these capacitors directly between the output transformer HT tap and the cathode returns of the output valve(s), but we can, and should, connect the bypasses between these points. See Fig. 3.7.

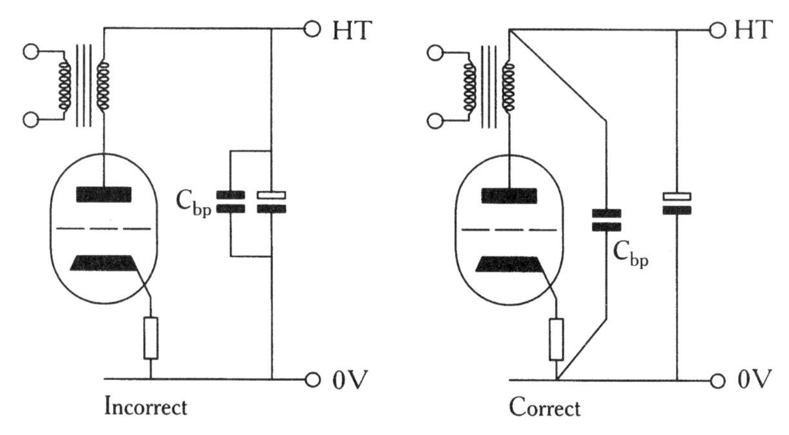

Fig. 3.7 *Connection of bypass capacitors*

A very useful assumption to make when considering the layout of *any* circuitry, is to treat each wire as an aerial with inductance, and to assume that the circuit is immersed in a strong radio frequency (RF) electromagnetic field which will induce currents into each and every wire. (This assumption is not very far from the truth.) It is for this reason, that a bypass capacitor that aims to make the composite capacitor a perfect short circuit at RF should be connected to its point in the circuit by *very* short wires. Thus, the smallest bypass should be nearest to the load, whilst the larger capacitors may be further away. Each capacitor should be connected using individual wires to the load, so that they are 'starred' to the load point, this will minimize ground noise. If this results in seemingly untidy wiring, so be it.

Magnetic components

Magnetic components include transformers and inductors. Transformers may be signal transformers, such as output transformers and moving-coil step-up transformers, or they may be mains transformers. Inductors may be the small signal inductors used in filters, or they may be the power chokes in an HT supply.

Magnetic components are the least perfect of 'passive' components (resistors, capacitors and inductors/transformers), and for this reason, many designers shun them. This is unfortunate, because it seriously restricts design choice.

Inductors

Inductors store energy in the form of a magnetic field. Any current passing through a wire will generate a magnetic field, and it must therefore possess inductance. We can deliberately increase this inductance by winding the wire into a coil, whilst placing the coil around an iron core will increase inductance still further.

It would be nice to be able to use a convenient formula, such as we had for the parallel plate capacitor, to allow us to calculate the inductance of a coil:

$$L \propto \frac{\mu_0 \mu_r A N^2}{l}$$

Where:
L = inductance
μ_0 = permeability of free space
μ_r = relative permeability of the core magnetic material
A = magnetic path cross-sectional area
l = magnetic path length
N = number of turns

Relative permeability is the magnetic analogue of relative permittivity that we met earlier, and has a value of 1 for air and ≈5500 for iron. The magnetic path length is the length through the core back to the starting point, and the cross-sectional area of the magnetic path is simply the cross-sectional area of the core, so it ought to be easy to derive a useful equation for calculating inductance.

Unfortunately, μ_r varies with flux density; the path length is easily affected by air gaps; and some of the flux may escape the core, thus reducing the cross-sectional area. We will look at each of these problems in more detail in a moment, but suffice to say that we cannot often accurately calculate the inductance of a coil. We are forced to make an informed guess, add a few turns, measure the inductance under operating conditions, and then remove turns until the desired inductance is achieved.

A curve that appears in virtually every discussion of magnetic materials is the *BH* curve. This is a plot of the relationship between applied magnetizing force and resulting magnetic flux. For our purposes, we need only note that μ_r is proportional to the gradient of the curve, and that as the gradient changes with level, so must μ_r. See Fig. 3.8.

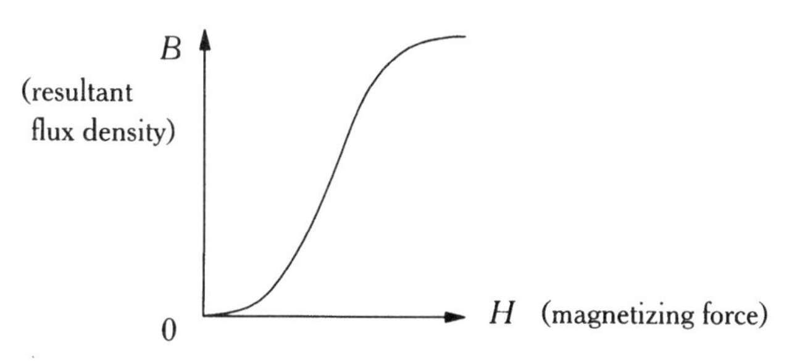

Fig. 3.8 *BH curve: non-constant slope implies non-constant* μ

Air-cored inductors

We can completely avoid the problem of non-constant μ_r with level by not using a magnetic material in the core. *Air-cored* inductors have constant inductance with applied signal level, and do not therefore cause distortion, making them popular in high quality loudspeaker crossover networks. It is now difficult to determine the magnetic path area since this theoretically extends to infinity, whilst the path length is similarly awkward to define. Nevertheless, formulae have been produced for various core geometries, and a particularly useful set of formulae for optimum (lowest) resistance air-cored copper wire coils based on a paper by A. N. Thiele[2] is given below:

$$R = 8.01 \times 10^{-3} \sqrt[5]{\frac{L^3}{d^8}}$$

$$N = 10.2 \sqrt[5]{\frac{L^2}{d^2}}$$

$$c = \frac{d\sqrt{N}}{0.841}$$

$$l = 0.188 \sqrt{Lc}$$

where: R = resistance in Ω
L = inductance in μH
d = diameter of wire in mm
N = number of turns
c = core radius (see Fig. 3.9)
l = length of wire in m

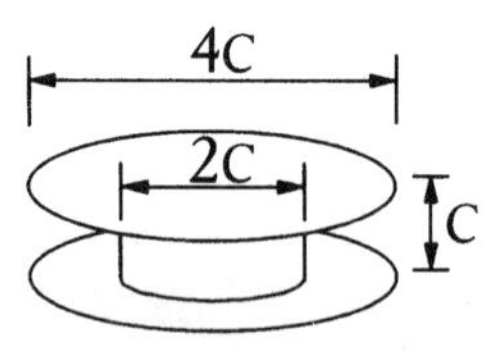

Fig. 3.9 *Relative bobbin dimensions for air-cored coils using above formulae*

The formulae are given in this form because wire is available only in a range of standard diameters, and the resistance of the coil is not usually critical. If the resistance is different from that desired, then a different wire diameter can be tried.

The simplest way of performing these calculations is on a computer, so a QBASIC program is given below:

```
CLS
L = 1
PRINT "This program designs air-cored copper wire coils
according to the Thiele"
PRINT "formulae. L is in microhenries, d (wire diameter) is
in mm."
PRINT
PRINT "To quit, input L = 0."
PRINT
DO WHILE L > 0
INPUT "L"; L
IF L = 0 THEN END
INPUT "d"; d
R = (((8.01 * 10 ^ (–3) * (L ^ 3 / d ^ 8) ^ (1 / 5)) * 100) \ 1) / 100
N = (((10.2 * (L ^ 2 / d ^ 2) ^ (1 / 5)) * 100) \ 1) / 100
c = (((d * N ^ (1 / 2) / .841) * 100) \ 1) / 100
Q = (((.1875 * (L * c) ^ (1 / 2)) * 100) \ 1) / 100
PRINT
PRINT "You need"; N; "turns on a core of"; 2 * c; "mm in
diameter,"; c; "mm thick."
PRINT "It will use"; Q; "metres of wire, and will have a resistance
of"; R; "Ohms."
PRINT
LOOP
```

Experimenting with this program will soon reveal that air-cored inductors have significant resistance, or that they are very large. This problem of resistance is common to all inductors, and is one of their imperfections. Air-cored inductors are not only useful for loudspeaker crossovers; they can also be used in the output filter of an audio Digital to Analogue Converter (DAC), where the resistance is far less of a problem.

It should be noted that because of practical considerations (winding efficiency, wire diameter, etc.) the above formulae do *not* give exact answers, and it is therefore wise to design 5% oversize and then remove turns whilst measuring the inductance with a component bridge.

Many component bridges use a 1 kHz internal oscillator. When measuring air-cored coils, the relatively high resistance may swamp the inductive component and the bridge will give misleading results. If it is possible to feed the bridge from an external source of AC, it should be fed with the highest frequency that the bridge manufacturer allows (typically 20 kHz), and this will allow sensible measurements to be taken.

Gapped cores

One way to reduce resistance without suffering gross distortion is to use a coil with a magnetic core that has an air gap. This *gapped* core will significantly increase inductance over an air-cored coil, but because the air gap is effectively a large magnetic resistance in the path of the magnetic flux, it swamps the variation in permeability of the much smaller magnetic resistance of the core, and inductance is more nearly constant. As the gap becomes larger, inductance falls, and if the gap is infinitely large, then we are back to an air-cored coil. This technique was used for many years in crossover inductors for BBC designed loudspeakers.

We may unintentionally make an inductor with a gapped core. Many of the ferrite cores used for small inductors are supplied as two mating halves which fit around the core once wound. Dust on the mating surfaces will cause an air gap, and if the cores are gently squeezed together whilst inductance is being measured, a significant increase in inductance may be seen.

If an inductor has to pass DC, it is essential that the DC current should not saturate the core, since this would drastically reduce the AC inductance. Iron-cored inductors passing DC are therefore invariably gapped in order to maintain their inductance up to their rated maximum current. They will therefore have significant leakage flux, which may interfere with nearby circuitry.

Self-capacitance

If a coil is made up of many turns of wire, and there is a potential between different turns, or layers of turns, then we must expect the inductor to have capacitance in parallel with its inductance. See Fig. 3.10.

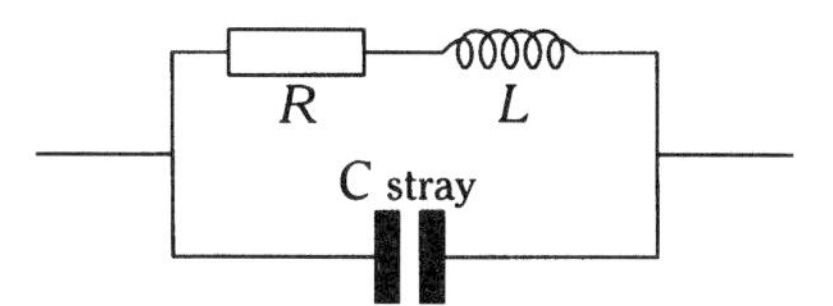

Fig. 3.10 *Equivalent circuit of practical inductor*

We now have our familiar parallel resonant circuit, which means that as we rise beyond the resonant frequency, the coil no longer looks like an inductor and behaves as a capacitor. The easiest way of finding this capacitance is to set up the test circuit in Fig. 3.11.

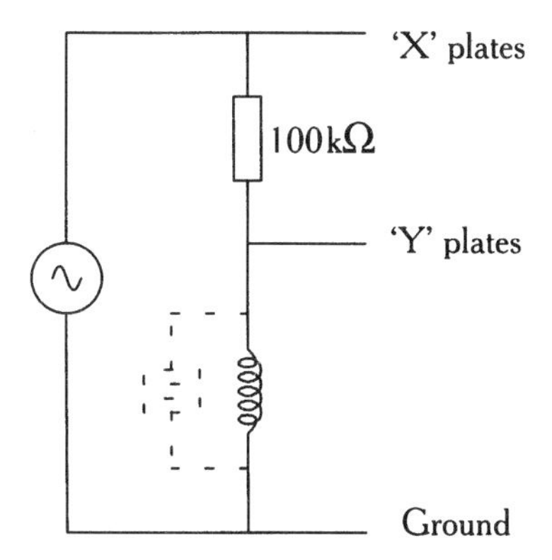

Fig. 3.11 *Using Lissajous figures to find self-resonant frequency of inductor*

The oscilloscope is used in XY mode and as the frequency of the oscillator is varied, the resulting Lissajous figure will change from an ellipse to a straight line; this is the resonant frequency of the inductor. If necessary, we can now calculate the shunt capacitance using:

$$C = \frac{1}{4\pi^2 f^2 L}$$

Power supply chokes for valve amplifiers are typically 10–15 H, 100–250 mA, and these chokes have a resonant frequency ≈3–12 kHz.

This means that the choke does not provide a barrier to the RF noise generated by rectification, or to RF incoming from the mains.

We will consider power supply chokes further in Chapter 4.

Transformers

In a perfect transformer, the primary winding produces a magnetic flux which is coupled to the secondary winding with no loss whatsoever. Practical transformers are somewhat different.

In a transformer, the losses are often divided into two distinct groups, *iron losses*, so called because they are due to imperfections of the core, and *copper losses*, due to imperfections of the windings.

Iron losses

As the core is successively magnetized and demagnetized through opposite polarities, work has to be done to change the alignment of the magnetic dipoles. This loss is known as *hysteresis* loss, and may be calculated by investigating the hysteresis curves for the particular core material used. Because it is the loss caused by changing the core magnetization through one complete cycle of the applied AC waveform, there will be more loss in a given time if more cycles of magnetization are traversed.

Hysteresis loss is therefore directly proportional to frequency, and can only be reduced by using a lower hysteresis core material.

Magnetic cores are metal, and therefore conduct electricity. As far as the primary winding is concerned, there is no distinction between an intentional secondary winding connected to a load and a conductive path parallel to the primary winding through the core. Conductive paths through the core cause *eddy* currents to flow, which, because they are short circuits, cause losses. To reduce these losses, the core can be constructed from a stack of *laminations* which have had their surfaces chemically treated to make them insulating. The ultimate approach to this problem is to make the core of iron dust particles whose surface has been treated, and then bond these with a ceramic to form a solid core known as a *ferrite dust* core.

Eddy current loss is proportional to f^2 because not only is the loss proportional to the number of traverses of the magnetization loop in a given time, but higher frequencies have smaller wavelengths and allow more loops of current to form within the core. Although thin steel laminations are satisfactory for audio frequencies, ferrites are necessary for RF frequencies, and at VHF, almost all core materials are excessively lossy, and air-cored transformers must be used.

Hysteresis and eddy current loss are often combined and known as magnetizing current in power transformers, and are responsible for the heating of the core despite the fact that no load may be connected at the time.

Not all of the flux from the primary will flow through the secondary winding, and this loss, combined with hysteresis and eddy current loss, is known as *leakage inductance* in audio transformers. Theoretically, leakage inductance (referred to the primary) is found by measuring the primary inductance with the secondary short circuited. In practice, leakage inductance is very difficult to measure accurately because a measurement at one frequency will be upset by the stray capacitances. Nevertheless, leakage inductance is an important theoretical concept, since it determines the high frequency operating limit of the transformer.

Leakage inductance is dependent on the size (q), the turns ratio N^2, and the geometry of the transformer (k), but is independent of μ_r:

$$L_{\text{leakage}} \propto qN^2k$$

For a given frequency, a higher power rating transformer will be larger than a lower power rating transformer, and will consequently have higher leakage inductance.

Since leakage inductance is proportional to N^2, we should always try to keep the turns ratio as low as possible. It is for this reason that paralleling output valves in a valve amplifier is such a good idea, since it reduces the turns ratio required.

Geometry can be improved in two fundamental ways, we can either improve the shape of the core, or we can improve our winding technique.

Standard transformers are made with E/I cores, where each lamination of the core is composed of an E shape and an I shape. On alternate laminations, the orientation of the shapes is often reversed to reduce the air gap at the joint. See Fig. 3.12.

Traditionally, superior cores were made as C cores. These were made by winding the core out of a continuous strip, which was then cut in half, and the resulting faces ground smooth. The coils were then wound, and

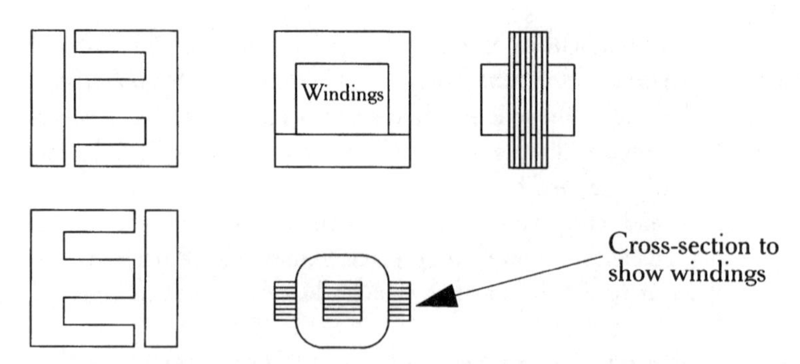

Fig. 3.12 *E/I core laminations arrangement to reduce leakage flux*

the cores were inserted so that the ground faces were perfectly aligned with minimal air gap, and steel straps were used to hold the assembly firmly together. See Fig. 3.13.

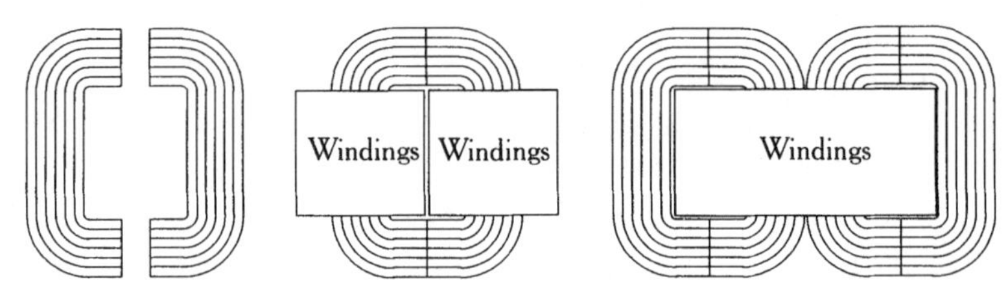

Fig. 3.13 C *core arrangement*

The C core was an expensive process, and inaccurate assembly could create an air gap, thus creating the very imperfection that the design intended to avoid. The more modern approach is to wind the core as a toroid, but not cut the core, and, using special coil winding machines, to wind the coils directly onto the core, resulting in a very low leakage core. See Fig. 3.14.

Incidentally, although toroids are thought of as being modern, the first transformer ever made was a toroid! (Michael Faraday, August 1831).

Both the C core and the toroid have the further advantage that the magnetic flux always flows in the same direction relative to the grain direction of the crystal structure of the core, whereas in the E/I core it has to flow across

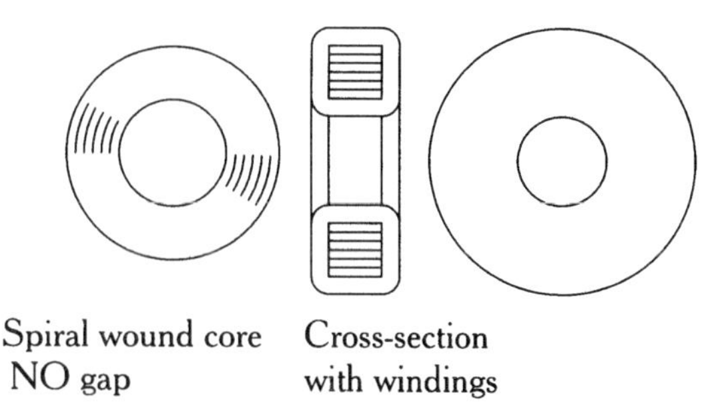

Fig. 3.14 *Toroidal core arrangement*

the grain in some parts of the core. This is significant because Grain Oriented Silicon Steel (GOSS) can tolerate a higher flux density before saturation in the direction of the grain, than across the grain. E/I cores can therefore only operate at flux densities below saturation across the grain, whereas C cores and toroids can operate at significantly higher flux densities, and core size can thereby be reduced.

The geometry of the coil can be improved by winding the primary and secondary out of many interleaving layers or sections, rather than winding one half of the bobbin with the primary and the other half with secondary. Increasing the number of sections improves the coupling between primary and secondary, thus reducing L_{leakage}, but probably increasing stray capacitance. See Fig. 3.15.

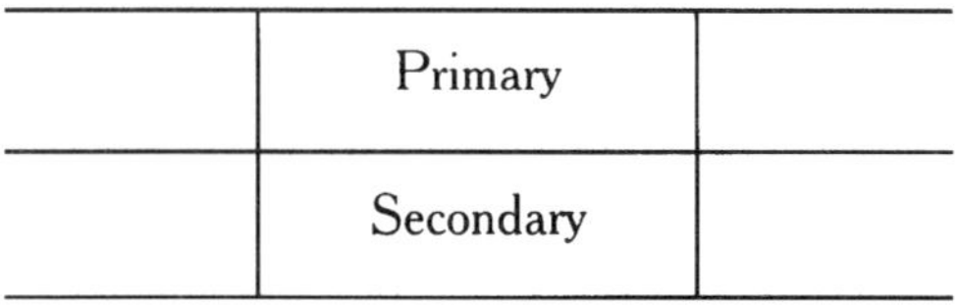

Primary

Secondary

Fig. 3.15 *Winding geometry giving good primary/secondary isolation but poor leakage inductance*

Although sectioning the windings is relatively easy on an E/I or C core, it is very difficult on a toroid; moreover, winding geometry on a toroid is quite poor, and so it is easy to lose the benefits of the improved core by having a poor coil. Toroidal mains transformers are notorious for their leakage flux at the point where the windings exit for this very reason.

An alternative technique for improving winding geometry is to use *bifilar* winding, where two wires are simultaneously wound side by side. If one of these wires is part of the primary, and the other is part of the secondary, this will promote excellent coupling between the windings, and leakage inductance will be significantly reduced. The technique is cheaper than sectioning, and providing the coil winding machine can cope, there is no reason to stop with two wires – three or four could be used.

Unfortunately, there are two snags to multifilar winding. First, the thin polyurethane insulation on the copper wire is easily damaged during winding, and may break down if we have >100 V between the windings, making it difficult to make a transformer capable of isolating the HT supply; neverthless, the seminal 50 W McIntosh amplifier used a multifilar output transformer and a 440 V HT supply. Second, the greatly increased capacitance between primary and secondary may resonate with the reduced leakage inductance to produce a lower resonant frequency than a sectioned transformer.

Multifilar winding is best used in small signal transformers with a very low turns ratio (ideally 1:1), such as balanced line output transformers used in studios.

DC magnetization

If a net DC current is allowed to flow in a transformer, it will shift the AC operating point on the *BH* curve and will cause significant distortion due to saturation on one half cycle. For this reason, output valve anode currents in push–pull amplifiers should be carefully balanced, and half-wave rectification should never be used on mains transformers.

Copper losses

Copper wire has resistance, and in a well-designed transformer, the losses due to resistance will be equal in primary and secondary, and will therefore be related by:

$$R_s = \frac{R_p}{N^2}$$

Where N is the primary to secondary turns ratio.

The total copper losses may in turn be traded against iron losses in a given transformer design, so that two transformers may have different proportions of iron to copper to achieve equal power ratings.

Electrostatic screens

Capacitance between primary and secondary sections is significant in audio transformers because it is multiplied by the turns ratio of the sections concerned in a manner similar to Miller effect in the triode valve. The solution to this problem is the same: an earthed electrostatic screen, usually made of foil, is interposed between the affected windings. We now have capacitance to earth, but the effect of this capacitance is minimal. It is most important that the two ends of the foil do not contact electrically, as this would constitute a shorted turn.

An electrostatic screen between primary and secondary is often fitted to mains transformers for rather different reasons. If the insulation were to break down between primary and secondary, mains voltage would be connected to the secondary, which would be a safety hazard. By interposing an electrostatic screen, the fault current flows directly to earth and blows the mains fuse, thus making the equipment safe.

The electrostatic screen prevents RF interference on the mains from being capacitively coupled to the following circuitry. In audio, the significance of RF interference cannot be over-emphasized, and this is sufficient incentive for using an electrostatic screen.

Magnetostriction

Valve amplifiers with output transformers tend to 'sing' audibly when operated at high powers. Occasionally, this is due to a loose lamination, but it is more likely to be due to *magnetostriction*, which is an effect whereby a magnetic material changes its length according to the strength of the magnetic field passing through it. Output transformers support quite strong magnetic fields, and so the effect can become noticeable. Since the magnetic field is varying, it causes vibration, but because magnetostriction is not polarity sensitive, in a push–pull amplifier the sound that will be heard is pure second harmonic distortion.

Magnetostriction is inversely proportional to μ_r, so a higher quality transformer is less likely to suffer from this (admittedly minor) problem.

Transformers and DC

Allowing an out-of-balance DC current to flow through a transformer will result in odd harmonic distortion unless the transformer has been specifically designed to cope with this current.

A traditional way of checking output valve DC balance in push–pull amplifiers is to measure the voltage between the anodes of the output valves, and adjust for zero volts. Zero voltage between anodes means equal voltage drops, and this implies equal currents with no out-of-balance current, *if the winding resistances are equal.* It is therefore essential to check these resistances before using this method.

Output transformers, feedback and loudspeakers

It is far more convenient to derive feedback from a dedicated feedback winding, or from the end of a tapped winding, because it means that the user can change the matching of the amplifier to the loudspeaker without having to adjust feedback. The Leak amplifiers were designed using this scheme, allowing a simple link to determine matching, but it means that the output transformer is not used optimally.

As an example, if the 4 Ω setting is chosen, then only half of the secondary winding is used, resulting in poorer leakage inductance. Worse, the feedback (which would ideally be applied at the output terminals) has to be coupled via the tapped secondary before being applied, and the coupling from one part of the secondary to another cannot be perfect. The optimum way to apply feedback is to derive it from the amplifier output terminals (or even the loudspeaker terminals). Ideally, transformer performance should be optimized by using as many of the secondary sections as possible.

Older transformers tend to have a pair of secondary sections which are connected in series for 15 Ω loudspeakers, and in parallel for 4 Ω. The sections are not necessarily from the same layers, so the sections have differing resistances and leakage inductances. Connected in series (15 Ω matching) this is not a problem, but when in parallel, the mismatched Thévenin sources will drive current into one another, which is undesirable. Better quality transformers have four secondary sections to cleanly give 1 Ω, 4 Ω and 16 Ω, but still have the same problem as before if configured for 8 Ω. Modern loudspeakers are nominally 8 Ω (often with dips that fall lower), or nearly resistive 4 Ω. Either way, it is better to treat all loudspeakers as 4 Ω – the slight loss of power is insignificant, but the boost in quality is worthwhile.

$16\,\Omega$ loudspeakers would not need such a low source resistance for optimum damping and would therefore be less upset by loudspeaker cable resistance. Transistor amplifiers could be designed more easily, valve amplifiers could have their secondary sections optimized, and the reduced turns ratio would further improve the transformer. However, any manufacturer who introduces a $16\,\Omega$ loudspeaker will have that loudspeaker branded as inefficient, because a given voltage would produce 3 dB less acoustic power than the $8\,\Omega$ design – see published comments about the BBC LS3/5a ($12\,\Omega$). So, we are stuck with $8\,\Omega$, and the trend is towards $4\,\Omega$.

Transformer models

Because real transformers are such complex devices, it is usual to devise simplified models that attempt to represent operation at low, mid and high frequencies.

At low frequencies, the transformer may be represented as a perfect transformer in parallel with the primary inductance of the real transformer, driven by the non-zero resistance of the source. See Fig. 3.16.

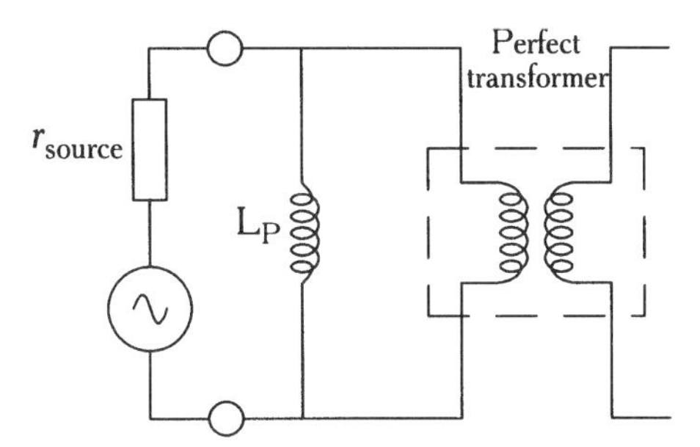

Fig. 3.16 *Transformer equivalent circuit at LF, showing effect of primary inductance*

The combination of source resistance and finite primary inductance creates a high-pass filter whose cut-off frequency is given by:

$$f_{-3\,dB} = \frac{R_s}{2\pi L_p}$$

With a given transformer, we will obtain better low frequency performance if we can reduce the source resistance. An EL34 pentode has $r_a = 15\,\mathrm{k\Omega}$, but the same EL34, used as a triode, has $r_a = 910\,\Omega$, and used as a cathode follower, $<100\,\Omega$.

Once we have decided on source resistance, we will want a high primary inductance, which can either be achieved by increasing primary turns, or by using a core material with a higher μ_r. A better core material is preferable because the bandwidth (in octaves) of the transformer is related by:

$$BW_{(\mathrm{octaves})} \propto \frac{L_{\mathrm{primary}}}{L_{\mathrm{leakage}}}$$

This ratio is dependent on the geometry of the transformer, and on μ_r, but not size or number of turns. All other things being equal, a core of higher μ_r will produce a transformer of greater bandwidth. Although in this instance we are suggesting that low frequency performance could be improved by using a material of higher μ_r, we could use the superior core material to improve high frequency performance. The same primary inductance could be achieved by winding fewer turns, which would then reduce leakage inductance, resulting in better high frequency performance.

At mid frequencies, we can consider the losses due to the resistance of the windings; it is usual to reflect the secondary circuit into the primary circuit. See Fig. 3.17.

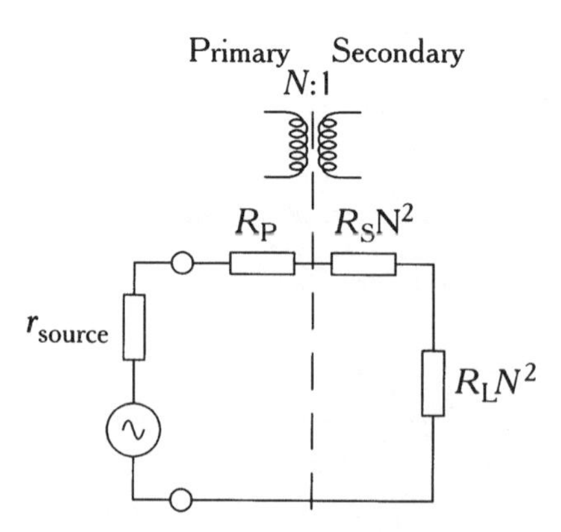

Fig. 3.17 *Transformer equivalent circuit at MF, showing effect of winding resistance*

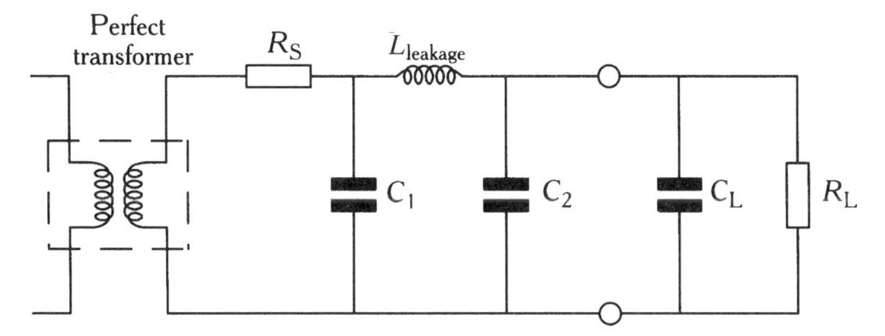

Fig. 3.18 *Transformer equivalent circuit at HF, and its similarity to classic 3rd order filter*

The high frequency model is much more complex. See Fig. 3.18.

In this model, the primary circuit has been reflected into the secondary, and the source resistance, primary resistance and secondary resistance have been lumped together. Interwinding capacitance has been lumped and introduced in two positions, and leakage inductance is also included. The resulting circuit is a classic low-pass filter having an ultimate roll-off of 18 dB/octave, and with suitable choice of component values, this model will accurately simulate a real transformer at high frequencies.

Since the model is a classic filter, we can use the rules that apply to these filters. The most important of these rules is that performance is critically dependent on terminating resistances. For a normal filter, these terminating resistances are the source and load resistance, but for the transformer we should also consider load capacitance.

In almost all cases, adding shunt capacitance worsens the performance of the transformer, whereas reducing load resistance may improve performance. When using a moving-coil cartridge step-up transformer, it is well worthwhile experimenting to find the optimum load resistance for the *transformer*, before adjusting the load on the cartridge.

Why should I use a transformer?

With all that has been said about the imperfections of magnetic components, it might be thought that they should be avoided at all costs, particularly since they are invariably expensive.

An output transformer can be used to load match a low impedance loudspeaker to the high resistance valve output stage, thereby greatly increasing efficiency. If multiple secondary windings are provided, it will allow matching to a range of impedances, rather than having to redesign the amplifier.

An input transformer, such as a moving-coil cartridge step-up transformer, can step up a small signal sufficiently so that it can be amplified by the following amplifier with minimum noise due to the amplifier. As a bonus, the primary can be left floating so that any hum induced into the connecting cable from the cartridge to the transformer will be rejected by the transformer. (See Chapter 6 for a fuller explanation of these benefits.)

With the possibility of multiple windings, a transformer may allow novel methods of feedback to be applied to a circuit, further improving its performance. This technique has frequently been exploited in power amplifiers.[3]

A transformer isolates the DC on the primary from the DC on the secondary. This is often essential!

General considerations in choosing transformers

These considerations only apply to audio transformers; power supply transformers will be dealt with in the next chapter.

Unless you are building a standard circuit, for which a transformer has already been designed, you will almost certainly need a custom designed transformer. It is therefore essential that you give the designer as many clues as possible in order that they may make the *right* compromises to suit your circuit:

- Is the transformer a power output transformer, or is it a small-signal transformer?
- What is the maximum signal level (mV) that will be applied to the primary at the lowest frequency of interest? Is this level constant with frequency? How much distortion can you tolerate at this frequency/level?
- What is the source resistance?
- What primary to secondary turns ratio is required?
- What shunt resistance and shunt capacitance will load the secondary? Can either of these be varied if necessary?
- What frequency range do you *need* the transformer to cover? Don't just say 5 Hz–500 kHz ± 0.1 dB, because it can't be done.
- Do you need an electrostatic screen?
- Do you want the transformer screened in a μ-metal can to reduce

electromagnetic hum?
- Are there any special requirements that the designer ought to know about?

If the answer to the first question was 'yes', then these additional questions should be answered, and an annotated circuit diagram of the output stage is ideal:

- Is the output stage Class A or Class AB?
- What is the quiescent DC current? What is the maximum DC current?
- What is the maximum output power, and what is the lowest frequency at which this is required, for a given distortion level?
- Is the output stage push–pull, or single ended?
- Are the output valves triodes or pentodes? Will you need 'ultralinear' taps?
- What primaries and secondaries do you need? What DC is superimposed on each?
- What form of physical mounting do you want? Open flanges, shrouds, or drop-through?

All these questions may seem rather off-putting, but if you already have a clear idea of what you want, it is much more likely that the finished article will meet your expectations.

Identifying an unknown output transformer

Occasionally, you may find an old output transformer that appears to be of good quality, but have no information as to what it is. Fortunately, it is quite easy to identify old output transformers using only a DVM, as they almost always conform to a pattern.

Before testing, all external links on the transformer should be noted and removed. The primary is invariably centre tapped to allow for push–pull operation, and there may be further tappings to allow for ultralinear operation. Typically, the DC resistance measured by a multimeter from end to end will be the maximum winding resistance that can be measured, and is commonly between 100 Ω and 300 Ω. Once you have found a resistance of this order, you have identified the A_1 and A_2 connections.

Best quality transformers have the primary wound symmetrically, with equal resistance from the HT centre tap to both A_1 and A_2, so the next step is to find the connection that has half the end to end resistance. However, cheaper transformers may not be wound quite so well, and may not have perfectly equal resistances.

The same gauge wire is invariably used for the entire primary of the transformer, so a tap that is 20% (maximum power configuration) from the HT centre tap to A_1 or A_2 will have 20% of the resistance from HT to A_1 or A_2. If it is from a higher quality amplifier, then 47% taps (minimum distortion configuration) are more likely.

The secondary is likely to be composed either of an even number of windings, or one tapped winding. Remember that in the valve heyday, loudspeakers were either 15 Ω (best quality loudspeakers), or 4 Ω, so output transformers were optimized for these impedances.

The most common configuration is to have two identical sections which were wired in series for 15 Ω, or in parallel for 4 Ω (actually 3.75 Ω). If, having identified the primary, you find two windings with ≤0.7 Ω DC resistance each, then you have a standard pattern transformer.

Better quality transformers took the above idea rather further, and had four equal sections comprising their secondary. Wired in series, they again matched 15 Ω, but with all four in parallel, they matched 1 Ω. This was not because loudspeakers with 1 Ω impedance were available (the technology for designing bad crossovers had not yet arrived), but because further sectioning of the transformer enabled a better quality component. This time, you are looking for four roughly equal DC resistances of ≈0.3 Ω. Bear in mind that contact resistance is very significant when making these measurements, so not only do you need a clean, firm contact, but also that even a 4½ digit DVM is inaccurate when measuring such low resistances, so a certain amount of guesswork is involved.

If after identifying the primary, all of the remaining wires appear to be connected together, then you have a tapped secondary, and the connections with the most resistance are between the 0 Ω and (probably) the 16 Ω tap. Provided that an 8 Ω tap is not fitted, the smallest DC resistance from *either* of these is the 4 Ω tap, and the 0 Ω is nearest to the 4 Ω tap. If an 8 Ω tap is suspected, then the tappings are best found using the AC method to be described later.

If some windings are still unaccounted for, these are likely to be feedback windings, perhaps to individual output valve cathodes, or for global feedback. Either way, these can be identified later, as the next step is to determine the turns ratio, and from this, the primary impedance of the transformer.

Although (when correctly carried out) the following test is not likely to be dangerous to the output transformer, it does expose LETHAL voltages, and if you are in ANY doubt as to your competence to perform the test, it should not be attempted.

Valve output transformers are designed to step down signals at hundreds of volts down to tens of volts at frequencies from 20 Hz to 20 kHz, so applying mains voltage from A_1 to A_2 is well within their capabilities. Provided that A_1 and A_2 have been correctly identified, there is no danger to the transformer in connecting the AC mains directly between A_1 and A_2, and measuring the AC voltage that appears across a secondary to determine the turns ratio.

The test should be done in the following manner:

- Fit a mains plug with the lowest value fuse you can find, 3 A will do, but 1 A is preferable.
- Wire the plug with a short length of three core flex. For obvious reasons, these leads are known as 'suicide leads', and should be kept locked away when not in use.
- Solder a solder tag to the 'earth' wire, and screw this to the metallic chassis of the transformer, remembering to use star washers to ensure good electrical continuity.
- Solder the 'line' wire to A_1, and the 'neutral' to A_2.
- Ensure that all links across secondary sections have been noted and removed.
- Set your DVM to its 'AC volts' range and connect it across one of the secondary sections.
- Ensure that you can see the meter clearly, plug the mains plug in, and switch on. If the meter does not *instantly* react, switch off. Assuming that a reading appears, allow the reading to stabilize, note the reading, switch off, and unplug the mains plug.
- Check the mains voltage by reconnecting your DVM across A_1 and A_2, and applying mains again. Note the reading.

We can now find the turns ratio using the following formula:

$$n = \frac{\text{mains voltage}}{\text{secondary voltage}}$$

This may not seem all that useful, but remember that impedances are related by n^2, so once we know n, we can determine primary impedance, since we already know the secondary impedance.

Example
Out of all its wires, the transformer has five wires that are all electrically connected when tested with the DVM on its resistance range. The largest resistance between two wires is found to be 236 Ω, so these wires are designated A_1 and A_2. Leaving one lead on A_1, another wire is found to give a resistance of 110 Ω, this is near enough to 118 Ω to be the centre tap, so this is designated HT. One lead is now moved to the HT tap, and the two remaining primary wires are tested, one reads 29 Ω, and the other reads 32 Ω, so these are likely to be 20% ultralinear taps. The one that reads the lowest resistance to A_1 is the $g_{2(V1)}$ tap, and the other is the $g_{2(V2)}$ tap. See Fig. 3.19.

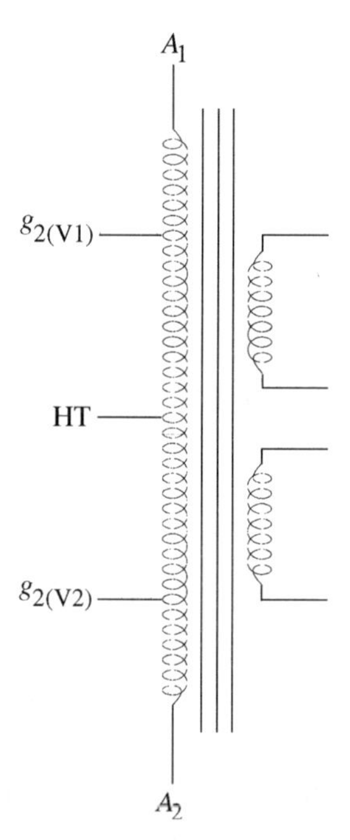

Fig. 3.19 *Identifying an unknown output transformer*

The secondary has only two sections, so each is likely to be a 4 Ω section. This is then confirmed, as one measures 0.6 Ω, and the other 0.8 Ω, which are typical values for 4 Ω windings.

When connected, mains voltage reads 252 V_{AC}, and the voltage across a single secondary was 5.60 V_{AC}. Substituting the values into the equation:

$$n = \frac{\text{mains voltage}}{\text{secondary voltage}} = \frac{252}{5.60} = 45$$

Impedances are changed by n^2, so the ratio of primary to secondary impedance is $45^2 = 2025$, and because the secondary voltage was measured across a $4\,\Omega$ section, the primary impedance must be $2025 \times 4 = 8100\,\Omega$. Unfortunately, testing with $240\,V_{RMS}$ at 50 Hz may drive the transformer close to saturation, causing errors, so we would round this value to $8\,k\Omega_{(a-a)}$.

We now need to 'phase' the secondary sections of the transformer, this is done by connecting one wire only between one section and another, to put the windings in series. If when AC mains is applied across the primary we measure double the original secondary voltage, then the two sections are aiding one another, and we have connected a 'phase' to a 'neutral', and we can designate the sections at the end of the linking wire '+', and the other '–'. However, if no voltage is seen, then we have connected like to like, and we can designate both as '+' or '–'.

If all identical sections have been identified and phased, any remaining windings can be measured and their turns ratio determined, either with reference to the secondaries, or to the primary – whichever seems the most convenient. At this point, annotated diagrams are usually the best way of identifying windings, and factors of two for secondary voltages are significant as they may indicate centre tapped sections, or $4\,\Omega$ and $16\,\Omega$ taps.

Uses and abuses of audio transformers

Transformers are among the most reliable of electronic components, often lasting 30 years or more, but they *can* be damaged. Transformers are made of wire that can fail if excessive current is passed, and insulation that can break down if too many volts are applied across it.

The most common way of destroying an output transformer is to drive the amplifier well into overload so that one output valve switches off completely whilst the other is hard on. The leakage inductance of the half of the transformer associated with the switched-off valve tries to maintain its current, and in doing so produces a high primary EMF causing the interwinding insulation to break down:

$$V = -L_{\text{leakage}} \frac{di}{dt}$$

Since $di/dt \approx \infty$, the EMF developed is far higher than HT voltage, and easily capable of punching through transformer interwinding insulation. If damp has been allowed to get at the transformer, then the (paper) insulation will already be slightly conductive, and the possibility of breakdown is increased.

The author has not managed to damage an output transformer (yet), even when driving amplifiers to their full voltage output with only a pair of electrostatic headphones to load the output stage, but the possibility should be considered.

Guitar amplifiers and arcs

Since the rate of change of current at overload is high, and output transformers for guitar amplifiers are deliberately poor, implying a large leakage inductance, a sufficiently high voltage can be developed to strike an arc externally, even though the transformer may be designed to survive the experience. The voltage needed to strike an arc depends partly on the cleanliness of the path, so a dirty (conductive) path will lower the voltage, and a carbonized trail from a previous arc will certainly reduce the voltage needed.

Although a high voltage is needed to strike an arc, once struck, it can be maintained by quite a low voltage. As an example, the xenon lamp used in a small cinema projector must be struck by a capacitive discharge of thousands of volts, yet may be maintained by only 26 V at 75 A. If an amplifier strikes an arc from the anode, it can only maintain the arc to a place that has a low resistance to ground, because a high resistance, such as a grid leak or cathode resistor, would limit the current, thus extinguishing the arc. The heater pins are connected directly to ground via the LT centre tap, so the most likely place for an external arc to strike is between anode and heater pins; the only limiting resistance being that of the HT supply.

If we know that the amplifier will be thrashed, then a possible solution (depending on the amplifier) is to insert a resistor between LT CT and 0 V HT, perhaps a 4.7 kΩ 6 W W/W, in order to extinguish the arc. However, floating the LT supply may now cause hum problems because of poor heater wiring (routeing, dressing, connection to chassis).

Other modes of destruction

Excessive current through an output valve may cause thermal runaway from grid emission, melting the internal valve structure, and drawing enough current from the power supply through the output transformer that the primary winding fails. The cure for this is simple, keep your valve amplifier on display, and if a valve anode starts glowing cherry red, switch off immediately. (Output stages in valve amplifiers very rarely have fuses partly because the non-linear resistance of the fuse can cause distortion, but mostly because a fuse would not blow sufficiently quickly to protect the output valves).

Small-signal transformers are usually damaged mechanically. They are fragile and have windings of very fine wire which is easily broken. Treat them with respect.

Transformers screened in μ-metal cans must not be dropped as the impact can work harden the μ-metal screen, reducing its effectiveness.

Thermionic valves

History

The thermionic valve was not so much invented as discovered, and was due to Edison's research into extending the longevity of incandescent filament lightbulbs. It had been observed that as the lightbulb neared the end of its useful life, the glass became discoloured and darkened. (This effect is not now generally visible on a domestic lightbulb, but it can be seen on non-quartz halogen torch bulbs and stage lamps.) The cause of the darkening was that the surface of the tungsten *filament* was being evaporated and deposited on the inner surface of the glass.

In an attempt to counter this effect, a *plate* was introduced into the (evacuated) *envelope*, and it was then noticed that if the plate was positively charged with respect to the filament, a current flowed across the vacuum. (Lightbulbs need a vacuum because the incandescent tungsten filament would otherwise oxidize so rapidly that it would burn.)

In 1904, Fleming went rather further, and connected a source of AC between the plate and the filament, and found that the current would only flow in one direction. The resulting device was the *thermionic* diode, whose name derives from the thermal energy required to produce ion flow, although strictly it is electron flow.

Although the diode had great curiosity value, it was of little practical use, and it wasn't until Lee de Forest invented the Audion valve, later known as the triode, that the devices could be used for radio.

Emission

All metals have free electrons within their crystal structure, so some of them must be at the surface of the metal, but they are bound there by the nuclear forces between them and the adjacent atoms. However, the atoms and electrons are constantly vibrating due to thermal energy, and if the metal is heated sufficiently, some electrons may gain sufficient kinetic energy to overcome the attractive forces of the atoms and escape.

The heated metal in the valve is the cathode, and when this is heated to a temperature determined by the work function of the metal, an electron cloud or space charge will form above the surface of the cathode. Eventually the cloud will have enough charge to prevent other electrons escaping from the surface because like charges repel, and an equilibrium is therefore reached.

If we now connect the plate, or anode, to the positive terminal of a battery, electrons will be attracted from the cloud, and will be accelerated through the vacuum to be captured by the anode. The electron cloud has now been depleted, and no longer repels electrons so strongly, so more electrons will escape the surface of the cathode to replenish the electron cloud.

Current cannot flow in the opposite direction because only the cathode can emit electrons, and only the positive anode can attract electrons.

Electron velocity

We mentioned that electrons were accelerated towards the anode, and this is quite literally true. At the instant that an electron leaves the cloud, it has almost zero velocity, but it is constantly accelerated by the electric field of the anode, and acquires energy proportional to the accelerating voltage:

$$E = eV = \tfrac{1}{2}mv^2$$

Rearranging and solving for velocity:

$$\text{velocity} = \sqrt{2V \cdot \frac{e}{m_e}}$$

The ratio e/m_e is known as the electron charge/mass ratio, and has an approximate value of 1.7588×10^{11} C/kg. If we apply 100 V between the anode and cathode, the electrons will collide with the anode with a velocity of $\approx 6 \times 10^6$ m/s, or 13 million miles per hour.

Using the previous equation, it would appear that 512 kV would be sufficient to accelerate the electrons to faster than light speed – which is an impossibility. The flaw is that the simple equation only accounts for the rest mass of the electron, but at relativistic (approaching the speed of light) velocities, the mass of the electron increases, thus requiring an infinite voltage to accelerate it to light speed. To account for this, the elegant equation given by Alley & Atwood[5] may be used:

$$\text{velocity} = c \cdot \sqrt{1 - \frac{1}{\left(1 + \frac{e}{m_e} \cdot \frac{V}{c^2}\right)^2}}$$

where c = velocity of light in a vacuum $\approx 2.998 \times 10^8$ m/s.

As an example of relativity at home, the display tube in a good quality colour television typically has a final anode voltage ≈25 kV, and therefore an electron collision velocity at the tube face of 202 million miles per hour, but the simple equation predicts a velocity 3.5% high.

Hospital X-ray machines deliberately fire high velocity electrons at a target, since X-rays are easily produced once the collision velocity exceeds 200 million miles per hour. Domestic display tubes do not therefore use a final anode voltage >25 kV, even though it would allow better focus and sharper pictures.

Note that the distance between the anode and cathode does not feature in either equation because an infinite distance would also allow an infinite time for acceleration, and even if the rate of acceleration was very low, the collision velocity would still be reached.

Many effects within valves can be understood by having an appreciation of the collision velocity of the electrons as they hit the anode.

Individual elements of the valve structure

The cathode

Early valves betrayed their lightbulb origins, and were *directly heated* using a tungsten filament that was also the cathode. (Tungsten was used in lightbulbs because, at 3410°C, it has the highest melting point of all conductors and could therefore withstand the high temperature necessary to generate light.)

Although it was not necessary for the filament to produce bright light, it was found that as cathode temperature was reduced, the number of electrons emitted fell drastically. The emitted current per unit area is:

$$I \propto T^2 e^{-k/T}$$

where T is the absolute temperature of the cathode, and k is a constant partly determined by the work function of the metal. (See Appendix for full Richardson/Dushmann equation.)

The emission efficiency of the cathode is important because significant power is dissipated in the filament, which not only increases the input power requirement of the equipment, but the heat also has to be lost without damaging any other components. We therefore want to maximize electron emission for a given filament power, and the history of the cathode is thus concerned with the developing chemistry of the cathode emissive surface.

The first improvement was to use a *thoriated* tungsten cathode which not only had a reduced work function, but could operate at 1700°C rather than 2700°C. This reduced temperature was significant because valves primarily lose heat by radiation, and by Stefan's law:

$$E = \sigma T^4$$

where: E = power per unit area
σ = Stefan's constant (5.67×10^{-8} JK^{-4} m^{-2}s^{-1})
T = absolute temperature

Thus 1700°C only requires one-fifth the heater power to overcome the losses due to radiation compared to 2700°C. For this reason, although the improved work function only doubled emission, the reduction in heater power by a factor of five meant that the total improvement in emission efficiency was a factor of ten. An unpleasant side-effect of thoriated tungsten cathodes is that they are very brittle, and so valves such as the 845 should be handled with extreme care, and not be subjected to mechanical shock.

The real improvement came with the oxide coated cathode, usually barium oxide, which operated at only 750°C, and was one hundred times as efficient as the pure tungsten cathode.

The emissive surface in an oxide coated cathode is only one molecule thick and consequently fragile. The vacuum in a valve is never perfect, and there will always be stray gas molecules between the anode and cathode. A cold cathode will prevent anode current, zero voltage will be dropped across the anode load resistor, and V_a rises to the full HT voltage. As the cathode warms from cold, sufficient electrons may be attracted towards the anode to collide with stray gas molecules and produce ions without significantly lowering V_a, thus repelling the ions towards the cathode with sufficient velocity to strip its surface. If this process of *cathode stripping* occurs sufficiently often, the cathode emissive coating will be significantly impaired, and valves with large envelopes in proportion to their (oxide coated) cathodes are particularly vulnerable, whereas thoriated tungsten cathodes (such as 845) are comparatively robust. The cure is to ensure that the vacuum is as nearly perfect as possible, and we will see how this is achieved when we look at the vacuum.

Another problem with oxide cathodes is cathode *poisoning*. If the cathode is kept at full operating temperature, but little or no current is drawn, a high resistance layer of barium orthosilicate is formed at the interface between the barium oxide emissive surface and the nickel cathode structure. This layer not only reduces emission, but more significantly it increases the noise resistance of the valve.

Poisoned cathodes may be gradually recovered by operating the valve at a high current. Another method, often used on the cathode ray display tube used in televisions, is known as *rejuvenation*, and this works by temporarily heating the cathode to a higher temperature, by increasing heater volts, and simultaneously drawing a large current. It should be realized that rejuvenation carries a risk of evaporating some of the cathode emissive surface, and contaminating the (nearby) control grid.

To avoid both cathode stripping *and* cathode poisoning, we can leave the heater supply permanently energized at 63% of full voltage, and only apply full voltage at full switch-on. There is nothing magical about the 63% proportion, it is simply a ratio that was found to extend the life, and reduce warm-up time, of (6.3 V) television camera tubes.

The heater

Traditionally, the filament was fed from a tapping on the mains transformer and AC was therefore injected directly into the signal circuitry, causing audible *hum* from the loudspeaker in audio amplifiers.

The filament had to be made of fine wire in order to give a sufficiently high resistance that it could be heated with a low current, and would not require excessively thick wires from the transformer. Because the wire was so fine, the mass of the filament was low, and the temperature of the filament was partly able to track the applied AC voltage. This then modulated cathode emission, and hum was also produced by this mechanism.

The solution to these problems was the *indirectly heated* cathode, consisting of a cylindrical emissive cathode surrounding the filament, but electrically isolated from it, which was then called the *heater*. In order to reduce the modulation in temperature, the cathode was thermally decoupled from the heater by increasing its mass.

Indirectly heated cathodes require five times as much heater power as directly heated cathodes, and take longer to reach operating temperature, but for small signal valves, the reduction in hum is invaluable, and the loss of efficiency can be tolerated.

Only the outer emitting surface of the cathode now needs to be oxide coated, and the heater can revert to pure tungsten. However, the pure tungsten heater is still capable of limited thermionic emission, and the cathode looks like an anode to the heater filament. If a current were to be allowed to flow from the heater to the cathode, then this would be summed with the intended anode current, and hum would result. Fortunately, this problem is easily cured by superimposing a small (10 V is sufficient) DC voltage on the heater supply, which then turns off the unintentional diode between the heater filament and the cathode.

Despite all these efforts to eliminate hum, the heater filament can still induce hum into the grid circuitry either by capacitive coupling, or by inducing a current into the wire of the grid itself. In a further effort to reduce hum, the heater filament of the EF86 is wound as a helix in order to cancel the magnetic field caused by the heater current.

The only way to *totally* eliminate hum is to use a DC heater supply with no AC content whatsoever, and this implies a *stabilized* supply, which has other benefits. Because cathode emission is so strongly temperature dependent, it is essential that the heater voltage is correct, and Mullard quote a maximum permissible heater voltage variation of ±5%, which is exceeded by the UK legal limit for mains voltage variation (+10%, –6%). A stabilized heater supply stabilizes the characteristics of the valve, and the elimination of thermal cycling of the cathode surface reduces low frequency noise.

As an aside, the author recently fitted an Automatic Voltage Regulator (AVR) to the mains supply for his AVO VCM163 valve tester to combat mains fluctuation, and found that the AVR worked hardest between 4 pm and 11 pm. Is it a coincidence that the sound of the hi-fi seems to improve after midnight?

All of the final generation of small-signal valves use indirectly heated cathodes, and directly heated pure tungsten cathodes are now only used for very high power valves.

Heater/cathode insulation

An indirectly heated cathode consists of a heater helix electrically insulated from the enclosing tubular cathode, but in thermal contact with it. No insulator is perfect, and they all deteriorate as temperature rises, which is unfortunate, since this particular insulator is red hot! All heater cathode insulation must therefore be electrically leaky, allowing leakage currents to flow between heater and cathode. Worse, the insulation is often contaminated, so we must expect this imperfection to produce $1/f$ noise. Irritatingly, one of the worst offenders for poor heater/cathode insulation is the otherwise excellent 12B4-A, so this valve must be screened to exclude those samples with poor (hot) insulation if noise is critical.

The heater/cathode electrical insulation is unavoidably also a thermal insulator, increasing the heater power required to raise the cathode to its operating temperature. This power could be reduced, and the valve made more efficient, by reducing the thickness of the heater/cathode insulation, and this is exactly what was done in the transition from the International Octal-based generation to the later B9A generation. Many observers have commented that octal valves sound better (hence the popularity of the 6SN7), and it seems highly likely that reduced heater/cathode leakage current is the cause.

Increasing the voltage across the heater/cathode insulation increases leakage currents. Although V_{k-h}(max.) is specified on data sheets as being anywhere from 90 V to 150 V, this is a very 'soft' limit, since it is usually given at an arbitrary leakage current; nevertheless, a sufficiently high voltage will punch through the insulation to rupture the heater. Heater failure due to heater/cathode insulation breakdown is uncommon, but is most likely in power amplifiers with output stages using cathode followers or distributed loads (such as the McIntosh design).

Heaters and their supplies

Because of the Richardson/Dushmann equation, electron emission, and therefore anode characteristics, are heavily dependent on cathode temperature. Provided that anode dissipation is sufficiently low, and does not further heat the cathode, cathode temperature is directly related to heater power dissipation.

It is usual to supply heaters in parallel from a constant voltage source (typically 6.3 V), or in series from a constant current source (typically 300 mA). If either type of supply drifts from its nominal value, undesirable changes in anode characteristics will occur. Although 6.3 V regulators are easily made, linear regulators become increasingly inefficient as load current rises, and switched mode regulators are inevitably electrically noisy. By comparison, a 300 mA constant current supply feeding a pure series heater chain is easily and efficiently implemented.

Although valve manufacturers sometimes specify either series or parallel heaters, is there actually any fundamental difference between the filaments of the two types, and could we use 6.3 V heaters (of equal current requirements) in a constant current chain?

Some 6.3 V valves were tested to see if there was any significant difference between the behaviour of their heaters. The valves were deliberately chosen to be as different as possible in order to magnify any difference between heaters – the 12AT7 was selected to be one whose heater flashed white at switch-on. See Table 3.1.

Within the limits of experimental error (which significantly worsened towards 1 V), the heater currents are in very close agreement, suggesting that parallel heater valves have essentially similar filaments. This is broadly to be expected, since tungsten is the only practical filament material.

Table 3.1 Percentage of normalized heater current against heater voltage for parallel (6.3 V) heaters

Heater voltage (V)	CV4024 12AT7 (0.30 A)	Mullard ECC83 (0.29 A)	Raytheon 5842 (0.30 A)	GE 6BX7 (1.45 A)	Mullard EL84 (0.79 A)	Mullard EL34 (1.475 A)	Mean	SD σ
6.30	100	100	100	100	100	100	100	0
6.00	97	97	100	97	96	97	97.3	1.37
5.50	92	93	93	93	92	93	92.7	0.52
5.00	87	90	90	88	87	87	88.2	1.47
4.50	83	83	83	83	82	85	83.2	0.98
4.00	77	78	80	78	77	78	78.0	1.10
3.50	70	66	73	72	72	73	71.0	2.68
3.00	67	60	67	66	67	67	65.7	2.80
2.50	58	62	60	61	60	60	60.2	1.33
2.00	52	55	53	54	51	53	53.0	1.41
1.50	42	45	47	45	43	44	44.3	1.75
1.00	30	34	37	37	32	34	34.0	2.76
0.50	20	21	20	26	19	20	21.0	2.53

Table 3.2 Percentage comparison of current heater valve (PL508) with mean current of voltage heater valves

Voltage	6.3	6.0	5.5	5.0	4.5	4.0	3.5	3.0	2.5	2.0	1.5	1.0	0.5
Av.	100	97	93	88	83	78	71	66	60	53	44	34	21
PL508	100	97	93	88	83	78	71	66	60	54	45	35	23

A PL508 was then tested set to its correct current (300 mA), the voltage at that current measured, and a series of measurements was taken, which were normalized to 6.3 V and compared to the mean currents from Table 3.1. See Table 3.2.

A somewhat improved measurement technique was available when the PL508 was tested; nevertheless, the correlation between this and the parallel heaters is remarkable. There appears to be no significant difference between the filaments in valves specified for series or parallel heaters, and provided that individual heaters consume their correct power, there seems no reason why we should not mix the two types at will. This technique will be used in the 5842 RIAA pre-amplifier in Chapter 6.

Having established that valve heaters are essentially the same, is the series or parallel chain best in terms of heater power regulation?

The heater current of an EL34 was investigated at ½% intervals within a ±5% nominal voltage range, and the results plotted as a graph. An extremely close fit to a straight line was observed, indicating that the heater behaves as a constant resistance over this very limited range. Since $P = I^2R$ and $P = V^2/R$, two conclusions emerge.

First, parallel chains should be constant voltage (Thévenin) regulated, and series chains should be constant current (Norton) regulated.

Second, we should not mix topologies – series/parallel heater chains cause errors, because each heater no longer sees a perfect Thévenin or Norton source. Because some double triodes, such as the 6SN7, may be observed to have their heaters internally wired in parallel, they should *only* be used with a 6.3 V Thévenin supply. (If necessary, this particular problem can be resolved with a pair of 6J5 single triodes.)

As a more insidious example, a double triode initially tested on a valve tester with 6.3 V parallel heaters, and found to have perfectly matched anode characteristics between sections, would be mismatched by configuring the heaters in series unless the heaters were also perfectly matched. Matching should closely replicate the proposed conditions of use.

The control grid

The control grid is wound from stiff, fine wire as a helix around the cathode, and is most effective close to the cathode surface, where the velocity of the electrons is low, than near the anode, by which time the electrons have acquired considerable momentum and are not easily repelled. For this reason, even in a valve having a succession of grids, the control grid is *always* the grid nearest to the cathode. The control grid for a high μ valve has a finer mesh and is closer to the cathode surface than in a low μ valve.

Because the control grid is so close to the cathode, a very small movement of the grid will have a significant effect on the flow of electrons, and this is the cause of valve microphony.

Although the control grid is normally a high resistance point, it can pass positive or negative grid current.

If we charge the grid positively with respect to the cathode, the grid will reduce the repulsive effect of the space charge on electron emission at the surface of the cathode, and will assist in pulling electrons away from the surface of the cathode. A much higher anode current will flow, but some electrons will now be diverted and flow into the grid circuit, resulting in positive grid current, which drastically reduces input resistance. This is why Class AB2 output stages, which operate with positive grid current, are invariably preceded by a power driver.

The grid is inevitably slightly contaminated with cathode emissive material during manufacture, and is close to the (hot) cathode. If the grid's temperature were allowed to rise, it would be capable of emitting electrons in its own right, so power valves cool the grid by winding the grid on thick copper axial supporting wires with heatsinks at their ends. Better quality valves plate the grid with gold because the increased work function of gold reduces grid emission, grid current, and therefore valve noise.

If the grid is allowed to emit electrons, negative grid current results, and depending on the value of the grid leak resistor and biasing, the potential of the grid may rise (*lowering* V_{gk}), causing an increase in anode current, and further heating the valve. The emissive material of the cathode then begins to evaporate, further contaminating the grid and increasing grid emission. At worst, the grid may become so hot that it slumps and touches the cathode, completing the destruction of the valve, but increased valve noise is inevitable even if the valve is not actually destroyed.

Other grids

In tetrodes and pentodes, some of the cathode current flows to the screen grid causing its temperature to rise. The screen grid is wound concentrically between the control grid and the anode, and a zirconium coating is applied to improve its radiation efficiency. Hot zirconium absorbs residual gas, so this grid assists the getter in maintaining a good vacuum.

Electrons are not attracted to the suppressor grid in a pentode, so this grid does not self-heat.

The anode

The anode is constantly bombarded by high velocity electrons. Although they have very little mass, their high velocity means that they possess considerable kinetic energy, which is converted into heat when they are stopped by the anode. An important specification for a valve is therefore the *anode* dissipation, because a hot anode heats the grid, causing grid emission. A secondary effect is that a hot anode releases gas, a phenomenon known as *outgassing*, which contaminates the vacuum.

The anode dissipates heat by radiation, and in order to maximize radiating area, anodes often have fins. The other way to improve the loss of heat by radiation is to colour the anode black, either by using a solid graphite anode, or by coating the surface of a conventional nickel anode with graphite.

The anode's surroundings must be cool and capable of absorbing radiant heat, otherwise they will emit or reflect heat back to the anode, which is perfectly coloured to absorb radiant heat, thus raising anode temperature. Chrome plated output transformers, etc. may look nice, but they raise anode temperature. The worst possible surroundings for a valve would be a concentric chrome plated cylinder, since this would reflect all radiant heat back to the anode.

The EF86 low noise pentode has an electrostatic screen surrounding its anode to reduce hum, and on some examples, is formed from a shiny metal sheet, but this should not be confused with the anode. The screen severely restricts anode cooling, but the *gm* of the EF86 is quite low anyway, and operating it at a high current (which would increase *gm*, but also increases P_a) would be pointless, so low noise circuits using the EF86 typically have a low P_a, and the electrostatic screen is not a problem.

Some of the high velocity electrons bombarding the anode will have sufficient energy to knock one, or more, electrons off the anode, such that the anode may actually emit more electrons than it receives. The relative level of this emission is determined by the Secondary Emission Ratio (SER) of the metal concerned. Nickel has a low SER, and together with its malleability is why it is commonly used for anodes and other valve electrodes.

The vacuum

The quality of the vacuum within the valve is critical because (initially uncharged) gas molecules in the valve will be struck by high velocity electrons on their way to the anode, possibly dislodging electrons to create ions. The positively charged ions are repelled from the anode, and strike the grid/cathode structure, to be immediately discharged by a balancing number of electrons flowing up from the external paths to ground. Since the formation of ions and their subsequent discharge by the grid/cathode structure is random, the process creates noise currents known as *ionization noise.*

Because valves are voltage operated devices ($I_a \propto V_{gk}$), the final effect of ionization noise depends not just on the noise current, but also on the value of the external resistance, in accordance with Ohm's law.

Low noise input stages use a high μ valve in order that subsequent stages do not degrade the noise performance, but high μ valves have a fine grid spacing, greatly increasing the probability of ions striking the grid rather than the cathode. Since R_g, the grid leak resistor, is invariably quite a high value, a significant noise voltage is developed and amplified. Since the grid effectively screens the cathode, very few ions strike the cathode, ionization current is greatly reduced, and even an undecoupled R_k has a low resistance to ground, further reducing the noise voltage developed in the cathode circuit. Ionization noise in high μ stages is thus dominated by grid ionization noise current, and can be minimized by a low grid impedance to ground at audio frequencies, so transformer input coupling will reduce the effects of ionization noise currents at low frequencies compared to capacitor coupling. (At very low frequencies, $Z_{sec.} \approx R_{DC(sec.)}$, which is quite low, whereas $X_C \approx \infty$, so capacitor coupling produces more $1/f$ noise.)

Stages following the intermediate stage tend to use low μ valves to preserve bandwidth, but low μ valves have a coarse grid spacing, strongly biasing the probability of ion strike in favour of the cathode. Low μ valves operate with high bias voltages (V_{gk}), requiring high values of R_k. The combination of an increased proportion of ionization current and high R_k means that low μ valves should not leave their cathodes undecoupled if the effects of ionization noise currents are to be minimized.

A good vacuum is referred to as being *hard* whilst a poor vacuum is *soft*; valves are therefore sometimes described as having 'gone soft'. During manufacture, the air in the valve is pumped out, but some air will remain that cannot be removed by pumps, and the remaining gas is removed by the getter.

The getter

The getter is a metal structure usually near the top of the valve, coated with a highly volatile powder (usually a barium compound similar to the cathode emissive surface). Once the valve has been sealed and as much gas as possible has been pumped out, the getter is heated and the powder explodes, consuming the remaining gas. The force of the explosion throws molten barium onto the inside of the envelope to give the familiar mirrored coating at the top of the valve. The explosion is initiated electrically, either by directly passing a large current through the getter's metallic supporting structure (metal cased valves), or by shaping the getter as a short circuited loop aerial and inducing a large current via an external RF field (glass cased valves).

Although some of the getter is deactivated by the explosion, the getter continues to consume gas molecules throughout the life of the valve because gas inevitably leaks into the valve vacuum, either via the seals where the leads leave the envelope, or by outgassing from a hot anode.

To be consumed by the getter, the gas molecules must touch the getter, and this is ensured by normal Brownian motion if the heater reaches operating temperature before HT is applied to the anode. Some valves improve the efficacy of the getter by bonding it to the anode, since the rate of chemical reactions doubles with every 10°C rise in temperature.

Soft valves can often be spotted by the gentle blue glow near the glass envelope which is due to the collision of ionized gas molecules with the glass. This should not be confused with the blue glow that can be seen within the electrode structure of valves such as the EL84, which is perfectly normal.

The mica wafers

The electrode structure, heatsinks and getter are supported and held rigidly in position by insulating mica wafers at the top and bottom of the anode structure. If this mica is not a perfect insulator, then paths for leakage currents will form, such as from the anode to the control grid, which would cause noise in a small-signal valve, but could cause destruction in a power valve.

When the getter is exploded, some of the molten metal may strike the mica wafer and make the mica slightly conductive. In order to lengthen these paths and increase their resistance, slots are cut in the wafer between the control grid and the anode.

Alternatively, the getter may be positioned such that it is less likely to spray onto the mica wafers, or the electrode supporting wafers may be screened by an additional mica wafer or metal plate. The designers of the GEC KT88 not only used all of the above techniques to reduce leakage, but also made the electrode supporting wafers undersize, so that they did not touch the (contaminated) envelope. See Fig. 3.20.

Fig. 3.20 *View of GEC KT88; note the measures taken to reduce leakage currents*

Even if the mica wafers have not been contaminated with conductive getter, mica is not a perfect insulator, and like all insulators, its resistance falls with increasing temperature.

The Sony C-800G studio condenser microphone uses a Peltier effect heat pump to cool the envelope of its pre-amplifier valve. Since the envelope is in contact with the mica wafers, they are also cooled, and it seems probable that the reduction in leakage currents through the wafers, together with reduced anode temperature and consequent outgassing is responsible for the reported improvement in noise.[6]

For a considerable time the author toyed with the idea of Peltier cooling the input valves of RIAA disc pre-amplifiers, but the idea has finally been dismissed. Anyone enjoying a cold drink on a hot summer's day will have noticed how quickly a layer of condensation forms on the outside of the glass. We do not want condensation forming on the envelope because water is conductive and would cause leakage currents to flow between the valve pins, increasing noise. Valve electrometers designed to measure minuscule currents (pA) hermetically sealed the input valve in a dry atmosphere, but this is not trivial. It seems simpler to sidestep the problem and completely avoid condensation by allowing the envelope to be a few degrees above room temperature (it will then evaporate any moisture). Thermally bonding the envelope to a large heatsink will easily allow the envelope to be cooled to room temperature +20°C with minimal cost and complexity.

Cooling the envelope to 40–50°C might seem an insignificant change, but low noise valves operate at high current (implying high P_a), causing a purely air-cooled envelope temperature to be far higher.

Not only does reduced envelope temperature reduce noise, it also improves valve life. To directly quote the Brimar Valves 'Components Group Mobile Exhibition' (Nov. 1959) manual:

The use of close-fitting screening cans of high thermal conductivity in intimate thermal contact with a large area of the bulb, in conjunction with an adequate heat sink can materially reduce the operating bulb temperature and very considerably improve the life of the valve.

Contrast this with a further comment from the same source:

The use of screening cans which are not in thermal contact with the valve may seriously interfere with the cooling of the valve.

Since valves in low noise stages need screening, are hot and expensive, it is well worth cooling them on a substantial (<2°C/W) heatsink which can also be used for cooling anode load resistors, regulators, etc. The ideal method is a snug fitting, black anodized aluminium P-clip covering the entire length of the valve, and bonded directly to the heatsink.

Valve sockets and noise

Out of curiosity, a few International Octal valve sockets from stock were tested for capacitance between adjacent pins. In balancing the bridge, not only was the capacitance determined, but also the '*d*' of that capacitance, which is inversely proportional to the leakage resistance, so a low value of '*d*' is desirable.

Type	***C* (pF)**	***d***	
White ceramic	Unglazed PCB	1.3	≈0.01
		1.2	
		1.4	
		1.3	
	Glazed chassis	1.4	
		1.4	
		1.4	
		1.3	
Phenolic	Black	1.7	0.08
		1.6	0.09
		1.6	0.095
	Brown	1.8	>0.1
		1.6	0.06
		1.7	>0.1
		1.6	>0.1
		1.9	>0.1

Although the ceramic sockets have slightly lower capacitance, the main difference is that phenolic sockets are typically ten times leakier, and will therefore increase noise currents from the anode into the grid. Although they could not be compared directly in this particular test, PTFE (PolyTetra-FluoroEthylene) sockets can be expected to have even lower leakage than ceramic, and are therefore thoroughly recommended.

PCB materials

Glass Reinforced Plastic (GRP) boards are rather less than ideal for valve audio because of leakage resistance. The leakage occurs because the epoxy resin does not always seal perfectly to the glass fibres and surface tension draws water vapour into the resulting gaps never to be released. Many years ago, the author built circuits on Synthetic Resin Bonded Paper (SRBP) and felt that they sounded better than the same circuit built on GRP, but at the time could not see any engineering reason why, and put it down to imagination. The crucial difference between GRP and SRBP is that SRBP is porous over its entire surface and not just at its edges. SRBP can therefore lose water vapour over its entire surface as it heats, whereas a GRP board can only lose water vapour at its edges. A warm SRBP board could therefore actually be less leaky (even though nominally a poorer material) than a warm GRP board. Because water is polar, the problem of dielectric leakage becomes even more acute at high frequencies, so microwave and >200 MHz oscilloscope designers have avoided the material for decades, preferring to use PTFE.

The glass envelope and the pins

The envelope maintains the vacuum within the valve; careless handling will crack the valve, and air will enter. The easiest way to crack the envelope is to bend the pins whilst attempting to insert the valve into a new socket. It is therefore good practice to plug old valves into the new sockets of a new amplifier, and test the amplifier before inserting the expensive new valves. This then spreads the fingers of the socket slightly, and the new valves can be inserted without fear of damage. A hidden problem is that repeated plugging and unplugging can create micro-fractures in the glass near the pins, allowing just sufficient gas to leak into the valve to subtly degrade the vacuum, and increase ionization noise (for this reason the British Standards Institution[7] warned against repeated testing of special quality valves). A terminally damaged envelope is easily spotted, because the mirror coating due to the getter turns white.

If the envelope is allowed to accumulate fluff, it will be thermally insulated, and the valve will run hotter, with all the dire consequences that entails. Valves should be clean and shiny to promote long life.

The pins are made of a special alloy that has the same coefficient of thermal expansion as glass in order that leaks do not occur at these seals. On some valves, these pins may be gold plated, but this gold plating will be quickly removed by repeated insertion and removal of the valve. Gold plated pins used to be a sign of quality (although Brimar did not bother to gold plate their excellent E88CC), but modern valve manufacturers cheerfully gold plate selected valves sourced by their standard production line, whereas traditional Special Quality valves were consciously designed/produced to be better, rather than selected from a standard production line.

References

1. Scroggie Iliffe, M. G. (1971) *Radio and Electronic Laboratory Handbook*, 8th ed.
2. Thiele, A. N. (1976) Air-cored Inductors for Audio. *Journal of the Audio Engineering Society*. June, p. 374. Vol. 24. No. 5.
3. Sowter, G. A. V. (1987) Soft Magnetic Materials for Audio Transformers: History, Production, and Applications. *Journal of the Audio Engineering Society*. October, p. 760. Vol. 35. No. 10.
4. McIntosh, F. H. & Gow, G. J. (1949) Description and Analysis of a New 50-watt Amplifier Circuit. *Audio Engineering*, December.
5. Alley, C. L. & Atwood, K. W. (1962) *Electronic Engineering*. Wiley.
6. Spencer-Allen, Keith (1992) Chill factor. *Studio Sound*, November, pp. 56–59.
7. The use of electronic valves. British Standard Code of Practice CP1005. BSI (1962).

Further reading

Mazda, F. F. (ed.) (1989) *Electronic Engineer's Reference Book*, 6th ed.

The Institute of Physics (1973) A Random Walk in Science. An anthology compiled by R. L. Weber and edited by E. Mendoza.

The Institute of Physics (1982) More Random Walks in Science. An anthology compiled by R. L. Weber.

Nelkon, M. & Parker, P. (1987) *Advanced Level Physics*, 6th ed. Heinemann Educational

Duncan, Tom (1994) *Advanced Physics*, 4th ed. John Murray

Williams, Tim (1991) *The Circuit Designer's Companion*. Butterworth-Heinemann.

Duncan, Ben. (1996) *High Performance Audio Power Amplifiers*. Newnes.

Eastman, A. V. (1949) *Fundamentals of Vacuum Tubes*, 3rd ed. McGraw-Hill.

4

Power supplies

A power supply is a device that converts one voltage to another, more convenient voltage, whilst delivering power.

Valve amplifiers need a DC High Tension (HT) supply and one or more heater, or Low Tension (LT) supplies, which may be AC or DC. Often, the supplies for the pre-amplifier and power amplifier will be derived from the same power supply, which is frequently integral to the power amplifier, but this need not be so.

In this chapter, we will identify the major blocks of a power supply, see how to design them, then design a pair of complete supplies.

The major blocks

There are two fundamental types of power supply, *linear* and *switchers*. See Fig. 4.1.

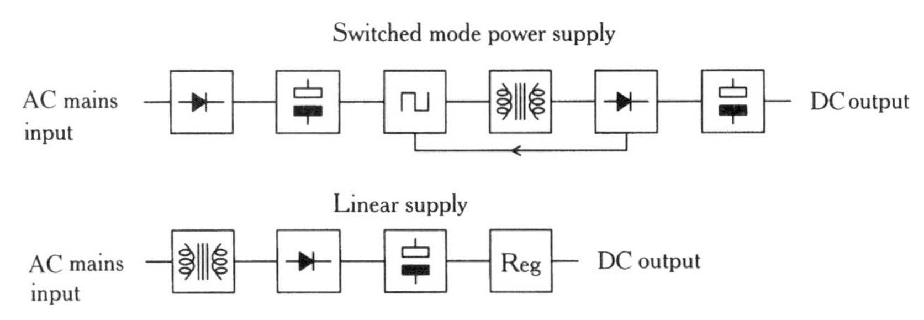

Fig. 4.1 *Comparison of linear and switcher power supplies*

In a switcher, the AC mains input is rectified, switched at a high frequency, typically >20 kHz, transformed, rectified, and smoothed; regulation is part of the switching element. Switchers are small, light and efficient. Their design is highly specialized, and unless very carefully designed, they generate copious RF noise. We will not mention switchers again in the company of valve amplifiers.

By contrast, a linear supply transforms the 50 Hz AC mains directly, requiring a bulky mains transformer. This is then rectified by valves or semiconductors, smoothed by large capacitors, and possibly even larger inductors, and then regulated if necessary. Linear supplies are heavy, inefficient, easily designed, and low noise. Valve amplifiers use lots of them, so we had better know how to design them.

Power supplies are designed from the output back to the input. Since they are designed after the amplification stages, it is tempting to think of them as an afterthought; indeed, some commercial products reflect this attitude. It is most important to realize that an amplifier is merely a modulator, and controls the flow of energy from the power supply to the load. If the power supply is poor, then the most beautifully designed amplifier will be junk.

Rectification and smoothing

Whilst we may not have a regulator on the output of the supply, we will always have rectification and smoothing. The two functions are inextricably bound together, and determine the specification of the mains transformer, so this is the starting point for design. Since we need to rectify the sinusoidal AC that leaves the transformer with maximum efficiency, we will only consider full-wave rectification, as not only is half-wave inefficient (because it only uses every other half cycle) but it causes DC to flow through the transformer, and even small DC currents can cause core saturation. A saturated core is lossy and generates leakage flux which may be induced into surrounding circuitry.

Choice of rectifiers/diodes

There are two forms of full-wave rectification; the centre-tapped rectifier, and the bridge rectifier: See Fig. 4.2.

The bridge rectifier is the usual topology in a modern circuit because it economizes on transformer windings. The centre-tapped rectifier was traditional in valve circuits because it economized on rectifiers (which were

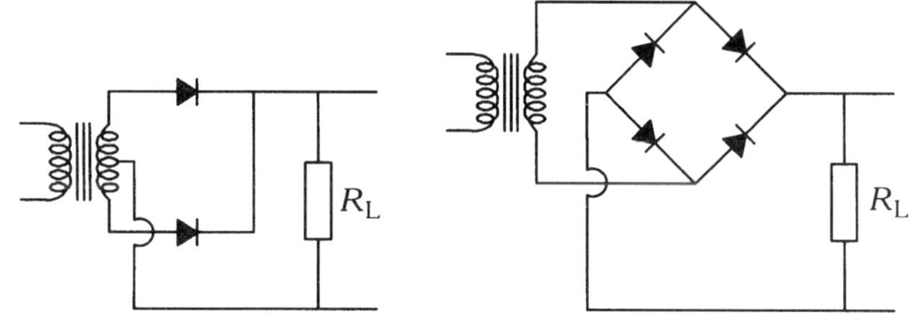

Fig. 4.2 *Full-wave rectification*

expensive), but it is still useful in low voltage/high current silicon circuits because of the lower voltage drop, only $1V_{be}$, compared to $2V_{be}$ for the bridge.

When we consider the HT supply, we have a choice between silicon or thermionic rectifiers such as the GZ34. Valve rectifiers are inefficient. Not only do they need a heater supply, but they also drop tens of HT volts across themselves and cause the supply to have a high output resistance. They are fragile in terms of ripple current (which we will investigate in a moment), and therefore there is a maximum value of capacitance that can be connected across their terminals (60 μF for GZ34). New Old Stock (NOS) Mullard GZ34s are now very scarce, and therefore expensive, whilst some modern GZ34s have been reported to be fragile at high voltages. However, the indirectly heated EZ81 is readily available, and ideal for pre-amplifiers and small mono power amplifiers.

Valve rectifiers have only one advantage over silicon, but this advantage may be sufficient to make us tolerate their foibles. They switch the HT on slowly, thus preventing cathode stripping of expensive signal valves and reducing inrush current into electrolytic capacitors. But as we will see later, there are other ways of dealing with these problems.

Whichever form of rectifier/diode we choose, we must ensure that it is capable of withstanding the stresses imposed upon it by the surrounding circuit. When considering low frequency rectification derived from the mains, we need only specify the voltage and current ratings. However, neither of these ratings are quite as straightforward as they might seem. See Fig. 4.3.

This is a centre-tapped rectifier using silicon diodes to rectify the 300–0–300 V secondary. The off-load voltage at the reservoir capacitor will be 424 V_{DC} (note that this is much higher than if a valve rectifier had been

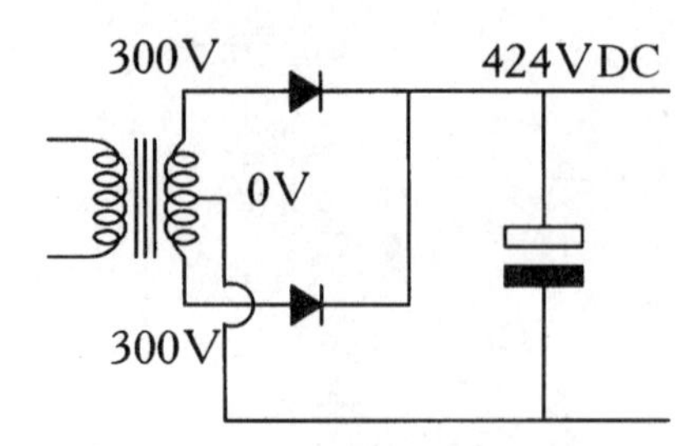

Fig. 4.3 *Effect of capacitor on rectifier ratings*

used – silicon diodes and valve rectifiers are *not* directly interchangeable). The voltage rating that concerns us is the reverse voltage rating, known as V_{RRM} (Reverse Repetitive Maximum), also known as PIV (Peak Inverse Volts). When a diode switches off, the input AC waveform at the anode continues to increase negatively until it reaches a peak of –424 V. Since the cathode of the diode is held at +424 V by the reservoir capacitor, the diode therefore has twice the peak AC voltage across it, and should be rated accordingly.

If the diode feeds a resistive load, then the current rating will simply be equal to the load current. When the diode feeds a reservoir capacitor, pulses of current many times greater than the DC load current flow. Fortunately, modern silicon rectifiers are designed with these peaks in mind, and it is usually sufficient to choose a rectifier current rating equal to the DC load current. We will look at this problem in more detail shortly.

Rectification and RF noise

Rectifiers are switches. Although the following argument implies a resistive load on the rectifier, the results are also valid for the load presented by a reservoir capacitor.

As the input AC waveform rises through 0 V, one or more diodes will switch on, and stay switched on until the waveform falls through 0 V, when the other diode, or diodes, switch on. All diodes need a minimum forward bias before they will switch on, even if it is only the 0.6 V required by silicon. This means that there is a dead zone symmetrically about 0 V when no diodes are conducting. The transformer, which is inductive, has been switched off, and it will try to maintain the current flow, and in doing so, will generate a voltage:

$$V = -L\frac{di}{dt}$$

Fortunately, there are usually plenty of stray capacitances within the transformer to prevent this voltage from rising very far, but if we are unlucky, the shock applied to the system may excite a resonance which will result in a damped train of oscillations. Using a search coil, the author once observed bursts of 200 kHz leaking from a mains transformer for this very reason. Happily, the cure for this problem is very simple; all we need do is to bypass each individual diode with a 10 nF film capacitor with a voltage rating equal to the diode voltage rating.

Whichever type of rectifier we use, we will still apply the same waveform to the succeeding circuit. This waveform, although it is of only one polarity, is not a smooth DC. The function of the smoothing element is to reduce the ripple, either to a satisfactory level, or to a level such that a regulator can cope with it.

The single reservoir capacitor approach

The simplest way of smoothing the output of the rectifier is to connect a reservoir capacitor across it, and feed the load from this reservoir. See Fig. 4.4.

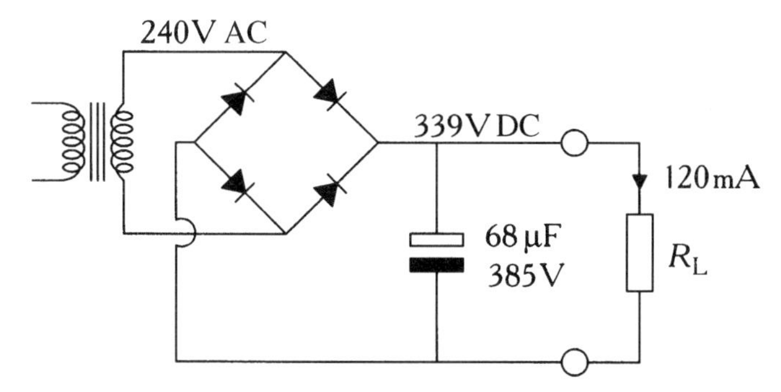

Fig. 4.4 *Power supply using reservoir capacitor*

Assuming no load current, the capacitor will be charged to the full peak value of the AC leaving the transformer, i.e. $V_{in} \times \sqrt{2}$.

Ripple voltage

The output of the rectifier tops up the charge in the capacitor every cycle, so that at the peak of the waveform, the capacitor is fully charged. The voltage from the transformer then falls away sharply, and the rectifier diodes switch off. Load current is now being supplied purely from the capacitor, which then discharges exponentially into the load until the transformer output voltage rises sufficiently to recharge the capacitor, and restart the cycle. See Fig. 4.5.

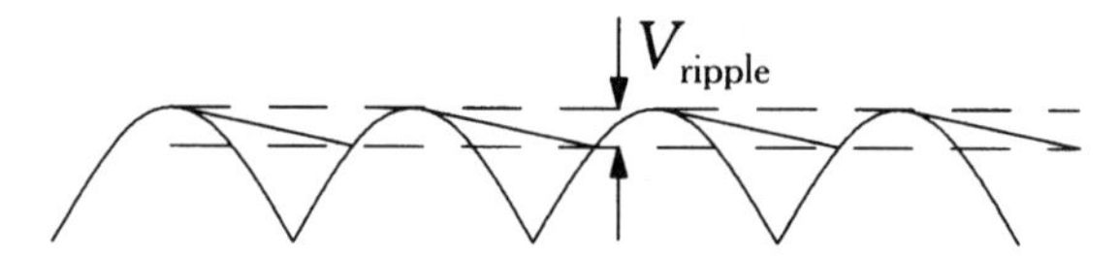

Fig. 4.5 *Ripple voltage across reservoir capacitor caused by charge/discharge cycle*

Although the capacitor discharges exponentially, for any practical value of reservoir capacitor, the discharge curve may be taken to be a straight line. Given this approximation, it is easy to calculate what the output ripple voltage will be.

The charge stored in a capacitor is:

$$Q = CV$$

The total charge, due to a current I, flowing for time t is:

$$Q = It$$

We can combine these equations:

$$CV = It$$

Rearranging:

$$V = \frac{It}{C}$$

This equation gives the voltage *change* on the capacitor due to the capacitor supplying current I, for time t. If mains frequency is 50 Hz, then each half cycle is 0.01 s. If we now make another approximation and say that the capacitor supplies current *all* of the time, then $t = 0.01$. We now have a useful equation:

$$V_{ripple} = \frac{0.01I}{C}$$

It might be thought that this equation is of little use, since two sweeping approximations were used to derive it, but as the reservoir capacitor is invariably an electrolytic capacitor, whose tolerance is likely to be ±20%, we will need a very inaccurate equation before it compares to the error introduced by component tolerances!

We can now calculate the ripple voltage at the output of our example circuit in Fig. 4.4, which had a 68 μF capacitor and a load current of 120 mA.

$$V_{ripple} = \frac{0.01 \times 0.12}{68 \times 10^{-6}}$$

$$= 18\,V$$

Which is about 5% of full voltage, a good design choice.

The previous method produces sensible results provided that the ripple voltage is between 5% and 20% of the total voltage (a good design would not allow ripple voltage to rise above this limit).

Ripple current and conduction angle

Now that we have looked at ripple voltage, we need to look at ripple current. This is the current that flows from the transformer into the capacitor to recharge it each half cycle. To do this, we need to find the conduction angle, which is the time for which the capacitor is charged. See Fig. 4.6.

To do this, we work backwards from the time that the capacitor is fully charged. We know what the ripple voltage is, so we can find the absolute voltage on the capacitor at the instant that the diodes switch on. The voltage at the output of the rectifier (ignoring the polarity) is:

$$v = V_{peak} \cos \omega t$$

At the instant that the diodes switch on, the capacitor voltage must be:

$$V_{peak} - V_{ripple} = V_{peak} \cos(\omega t)$$

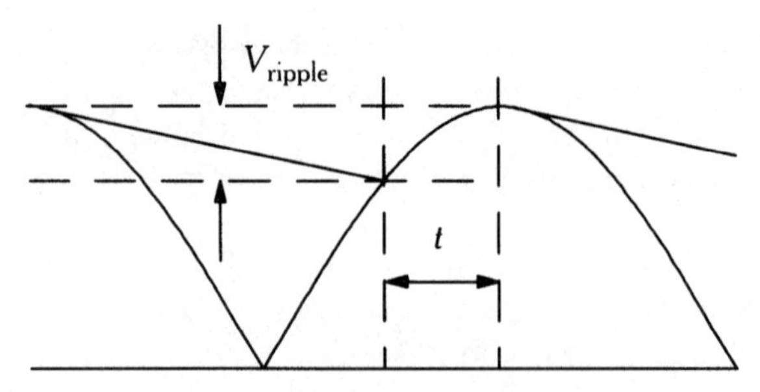

Fig. 4.6 *Determination of conduction angle from ripple voltage*

Rearranging:

$$\frac{V_{peak} - V_{ripple}}{V_{peak}} = \cos(\omega t)$$

$$\omega t = \cos^{-1}\left(\frac{V_{peak} - V_{ripple}}{V_{peak}}\right)$$

$$t = \frac{1}{2\pi f} \cdot \cos^{-1}\left(\frac{V_{peak} - V_{ripple}}{V_{peak}}\right)$$

If we now put some figures into this equation from our earlier example (Fig. 4.4), *remembering to work in radians and not degrees*:

$$t = \frac{1}{2 \times \pi \times 50} \cdot \cos^{-1}\left(\frac{340 - 18}{340}\right)$$

$$= 1 \text{ ms}$$

The capacitor only draws current from the mains transformer for 1 ms in every 10 ms, or 10% of the time. We should therefore expect this ripple current to consist of short, high current pulses. See Fig. 4.7.

We can now find the ripple current using the relationship:

$$I = C\frac{dV}{dt}$$

We need an expression for $\frac{dV}{dt}$:

$$V = V_{peak}\cos(\omega t)$$

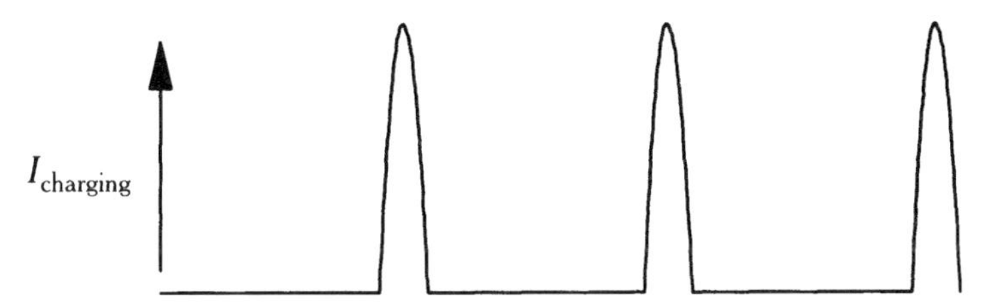

Fig. 4.7 *Ripple current waveform*

Differentiating:

$$\frac{dV}{dt} = -\omega V_{peak} \sin(\omega t)$$

And so:

$$I_{ripple} = -\omega C V_{peak} \sin(\omega t)$$

If we now put some values into this equation:

$$I_{ripple} = 2 \times \pi \times 50 \times 68 \times 10^{-6} \times 340 \times \sin(2 \times \pi \times 50 \times 1 \times 10^{-3})$$
$$= 2.2\,\text{A}$$

Which is considerably greater than the 120 mA load current!

Quick check: charge is equal to current multiplied by time, which would be *area* on a graph of current against time. If the capacitor has to charge in a tenth of the time that it took to discharge, then it is reasonable to suppose that it will require ten times the current. This brings us to 1.2 A. However, we observed earlier that the shape of the charging pulse is not rectangular, and because the area under this pulse is smaller than a rectangle of equivalent height and width, this accounts for the final difference in the two answers.

Summarizing: the answer is unexpectedly large, but believable.

These high current pulses are composed of a large number of harmonics of mains frequency that extend into the low RF region. Unchecked, these may be fed back into the mains to cause interference to other equipment.

The previous model found the maximum possible ripple current, and it is worthwhile ensuring that the rectifier and capacitor can cope with these current pulses by checking its *peak* current rating. In practice, ripple current will be limited by:

- Series resistance made up of: diode resistance, capacitor ESR, wiring resistance, transformer winding resistance (secondary and reflected primary).
- Transformer core saturation.

Transformer core saturation

Toroidal transformers are more susceptible to core saturation as a direct result of their more nearly perfect design. Transformer cores are made of GOSS (Grain Oriented Silicon Steel), which has the advantage of allowing a higher flux density in the direction of the grain. Traditional E/I cores are unable to take full advantage of this, since there is always a region where the flux is at right angles to the grain. Toroids, however, have all of their flux aligned with the grain, and can operate at a flux density much closer to saturation; this is the main reason for the reduced size of toroids. (A higher operating flux density results in a smaller core.) Consequently, toroids saturate sharply, whereas E/I cores have a much gentler limit.

Transformer core saturation is important because it causes a leakage field of magnetic flux which induces currents in nearby circuitry. Even worse, this saturation happens cyclically (100 Hz or 120 Hz) and so produces *bursts* of interference with harmonics extending to radio frequencies. Sharper saturation will produce a greater proportion of higher harmonics in the leakage field.

This is not merely an apocryphal tale of woe. The author tore his hair out searching for the source of (video) hum in a picture monitor, only to find that the cause was a saturating mains toroid inducing hum directly into the neck of the picture tube.

Choosing the reservoir capacitor and transformer

If we have designed our supply to have a ripple voltage of 5% of supply voltage, then for 90% of the time the transformer is disconnected, and the output resistance of the power supply is determined purely by the capacitor ESR and associated wiring resistance. This is why changing reservoir capacitors from general purpose types to high ripple current types produces a noticeable effect on the sound of an amplifier, they have a lower ESR (but a higher price).

The transformer/rectifier/capacitor combination is a non-linear system. This makes its behaviour considerably more complex than the ideal Thévenin source, and so we need to investigate it over different periods of time.

In the short term (less than one charging cycle) the output resistance of the supply is equal to capacitor ESR plus wiring resistances. This will be true even for very high current transient demands, which may appear in each and every charging cycle, provided that they do not significantly alter the charge on the capacitor. All that is required is that the capacitor should be able to source these transient currents. To be able to do this, the capacitor needs a low ESR, not just at mains frequencies, but also up to 40 kHz, because a Class B power output stage will cause a rectified (and therefore frequency doubled) version of the audio signal to appear on the power supply rails. (See Chapter 5 for explanation of Class B.) We could cope with this requirement by using an electrolytic capacitor designed for use in switch mode power supplies as the main reservoir, and then bypassing it with several smaller value capacitors. See Fig. 4.8.

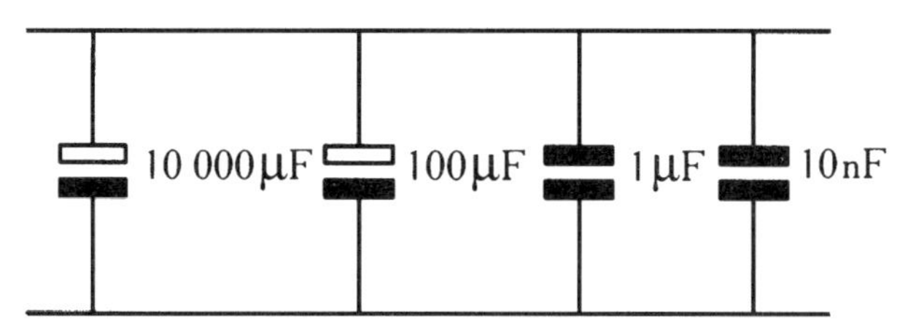

Fig. 4.8 *Use of bypass capacitors to simulate 'perfect' capacitor*

A power amplifier may significantly deplete the charge in the reservoir capacitor, and cause output voltage to fall, either by drawing a sustained high current, due to a continuous full power sine wave test, or by reproducing a short, but loud, sound – such as a bass drum.

Supplying a constant load is relatively easy, because we know exactly how much current will be drawn, and we simply design for that current. If the ripple voltage for a sensible ripple current is higher than we would like, then we simply add a regulator to remove it.

The difficulties start when we want to supply a changing load. It might seem that if the power amplifier is rated at 100 W continuous into 8 Ω, then all we have to do is to calculate what load current that implies, and design for that current. The drawbacks of this approach are more easily demonstrated using a transistor amplifier, where the load is directly coupled to the output stage and power supply:

$$W = I^2R$$

Therefore, for a sine wave:

$$I_{\text{RMS}} = \sqrt{\frac{W}{R}}$$

$$= \sqrt{\frac{100}{8}}$$

$$= 3.5\,\text{A}$$

But we have to supply the *peak* current, which is $\sqrt{2}$ greater, at 5 A. This amplifier would typically have ±50 V rails, and so a power supply capable of delivering ±50 V at 5 A is implied. We therefore need a 500 W power supply to feed a 100 W (mono) amplifier!

This is a very large and expensive power supply, and we would need some very good reasons for pursuing such an approach.

The key to the problem lies in the class of the output stage. (See Chapter 5.) If the output stage operates in pure Class A, then there will be a quiescent current equal to the peak current required at maximum power output, in this case 5 A. If we really draw a constant 5 A from the power supply, then we genuinely do need a 500 W power supply. See Fig. 4.9.

The reservoir capacitor value was easy to determine using our earlier formula and 5% ripple voltage criterion, but the transformer is quite a different matter. It is possible to determine the requirements of the transformer exactly, using the graphs originally devised by O. H. Schade.[1] In practice, the required transformer information is not often available, so a practical rule of thumb is to make the VA rating of the transformer equal to, or greater than, the required output power.

If the amplifier output stage is Class B, then on the crests of the sine wave at full output power, it will still supply 5 A to the load. But at other points in the cycle, the required current will be much lower. The effect of the reservoir capacitor is to average the fluctuating current demand, and for a sine wave:

$$I_{\text{average}} = 0.637 I_{\text{peak}}$$

The average supply current is 3.2 A, and a 350 VA transformer would therefore be chosen.

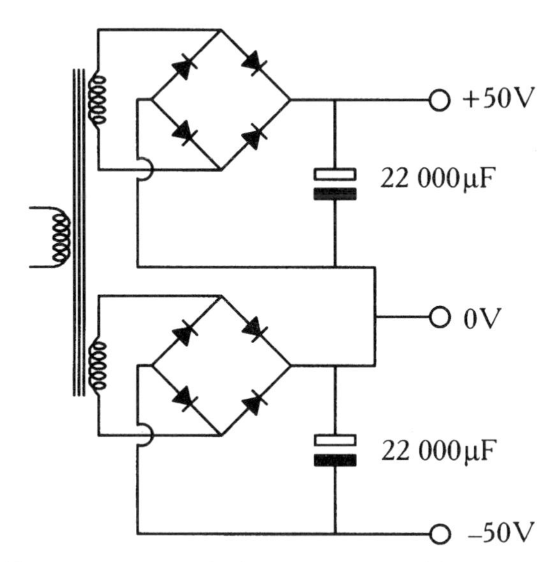

Fig. 4.9 *Typical power supply for transistor amplifier*

We could further argue that the amplifier will not operate at full power all the time, and that the short-term musical peaks requiring maximum output power do not last long. A smaller transformer could therefore be used, since the reservoir capacitor can supply the peak currents. This is a very seductive argument, and many commercial amplifier manufacturers have been persuaded by it, since £1 extra on component cost generally adds £4 to the retail price.

We do not have to work to such tight commercial considerations and, within reason, the bigger the mains transformer, the better.

The choke input power supply

Choke input power supplies were very popular in the heyday of valve amplifiers for the simple reason that large value capacitors were not available, and chokes had to be used for smoothing. See Fig. 4.10.

The great advantage of a choke input power supply is that it draws a very nearly continuous current from the mains transformer rather than a series of high current pulses. To understand why this is so, we need to consider the output waveform of the rectifier. See Fig. 4.11.

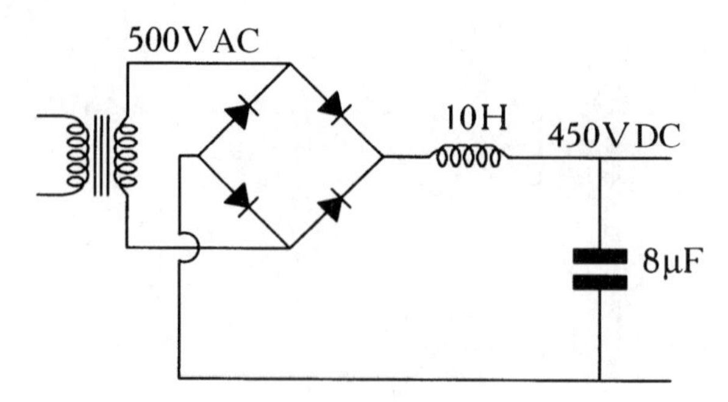

Fig. 4.10 *Choke input power supply*

Fig. 4.11 *Full-wave rectified AC sine wave*

This waveform is a full-wave rectified sine wave, but because it has undergone a non-linear process (rectification), the frequencies present in this waveform are not the same as went into the rectifier. Fourier analysis reveals that the frequencies present in a full-wave rectified sine wave are:

$$v = \frac{2\sqrt{2}V_{\text{in (RMS)}}}{\pi}\left(1 + \sum_{n=1}^{\infty}(-1)^{n+1}\frac{\cos 2n\omega t}{4n^2 - 1}\right)$$

Note that $V_{\text{in (RMS)}}$ is the voltage *before* rectification.

This is a mathematical way of presenting the infinite series, but for our purposes it is much easier to present the information as follows:

$$v = V_{\text{in (RMS)}}\,[0.9 + 0.6(2f) - 0.12(4f) + 0.05(6f) - 0.03(8f)\ldots]$$

This shows us that a full-wave rectified sine wave is made up of a DC component corresponding to $0.9\,V_{\text{in (RMS)}}$, plus a series of decaying even harmonics of the input frequency (f) *before* rectification. The choke has such a high reactance to these AC terms that only the DC component reaches the load. The output voltage of a choke input power supply is therefore $0.9\,V_{\text{in (RMS)}}$, rather than $\sqrt{2}V_{\text{in (RMS)}}$ for the capacitor input supply.

Minimum load current for a choke input supply

Unfortunately, choke input power supplies require that a minimum load current be drawn before they operate correctly. If less than this current is drawn, the circuit reverts to pulse charging of the capacitor, and the output voltage rises to a maximum of $\sqrt{2}V_{\text{in (RMS)}}$. The absolute minimum current that should be drawn is:

$$I_{\text{min.}} = \frac{2\sqrt{2}V_{\text{in (RMS)}}}{6\pi^2 fL}$$

In practice, the inductance of the choke depends partly on the current through it (*BH* curve), so it is wise to draw rather more current than this, and a handy approximation (appropriate for 50 Hz or 60 Hz mains) is:

$$I_{\text{min. (mA)}} = \frac{V_{\text{in (RMS)}}}{L_{\text{(H)}}}$$

The choke input power supply invariably feeds a capacitor, and the minimum current requirement is therefore important, since insufficient current could cause the voltage across the capacitor to rise to ≈157% of nominal voltage, which might destroy it. The traditional way of dealing with this problem was to use a 'swinging' choke, whose inductance rose at low currents, but these were not readily available after the 1960s.

Once the minimum current has been exceeded, the output ripple is constant with load current, and the AC components of the rectified sine wave are attenuated by a factor of:

$$\text{attenuation} = \frac{1}{6\sqrt{2}\omega^2 LC}$$

where $\omega = 2\pi f$.

The second harmonic of the mains (100 Hz) is the only significant term, since not only is the next harmonic attenuated by a far greater factor, but it is at a lower level before filtering.

Current rating of the choke

Although it might initially seem that the choke and mains transformer need only have a current rating equal to the maximum DC load current, they actually have to support a somewhat higher current, and it is particularly important that the choke is correctly rated. Remember that the choke generates magnetic flux in a core proportional to the current passing through the coil, but if too much magnetizing force is applied, the core will saturate, causing inductance to fall to zero.

Since the output of the rectifier comprises a DC component and an AC component, it is the summation of these components that determines the current rating of the choke. The DC component is easily determined, but the AC component requires a little more thought.

Because the choke is followed by a capacitor, which is a short circuit to AC, the entire AC component leaving the rectifier is developed across the reactance of the choke, causing an AC current to flow. Once we know the AC voltage across the choke, we can easily calculate the current.

As previously mentioned, the AC component is dominated by the second harmonic, so we can simplify the calculation to deal exclusively with this component.

The instantaneous AC voltage across the choke is therefore:

$$v = 0.6V_{\text{in (RMS)}} \cos(2\pi ft)$$

where f is the second harmonic of mains frequency.

The reactance of the choke is:

$$X_L = 2\pi fL$$

Using Ohm's law to combine the two equations, the instantaneous current through the choke is:

$$i_{\text{AC}} = \frac{0.6V_{\text{in (RMS)}} \cos(2\pi ft)}{2\pi fL}$$

We are only concerned with the maximum current, which occurs when $\cos(2\pi ft) = 1$, so this factor can be removed, leaving:

$$i_{\text{AC(peak)}} = \frac{0.6V_{\text{in (RMS)}}}{2\pi fL} \quad \text{(second harmonic only)}$$

It was stated that only the second harmonic was significant, but this assumption should now be examined. Referring to the Fourier series, the fourth harmonic is 20% (0.12/0.6) of the voltage of the second. The doubled reactance of the choke at the fourth harmonic halves the choke current, resulting in a fourth harmonic current that is only 10% of the second, so the approximation is fair, but there is room for improvement.

The sum of the AC currents drawn by each of the Fourier terms, up to and including the eighth harmonic, was investigated graphically on a computer to find the maximum positive peak. (The negative peak is irrelevant since when added to the DC load current, it reduces the total peak current.) The result of this exercise modified the equation to:

$$i_{\text{AC (positive peak)}} = \frac{0.544V_{\text{in (RMS)}}}{2\pi fL} = \frac{V_{\text{in (RMS)}}}{1155L} \text{ (50 Hz)}$$

$$= \frac{V_{\text{in (RMS)}}}{1386L} \text{ (60 Hz)}$$

But the total peak current flowing through the choke is the sum of the AC peak current and the DC load current:

$$i_{\text{total peak current}} = I_{\text{DC}} + i_{\text{AC (peak)}}$$

As an example, a Class A power amplifier using a pair of push–pull 845 valves requires a raw HT of 1100 V at 218 mA, and a 10 H 350 mA choke is available, but is this adequate? The transformer supplying the choke input filter has an output voltage of $1224V_{\text{RMS}}$, using the previously derived equation and assuming 50 Hz mains:

$$i_{\text{AC (positive peak)}} = \frac{V_{\text{in (RMS)}}}{1155L} = \frac{1224}{1155 \times 10} = 106 \text{ mA}$$

$$i_{\text{total peak current}} = I_{\text{DC}} + i_{\text{AC (peak)}} = 218 \text{ mA}$$

$$= 106 \text{ mA} = 324 \text{ mA}$$

The total peak current is 324 mA, so the 350 mA rated choke is just sufficient, but the example shows that choke AC current can be surprisingly high, particularly when high HT voltages are contemplated.

Transformer current rating for a choke input supply

The peak choke current must be supplied by the transformer, so the transformer should be rated appropriately. However, since transformer ratings assume resistive loads and sine waves, their current ratings are RMS sine wave, and they can deliver a peak current of $\sqrt{2}$ this value, so the previous example would require a transformer with an RMS sine wave current rating of 229 mA. This is sufficiently close (5% error) to the DC load current, that a common approximation is to assume that the transformer should have an AC_{RMS} current rating equal to the DC load current.

Choke input supply problems

Choke input power supplies are not perfect, and have two main problems.

Current spikes and snubbers

Although we said earlier that the choke input power supply drew a constant current from the mains transformer, this cannot be exactly true. Since the rectifier diodes require a certain voltage across them before they switch on (irrespective of whether they are thermionic or semiconductor), there must be a time, as the input waveform crosses through zero volts, when neither diode is switched on. The current drawn from the transformer is therefore not quite continuous, and must momentarily fall to zero. The choke will try to maintain current, and in doing so, will develop a voltage:

$$V = -L\frac{di}{dt}$$

A snubber network should therefore be fitted to the choke to prevent these spikes. (Curiously, unsuppressed spikes tend to destroy the interwinding insulation of the mains transformer rather than the choke.)

Traditionally, a resistor/capacitor snubber network was connected directly across the choke, but this reduces high frequency filtering because the snubber bypasses the choke. See Fig. 4.12a.

A method that actually improves high frequency filtering is to fit back to back capacitors across the choke, with their centre tap connected to 0 V, and use the internal resistance of the choke as the snubbing resistance. Ideally, C_1 should be chosen so as to resonate with the leakage inductance of the mains transformer at precisely the same frequency as the self-resonance of the choke, but this seems not to be critical, and curiously 220 nF is often a practical value, even for LT supplies. See Fig. 4.12b.

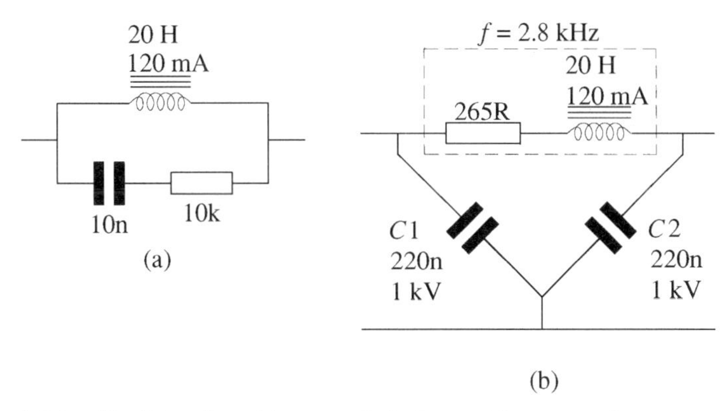

Fig. 4.12 *Traditional and improved choke snubber networks*

As mentioned previously, the entire AC component at the output of the rectifier is across the choke. In Chapter 3 we observed that output transformers could 'sing' due to loose laminations or magnetostriction, and the same is true here. The choke is likely to buzz at twice mains frequency, and if it has any loose parts, such as a loose screening can, it will rattle. Loudly.

Even worse, the choke is bolted to a resonant sounding board (the chassis) which will amplify the buzz. If the choke buzzes, then this is a very serious problem, and the choke will need to be acoustically isolated from the chassis, possibly with soft rubber grommets, but preferably with metal springs. The chassis material can be damped with bitumastic pads (as sold for deadening car body panels), but the amplifier may still need to sit on vibration isolating feet, to avoid exciting the surface on which it sits.

The author once built a stereo 10W EL84 amplifier using a choke input power supply, and had to resort to all the previous ruses to defeat the buzz. On reflection, it seems likely that the buzz was caused because the choke was very close to saturation, as it was rated at 250 mA, but was required to support a peak current of 240 mA. By contrast, the prototype electrostatic 'Beast' (described in Chapter 5) has a choke input HT supply which is silent, but the choke input LT supply (which currently uses a rather inferior choke) buzzes.

It is well worthwhile considering using a choke input power supply if you are designing from scratch, and are prepared to deal with the vibration problem, but don't even think of fitting it as a modification to an existing circuit!

Broad-band response of practical *LC* filters

So far, our investigation of choke input filters has focused on their behaviour at low frequencies, but we now need to broaden our outlook to include behaviour from DC to low radio frequencies. To attenuate low (≥100 Hz) frequencies, a large inductance is required, but this inevitably has internal shunt capacitance; conversely, the capacitor has series inductance, and these hidden components mean that any practical LC filter has a frequency response that may be divided into four main regions. See Fig. 4.13.

Region 1: This is the only region we can control, so it is well worthy of investigation. Apart from losses due to DC resistance, the low-pass filter does not attenuate frequencies below the LF resonance:

$$f_{res.} = \frac{1}{2\pi\sqrt{LC}}$$

We aim to position the (hopefully) subsonic resonance as low as possible by making L and C large, because every octave by which we can lower $f_{res.}$ produces an additional 12 dB of filtering.

If the resonance has $Q > 0.707$, it will cause an undesirable peak in the response of the filter, so it is useful to check Q:

$$Q = \frac{1}{R}\sqrt{\frac{L}{C}}$$

where: L = inductance of choke
R = internal resistance of choke
C = capacitance of smoothing capacitor

Strictly, the load resistance across the capacitor also damps the resonance, and this may be accounted for by adding a notional series resistance to the choke:

$$R_{notional} = \frac{L}{CR_{load}}$$

However, for most practical applications the effect of load resistance is negligible. Ideally, the resonance should be critically damped ($Q = 0.5$), which is achieved by adding external series resistance to the choke.

Example: a traditional filter might have used a 15 H choke with 260 Ω internal resistance coupled to an 8 μF paper/oil capacitor, resulting in

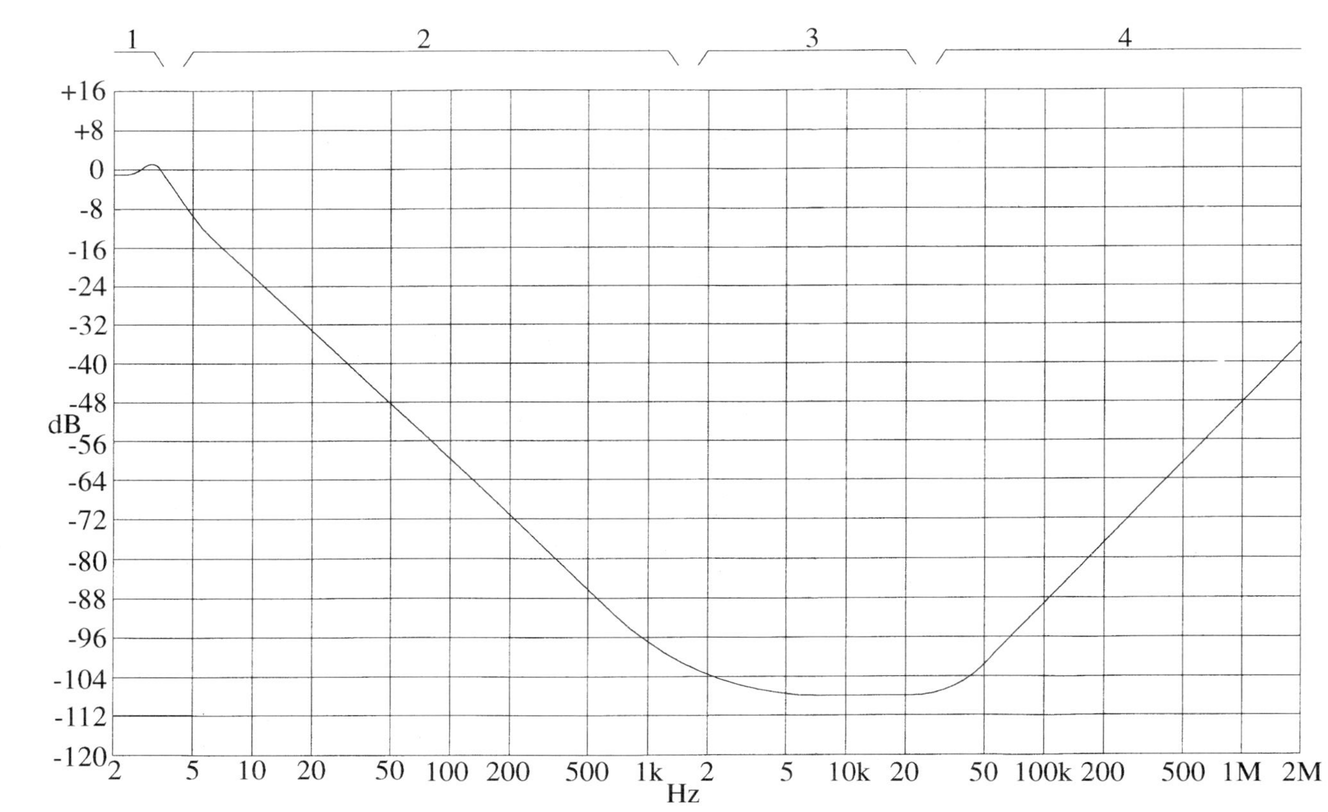

Fig. 4.13 *Response of LC filter (20H 50 mA, 120 μF 400 V polypropylene)*

$f_{res.}$ = 14.5 Hz, Q = 5.27. This Q is too large, and $f_{res.}$ is too near the audio band, but the additional 2.48 kΩ series resistance required to achieve critical damping would be wasteful of HT voltage. A better approach would be to replace the 8 μF capacitor with a 120 μF polypropylene, since this would give $f_{res.}$ = 3.75 Hz, Q = 1.36, and this Q might be tolerated, but only 447 Ω of additional series resistance would be required to achieve Q = 0.5.

Region 2: The reactance of the choke doubles for each octave rise in frequency, whilst the reactance of the reservoir capacitor halves, producing the familiar 12 dB/octave slope.

Region 3: The shunt capacitance of the choke begins to take effect. Once the reactance of the shunt capacitance is equal to the inductive reactance of the choke, the choke resonates, so this region may be defined as beginning at choke self-resonance (≈3–15 kHz for an HT choke). Above this self-resonant frequency, the shunt capacitance forms a potential divider with the smoothing capacitor whose loss is constant.

$$\text{Loss(dB)} \approx 20\log\left(\frac{C_{choke}}{C_{smoothing}}\right)$$

Region 4: The series inductance of the reservoir capacitor becomes significant, and this forms a hidden high-pass filter in conjunction with the shunt capacitance of the choke, and the output noise of the practical filter rises at 12 dB/octave.

Wide-band filtering

Power amplifiers have high signal voltages, and are relatively tolerant of RF noise on the HT supply. Pre-amplifiers have lower signal voltages, and stages may be designed to optimize noise performance at the expense of headroom, so there is a danger that RF noise may overload a stage, resulting in the noise being demodulated and translated into the audio band. For this reason, RF noise should be taken particularly seriously in pre-amplifiers, and wide-band filtering becomes desirable.

The best strategy for wide-band filtering is to use a cascade of different filters, with an LF frequency filter first. We should attempt to filter low frequencies first for two reasons:

- The output of the rectifier contains large amplitudes at low frequencies, for which an LF choke is designed, but which would saturate an RF choke.

- As frequency increases, stray inductance and capacitance become more significant. This means that even 100 mm of wire has inductance, and could be used as a choke at UHF, but it is also an aerial whose length should be minimized near the load.

Typically, we might start with a 15 H choke input PSU and 120 μF reservoir capacitor to reduce ripple to a few volts. If required, a practical second stage might comprise a low RF filter (perhaps 100 mH on a large ferrite toroid, with 10 nF to ground) close to the load. If fitted, any regulator should ideally be connected between this and any final stage of filtering, in order to minimize its exposure to RF. In critical situations, a final stage might use a VHF ferrite bead followed by a feedthrough capacitor directly adjacent to the anode load resistor, thus ensuring that the final high frequency filtering is performed as close to the load as possible, and minimizing the length of aerial.

Regulators

The best way of improving a power supply is to use a voltage regulator. A voltage regulator is a real world approximation of a Thévenin source; it has a fixed output voltage, and an output resistance that approaches zero. A true Thévenin source implies infinite current capacity, whereas the supply that feeds our regulator will have limited current capacity. It is therefore important to realize that the regulator will only simulate a Thévenin source over a limited range of operation, and we must therefore ensure that we remain within this range under *all* possible operating conditions.

Voltage regulators are based on the potential divider. Either the upper or the lower leg of the divider is made controllable in some way, and by this means, the output voltage can be varied. See Fig. 4.14.

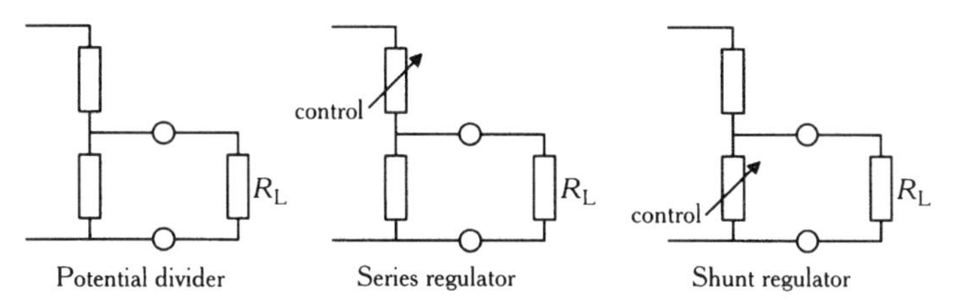

Fig. 4.14 *Relationship between voltage regulators and the potential divider*

If the upper element is made controllable, then the regulator is known as a *series* regulator, because the variable element is in *series* with the load. If the lower leg of the divider is made controllable, then the regulator is known as a *shunt* regulator, because the variable element is *shunted* by the load. Shunt regulators are inefficient compared to series regulators, and the design has to be carefully tailored to match their load.

The fundamental elements of a series voltage regulator are shown in Fig. 4.15.

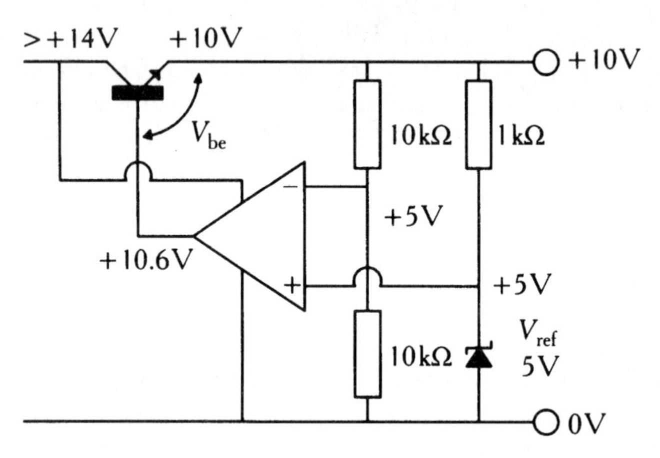

Fig. 4.15 *Series regulator*

This circuit is shown using semiconductors, but a valve version could equally well be built. The error amplifier amplifies the difference between the reference voltage and a fraction of the output voltage, and controls the series pass transistor such that a stable output voltage is achieved.

The circuit depends for its operation on negative feedback. We saw in Chapter 1 that when feedback is applied, input and output resistances change by the feedback factor $(1 + \beta A_0)$. Voltage regulators rely on reducing the output resistance of the system by the ratio of the feedback factor.

Suppose initially that the regulator is working, and that there is 10 V at the output. By potential divider action, there must be 5 V on the inverting input of the operational amplifier. The voltage reference is holding the non-inverting input at 5 V. The series pass transistor is an emitter follower fed by the error amplifier, and has 10 V on its emitter, so the base must be at 10.6 V.

Suppose now that the output voltage falls for some reason. The voltage at the midpoint of the potential divider now falls, but the voltage reference maintains 5 V. The error amplifier now has a higher voltage on its non-inverting input than on its inverting input, and its output voltage must rise. If the voltage on the base of the transistor rises, its emitter voltage must also rise. The circuit therefore opposes the reduction in output voltage.

Since the same argument works in reverse for a rise in output voltage, it follows that the circuit is stable and that the output voltage is determined by the combination of the potential divider and the reference voltage. If we redraw the regulator, we can easily see that it is simply an amplifier whose gain is set by the potential divider, and that it amplifies the reference voltage. See Fig. 4.16.

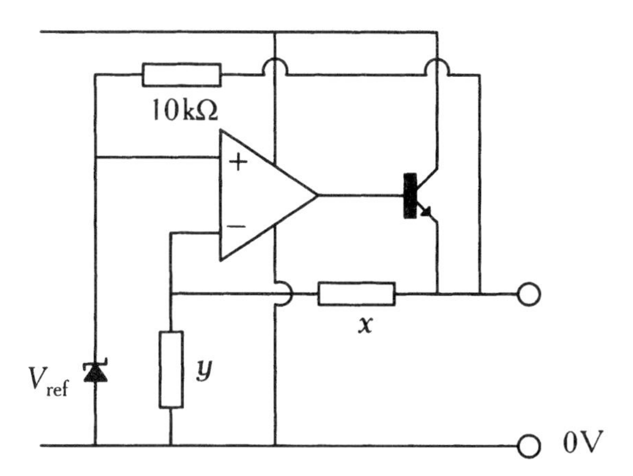

Fig. 4.16 *Series regulator redrawn to show kinship to non-inverting amplifier*

By inspection, the output voltage is therefore:

$$V_{out} = \frac{x + y}{y} \cdot V_{ref}$$

Since the error amplifier simply amplifies the reference voltage, any noise on the reference will also be amplified, and it is therefore good practice to feed it from as clean a supply as possible. Although the argument seems like a snake chasing its own tail, if we feed the reference from the output of the supply (which is clean), then the reference will be clean, and the output of the supply will also be clean. It might be thought that supplying the current for the reference voltage from the output voltage might cause instability, but in practice this is not a problem.

It should be noted that the regulator needs an input voltage higher than the output voltage. The minimum allowable difference between these voltages before the regulator fails to operate correctly is known as the drop-out voltage (because the regulator 'drops out' of regulation). With this particular design it is only a few volts, but drop-out voltage for a valve version could be 40 V or more.

The two transistor voltage regulator

The two transistor series regulator is a very common circuit. See Fig. 4.17.

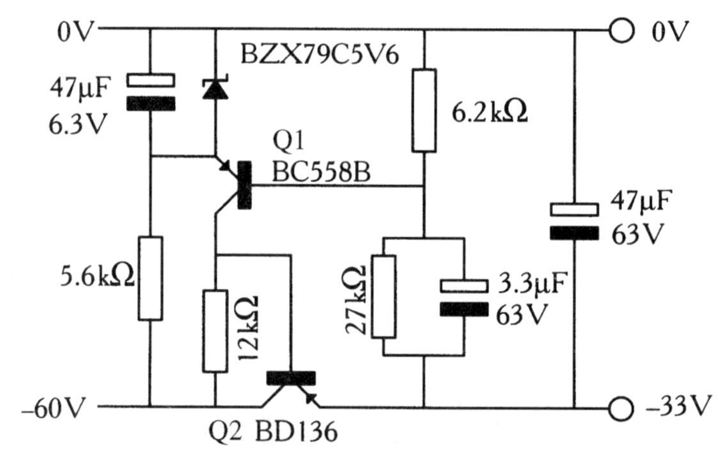

Fig. 4.17 *Basic two-transistor negative regulator*

This circuit is often used because of its extreme cheapness, but despite that, its performance is really quite good. Q2, the series pass transistor, is fed from the collector of Q1, a common emitter amplifier. The emitter of Q1 is held at a constant voltage by the voltage reference, whilst its base is fed a fraction of the output voltage by the potential divider. If the output voltage should rise, Q1 will turn on harder, drawing more current; its collector voltage and the base voltage of Q2 will therefore fall, causing the emitter voltage of Q2 to fall, and output voltage to fall, thus counteracting the initial error.

This circuit is ideal for use as a bias voltage regulator in a power amplifier, because we often find that we need to drop more volts across the regulator than an IC regulator would tolerate.

As presented, the circuit can only supply 50 mA of output current, because the base current for Q2 is stolen from the collector current of Q1. If we increased the collector current of Q1, Q2 could steal more, and output current could be increased, but a better solution would be to replace Q2 with a Darlington transistor, which would need less base current.

The Zener diode passes 12 mA, which is quite sufficient to ensure that it operates correctly, and has a stable output voltage. Because Zeners produce noise, it is usual to bypass them with a capacitor. (A better solution might be to replace the Zener with a red LED (1.6 V @ 15 mA) and adjust circuit values.)

The capacitor across the upper resistor of the potential divider is significant. The purpose of this capacitor is to increase the amount of negative feedback available at AC, and thereby reduce output ripple and noise. The value of this capacitor is critical, because if it is too large it slugs the response of the regulator to changes in load current. The maximum value for this capacitor is found by first calculating the AC Thévenin resistance that it sees:

$$r_{\text{Thévenin}} = \frac{1}{\dfrac{1}{h_{ie}} + \dfrac{1}{x} + \dfrac{1}{y}}$$

Remembering that:

$$h_{ie} = \frac{h_{fe}}{gm}$$

And also that:

$$gm = 35I_c$$

We find that for this circuit, with $h_{fe} = 200$, and $I_c = 12$ mA, $h_{ie} \approx 500\,\Omega$. So the Thévenin resistance seen by the capacitor is $\approx 450\,\Omega$.

The lowest ripple frequency to be attenuated is at 100 Hz (120 Hz USA), so we would like the capacitor to have a significant effect on this. The potential divider chain and capacitor is a step equalizer whose effect on the regulator is similar to that used for the RIAA 3180/318 μs pairing in Chapter 6. A good initial choice of capacitor value is to make the reactance of the capacitor at the lowest ripple frequency equal to the Thévenin resistance at the tapping of the potential divider. This value of capacitor is a compromise between ripple reduction and regulator LF transient response, and an infinitely large capacitor could only improve ripple reduction by a further 3 dB:

$$C = \frac{1}{2 \times \pi \times 100 \times 450}$$

$$= 3.5\,\mu\text{F}$$

The nearest value is 3.3 μF, and this is the maximum value that we should use, since larger values degrade LF transient response. This is quite a small capacitor, and the author has seen many similar circuits with oversized capacitors, and, indeed, built one himself. The subjective effect of the oversize capacitor was to create a bass boom that was incorrectly thought to be due to room acoustics. You may wish to experiment with still smaller values.

The regulator also has a capacitor across its output. Because the gain of the error amplifier falls with frequency due to Miller effect and stray capacitances, the amount of gain available for reducing output resistance falls. If $(1 + \beta A_0)$ has fallen, then the output resistance must rise, and the effect is that output resistance rises with frequency. A perfect Thévenin source in series with an inductor would look identical, and for this reason the output of regulators is often described as being inductive at high frequencies. The capacitor maintains a low output impedance at high frequencies.

A variable bias voltage regulator

We often need a bias voltage regulator to be variable between certain limits. In this example, we will look at a grid bias regulator needed for an 845 directly heated triode. Perusal of RCA anode characteristics (circa 1933) indicated a grid bias voltage of –125 V, but modern valves do not seem to exactly match the original curves, and we need to be able to equalize anode currents in this (push–pull) output stage in order to avoid saturating the output transformer with an out-of-balance DC current and causing distortion. A range of ±25 V either side of the nominal –125 V seems reasonable, but how can we design a regulator to fulfil this requirement exactly?

Fortunately, since it supplies a part of the circuit where the signal voltages are very high (up to 90 V_{RMS}), the regulator need not necessarily have an impeccable noise performance, and Zener diodes are perfectly acceptable. See Fig. 4.18.

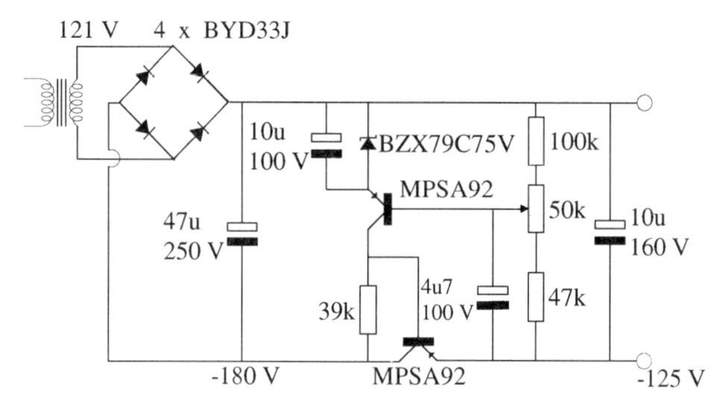

Fig. 4.18 *Adjustable –125 V bias regulator*

The higher Zener voltage chosen, the better regulation the finished circuit will have, but we must still allow a reasonable voltage between the collector and emitter of the control transistor. In practice, a Zener voltage of about half the maximum output voltage is usually a good choice, and 75 V Zeners are readily available.

The Zener will hold the emitter of the transistor at –75 V, and V_{be} = 0.6 V, so the base of the transistor will be held at a fixed potential of –75.6 V. Since the base of the transistor is connected to the wiper of the potential divider, the wiper must also be held at –75.6 V, no matter what output voltage is set. We can now calculate the required attenuation of the potential divider for the two extreme design cases:

$$\frac{100\,\text{V}}{75\,\text{V}} = 1.323$$

$$\frac{150\,\text{V}}{75.6\,\text{V}} = 1.984$$

By choosing a convenient value for the variable resistor in the middle of the potential divider, we now have enough information to calculate the required resistors either side. A low value of variable resistor would require a large current to flow in the potential divider, whereas too high a value will cause errors due to the (small) base current drawn by the transistor. A good engineering principle is that the potential divider chain should pass roughly ten times the expected base current, so a 50 kΩ variable resistor was a convenient standard value for this example.

When the wiper of the variable resistor is set to produce the largest voltage at the regulator output, it is connected directly to the grounded resistor (x), and vice versa. Using the standard potential divider equation, for –150 V:

$$\frac{x + y + 50}{x} = 1.984$$

And similarly for –100 V:

$$\frac{x + y + 50}{x + 50} = 1.323$$

We now have two equations that can be solved, either simultaneously, or by substitution, to give the values of the fixed resistors x and y. In this particular case, the values fell out very conveniently to give, $x = 100\,\text{k}\Omega$ and $y = 47\,\text{k}\Omega$.

The 317 IC voltage regulator

Although the two transistor regulator is the ideal choice for a bias regulator, because of its high voltage drop capability, once we need higher currents at lower voltages, its deficiencies become apparent.

It is perfectly possible to build a voltage regulator using a handful of components including an operational amplifier, a voltage reference, various transistors, resistors and capacitors. With care, the circuit can be made to work almost as well as an IC regulator, and only cost about three times as much. We need not feel guilty about using IC regulators.

The 317 is a standard device that is made by all the major IC manufacturers,[2] and they all have virtually identical performance. Linear Technology[3] make an upgraded version of the 317, the LT317, but the *only* difference is that the guaranteed tolerance of the voltage reference is tighter. A commercial design could therefore set its output voltage using fixed resistors rather than a variable resistor, thus saving money, because not only are variable resistors expensive to buy, but they then have to be adjusted. We do not have to worry about such considerations, and so the standard 317 is fine.

The 317 incorporates all of the fundamental elements of a series regulator in one three terminal package, and we only need to add an external potential divider to produce an adjustable regulator. See Fig. 4.19.

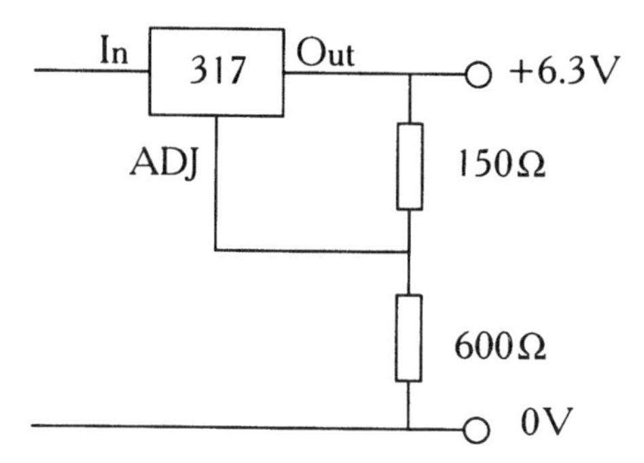

Fig. 4.19 *Basic 317 regulator circuit*

One end of the voltage reference is connected to the OUT terminal whilst the other is an input to the error amplifier. The other input of the error amplifier is the ADJ terminal. The 317 therefore strives to maintain a voltage equal to its reference voltage (1.25 V) between the OUT and ADJ terminals. All we have to do is to set our potential divider so that the voltage at the tap is $V_{out} - 1.25\,V$, and the 317 will do the rest.

In data sheets for the 317, you will invariably find that the upper resistor of the potential divider is 240 Ω. The reason for this is that the 317 requires 5 mA of current to be drawn before it can regulate. If the potential divider passes 5 mA, then this ensures that the device is able to regulate even if there is no external load.

The 317 also sources ≈50 μA of bias current from the ADJ pin, which flows down the lower leg of the potential divider. Normally, this is negligible, but if you are designing a high voltage regulator, and choose a lower potential divider current, this will need to be taken into account.

As an example, the manufacturer's data sheets show a regulator bypassing the ADJ pin to ground with a 10 μF electrolytic, from which they show an improved ripple rejection from 60 dB to 80 dB at 100 Hz. This is directly equivalent to the capacitor that we added in the two transistor regulator, but because the reference voltage is tied to V_{out}, rather than ground, the potential divider bypass capacitor connects to ground, rather than V_{out}.

We could therefore use the method derived earlier to see if we think that this is the optimum choice of capacitor. The ADJ pin is an input to an operational amplifier, so we can treat it as infinite input resistance, and we are only concerned with the external resistor values. If we were to use an upper resistor of 240 Ω, and a 3.9 kΩ lower resistor to set an output voltage of 22 V, then the optimum value of capacitor would be 7 μF, so a 10 μF electrolytic is a reasonable choice, although the author would probably prefer 6.8 μF, or less.

Just like the two transistor regulator, the output of the 317 is inductive, and the manufacturer's output impedance curves suggest that the output impedance is equivalent to ≈2.2 μH in series with 2.7 μΩ, so they recommend a 1 μF tantalum bead output bypass capacitor, as shown in the equivalent circuit. See Fig. 4.20.

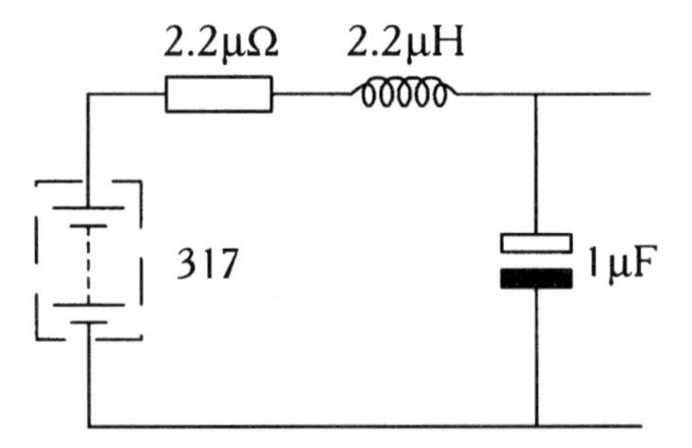

Fig. 4.20 *AC Thévenin equivalent of 317 plus 1 μF bypass capacitor*

If we assume that the tantalum bead capacitor is a perfect capacitor (!), then we have an underdamped resonant circuit, with a Q of:

$$Q = \frac{1}{R}\sqrt{\frac{L}{C}}$$

Theoretically, Q = 550 000! Stray resistance will reduce this considerably, but it will not reduce it to Q = 0.5, which would be critically damped. This would not matter greatly, because we would be unable to excite the circuit from the output (any external excitation would be short circuited by the capacitor). If we now concede that the capacitor is not perfect, we may be unlucky enough to be able to excite the resonance, and the circuit could become unstable. Using the formula, $3\,\Omega$ would critically damp the resonance, and the manufacturer recommends $2.7\,\Omega$ in series with the tantalum capacitor.

LT supplies and common-mode noise

Older pre-amplifiers had AC heaters and suffered hum. Modern pre-amplifiers have DC heaters, but because this high current (1–2 A) is awkward to smooth adequately purely passively, they invariably add a regulator to reduce hum to mV levels. The fact that the heater voltage is now stabilized at 6.3 V is a bonus.

Almost every effort within the *design* of audio circuitry to increase immunity to HT noise increases sensitivity to LT noise because it increases r_k (see Chapter 2), but heater supplies are somewhat different to HT supplies, and cause considerable confusion. LT supply ripple may be defined as differential mode noise, since it is the difference in voltage between one heater pin and another. Provided that differential noise is reasonably low, the considerable thermal inertia of an indirectly heated cathode will filter it out completely, and differential RF noise is irrelevant.

Nevertheless, there has recently been a trend to using exotic (and expensive) heater regulators in the belief that this allows the finished amplifier to sound better. If valve heaters were designed to be supplied by $6.3\,V_{RMS}$ of AC ripple, it hardly seems likely that they will be materially affected by differences between heater regulators producing perhaps $<10\,mV_{pk-pk}$ ripple, since this is essentially perfection anyway, so why are there reports of audible improvements caused by changing heater regulators?

Valves may not be particularly sensitive to differential-mode noise on LT supplies, but they certainly are sensitive to common-mode noise. In the context of an LT supply, common-mode noise leaves the potential difference between the ends of the heater constant, but *both* voltages are bouncing up and down, and if we view the heater as a single conductor, its voltage with respect to the cathode is changing, rendering it perfectly capable of inducing noise into the cathode.

Common-mode heater noise is a problem for small-signal valves because the noise current is capacitively coupled from the heater helix directly to the enclosing cathode. Unless decoupled by a cathode capacitor, the noise current develops a noise voltage at the cathode primarily determined by r_k, which is then added to the wanted signal and amplified by the valve. Cathode followers have a low (but non-zero) r_k, and differential pairs have quite a high r_k, so it is clear that we need to be able to prevent common-mode noise from reaching the heater supply.

Because regulators are designed to address differential-mode noise they are normally ineffective against common-mode noise, although their support circuitry may inadvertently assist in rejecting common-mode noise. Useful common-mode filtering is only gained by employing RF techniques involving series RF chokes and shunt capacitors to chassis. See Fig. 4.21.

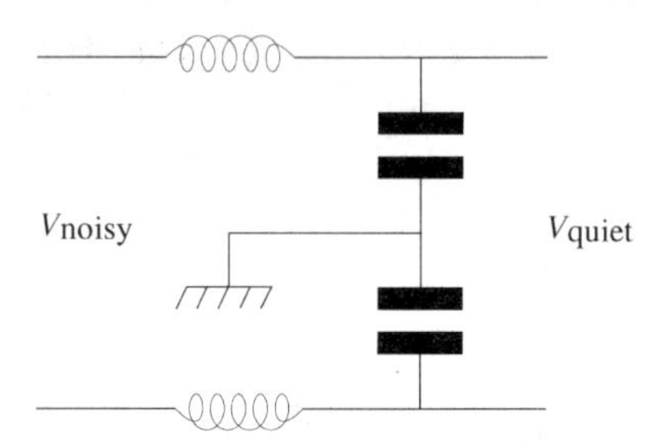

Fig. 4.21 *Filtering common mode noise*

Because the heater is capacitively coupled to the cathode, and the reactance of C_{h-k} is low at RF, the cathode is likely to be particularly sensitive to RF for two reasons:

- In a single ended stage, the cathode will be (should be) decoupled to ground using a large capacitor. But the capacitor is not connected directly onto the (hot) cathode, so the inductance in the connecting wires reduces its effectiveness at RF. Additionally, the (inevitably electrolytic) capacitor itself has some inductance.
- In a differential pair, the cathode unavoidably has quite a high resistance to ground, and cannot form a useful *CR* filter in conjunction with C_{h-k}. We are forced to rely on the (usually quite poor) RF balance of the differential pair to reject RF, so a differential pair is likely to be more sensitive to heater borne RF than a single ended stage.

Sources of common-mode LT noise

Common-mode noise currents are capacitively coupled to the LT winding from any adjacent winding on the transformer, with high voltage windings being the worst offenders. Although there is often an electrostatic screen around the mains winding, it is most unusual to find a screen around the HT winding. This is significant because the HT rectifier diodes generate copious RF noise when they switch, which is easily coupled via the interwinding capacitance. As an example, a recent version of the Bevois Valley amplifier (see Chapter 6) used a Leak Stereo 20 as the donor chassis (retaining the GZ34 rectifier), but HT switching spikes were easily observable on the 6.3 V_{AC} LT supply when the amplifier was first tested.

Typical mains transformers have an interwinding capacitance of ≈1 nF between adjacent windings, and so the HT supply should ideally use a separate transformer. However, the problem may be alleviated if this is not possible. The first step in the cure is to connect a small (≈10 nF is fine) capacitor from each side of the heater supply to chassis using extremely short leads. The two capacitors are in parallel so far as common-mode noise is concerned, so their 20 nF forms a potential divider in combination with the 1 nF interwinding capacitance. See Fig. 4.22.

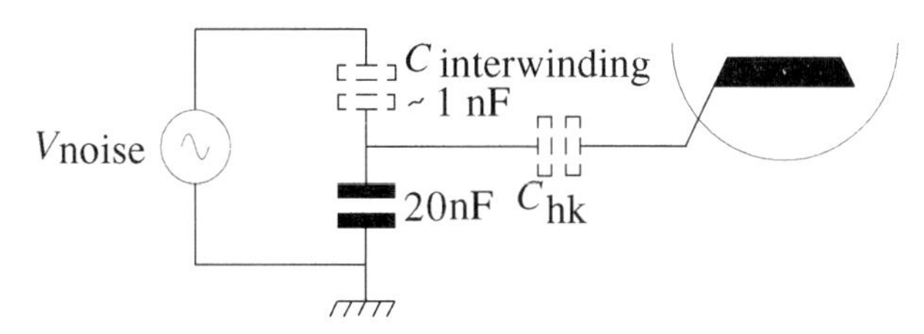

Fig. 4.22 *Potential divider formed by interwinding capacitance and heater bypass capacitors*

Because the reactance of a capacitor is inversely proportional to its capacitance, the equation for the loss through this potential divider is found using the *other* capacitor value as the numerator:

$$\frac{V_{out}}{V_{in}} = \frac{1n}{20n + 1n} = 0.048 = -26\,\text{dB}$$

Notice that a potential divider formed of only one type of component (*R* or C or *L*) attenuates all frequencies equally, so RF noise at *all* frequencies is attenuated now by 26 dB, but we can do better than this. If we were to add equal inductance to each leg of the heater supply, we could create an *LC* filter which would improve attenuation. Since we are trying to filter common-mode noise, rather than differential mode, we can wind a bifilar choke on a small ferrite core without worrying about core saturation, because the DC magnetization of the core caused by the heater current will cancel.

The 317 as an HT regulator

Because the 317 is a floating regulator, there is no reason why it should not be used to regulate a 400 V HT supply. However, because the 317 can only tolerate 37 V from input to output, support circuitry is needed to protect it.[4] See Fig. 4.23.

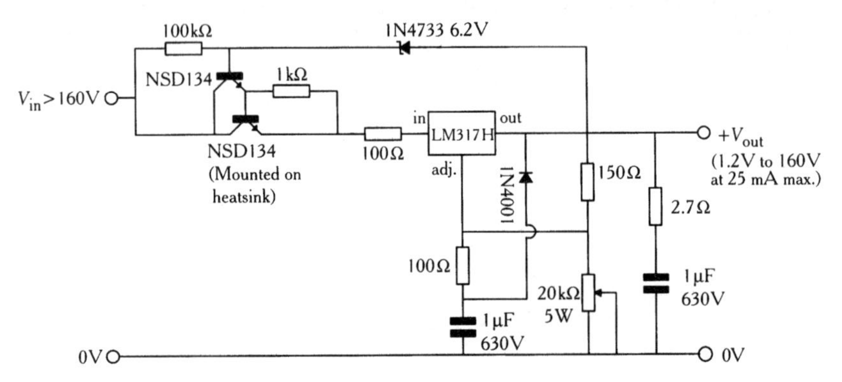

Fig. 4.23 *Basic high voltage regulator (Reprinted by kind permission of National Semiconductor)*

The 317 is preceded by a high voltage transistor Darlington pair whose sole aim in life is to maintain 6.2 V between the IN and OUT terminals of the 317; longevity of the 317 is thus ensured. The Darlington pair can easily cope with variations in mains voltage, but it should not be thought that this circuit is proof against a short circuit when used at typical valve voltages.

Accidentally short circuiting a regulator of this type with a 'scope probe results in an almighty bang and the silicon is destroyed. The author *knows*.

The lower arm of the potential divider is bypassed, but has a resistor in series with the capacitor to improve LF transient response by raising the lower $f_{-3\,dB}$ frequency of the step equalizer, and a diode has been added, allegedly to discharge the capacitor in the event of an output short circuit (although the author's experience is that it doesn't actually help).

This fundamental circuit will be investigated further when we design a complete HT and LT power supply.

The valve voltage regulator

Valve voltage regulators have always been very rare, and we will now see why. See Fig. 4.24.

The circuit is very similar to the two transistor regulator, it simply has valves and higher voltages. The Zener diode has been replaced by a neon stabilizer which burns at 85 V; this holds the cathode of the EF86 stable, and the grid is fed from the potential divider. The series pass element is a 6080 double triode ($P_{a(max.)} = 13$ W) which was specifically designed for use in series regulators and can pass high currents at low anode voltages.

A valve rectifier is used, and in deference to its limited ripple current capacity, an 8 μF paper/oil capacitor has been chosen for the reservoir, although a higher value (<60 μF) polypropylene would be a physically practical value. This results in considerable ripple voltage which is filtered by the following LC combination.

The performance of the regulator is slightly compromised by feeding g_2 of the EF86 from the raw supply, but if fed from the regulated supply, there is a danger that the circuit might not switch on, and might simply sulk. The gain of the EF86 is ≈100, and above ≈100 Hz this gain is available for reducing the output resistance of the 6080, whose $gm \approx 7$ mA/V, so $r_k \approx 200\,\Omega$ (including the effect of the external 100 Ω R_a). The regulator therefore achieves an output resistance of ≈2 Ω. The EF86 is noisy (2 μV), but this noise is swamped by the noise from the 85A2, which is even noisier at 60 μV.

Because each valve has its cathode floating above ground, three separate heater supplies are needed (the 85A2 is a cold cathode valve). The EF86 could be supplied by a grounded heater supply, but this is putting quite a strain on the heater to cathode insulation. One way of obtaining the remaining heater supply if a piece of classic equipment is being modified is to remove the GZ34 rectifier and to supply the (6.3 V heater) 6080 from the (5 V heater) GZ34 winding. The heater is thus considerably underrun, but it seems to work, provided not much current is asked of the regulator. Neither of these techniques is recommended.

The author once built a rather more sophisticated 420 V valve regulator with floating heater supplies and an error amplifier made of two cascaded ECC81 differential pairs. This had a commensurately lower output resistance, and the measured hum and noise was 400 μV_{pk-pk}. It was big. It

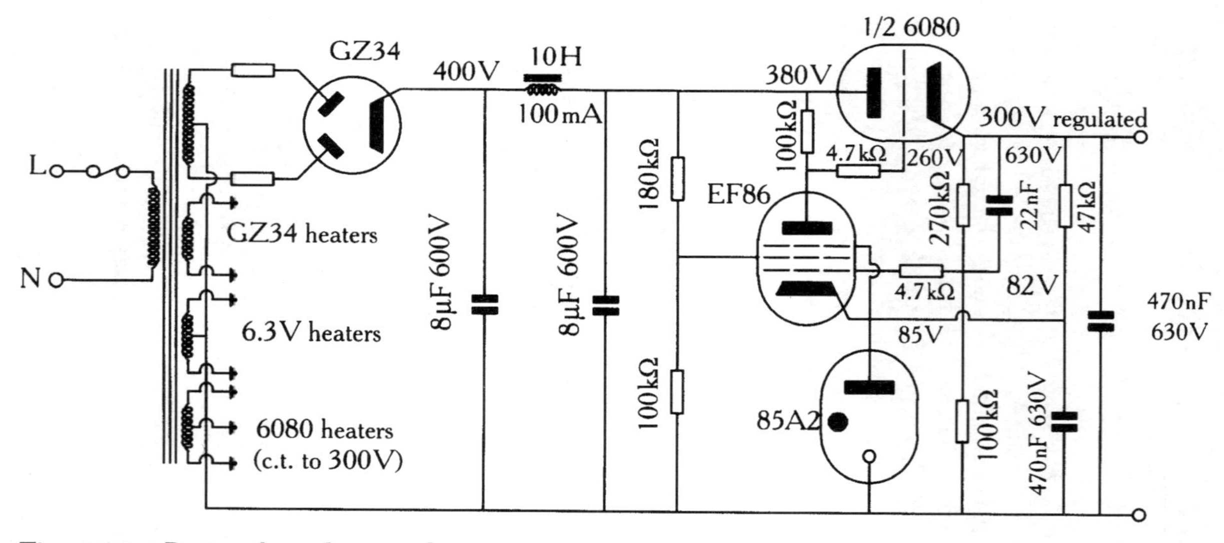

Fig. 4.24 *Basic valve voltage regulator*

was heavy. It drifted. Viewed on an AC coupled oscilloscope, the output voltage gently perambulated around 420 V due to heater voltage variation (unregulated AC heater supplies). The 317 regulator is rock steady.

Power Supply Rejection Ratio (PSRR) to individual stages and stability

Although individual stages may be designed and interconnected to form an audio system, each stage requires HT, which will ultimately always be derived from a common source. No practical source has zero output resistance, although the AC mains is quite a good approximation.

The issue of a common power supply with non-zero output resistance is crucial because it implies that as a given audio stage draws a varying supply current (in sympathy with the audio signal), a voltage will be developed across the source resistance of the supply. Although attenuated by individual stage PSRR, this voltage is now an *input* to all other stages, and if gain between stages is high (as in an RIAA stage) whilst PSRR is low, then loop gain via the power supply may be greater than unity, resulting in oscillation.

Traditional interstage filtering used a shunt capacitor to define source resistance (strictly impedance), resulting in increased source impedance at low frequencies since:

$$Z_{source} = \frac{1}{2\pi fC}$$

Low frequency (≈1 Hz) instability was known classically as *motorboating*, but marginal stability probably went unnoticed much of the time, because loudspeakers of that time had very stiff suspensions and would filter this out.

Modern designs use regulators giving excellent Z_{source} down to DC, but because the error amplifier must have a response falling with frequency in order to maintain its own stability, Z_{source} is inductive and rises with frequency, and so HF instability is a possibility.

Summarizing, any practical common supply will always have a non-zero output impedance, and system stability can only be maintained if individual stages have sufficient PSRR *to that common supply.*

To aid clarity, it is useful to define two new terms:

- Intrinsic PSRR: the PSRR due to the topology of an individual stage.
- Common supply PSRR: intrinsic PSRR plus any added PSRR (by whatever means) to the common supply point.

Any common cathode stage possesses intrinsic PSRR by virtue of the potential divider formed by r_a and R_L, but an E88CC operated such that $r_a = 5\,k\Omega$, and $R_L = 47\,k\Omega$ only results in an intrinsic PSRR (referred to the output) of 20 dB. Using the same valve as a μ-follower could improve this to 44 dB, and a differential pair might manage 62 dB, but used as a cascode, the 20 dB figure would be degraded to zero ($r_a \approx \infty$).

Any given stage may have its common supply PSRR increased by an arbitrary amount using individual filtering or regulation, and, apart from expense, it does not matter whether the common supply PSRR is made up mostly from intrinsic PSRR, or added PSRR via filters or regulators. Extreme methods might even include individual mains transformers and supplies for each stage, to increase common supply PSRR to the AC mains (the common supply point), whilst the use of a dedicated spur from the electricity supply company cable head would also be a means of reducing Z_{source}. The ultimate, mindboggling, solution would be to use individual batteries for each stage.

One method for increasing PSRR to each stage would be to use an individual regulator for each stage, but even 317 regulators are expensive if we use enough of them. A considerably cheaper method would be to design as many stages as possible to need the same HT voltage, and then to isolate the stages by adding the (high) intrinsic PSRR of an op-amp to the stage's intrinsic PSRR by supplying each stage via a voltage follower op-amp. See Fig. 4.25.

The idea is that each buffer has as its input an *RC* filtered supply, and the high current gain of the buffer allows a low output resistance to drive the load, whilst the high input resistance provides a very light load on the *RC* filter. A Junction Field Effect Transistor (JFET) would be ideal as the input buffer because of its very high input resistance. Unfortunately, JFETs tend to have a high output resistance and appalling device variation, so a bipolar emitter follower could be added to form what is occasionally known as a FETlington (FET/Darlington). This device will then need a bias network, and is rapidly becoming complicated, so why not use a dual JFET input op-amp as a voltage follower?

The Burr-Brown OPA2134 is a splendid JFET input, bipolar output, dual op-amp capable of sourcing up to 40 mA into the load. Its only disadvantage for our application is that it achieves its excellent specifications by consuming 4 mA per op-amp, which would be trivial if supplied from a dedicated ±15 V supply, but we are more likely to derive this current from the HT supply.

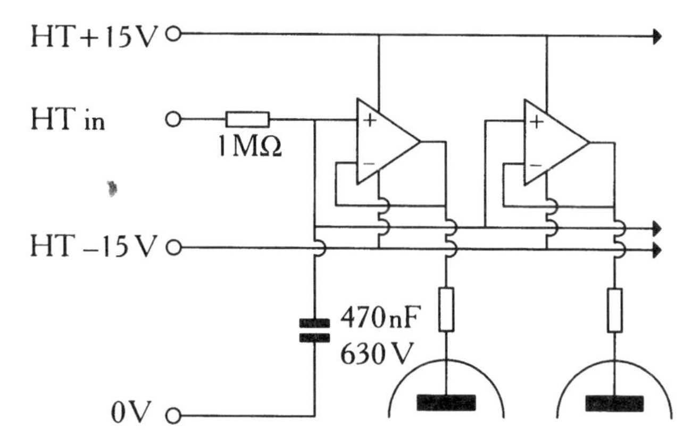

Fig. 4.25 *Use of op-amp buffers in HT supply to isolate stages*

When evaluated for a 20 kHz bandwidth, the noise of a JFET input op-amp is almost comparable to an EF86. In comparison with bipolar input op-amps, JFET input op-amps tend to have a high $1/f$ corner frequency, which means that the noise rises at 6 dB/octave below ≈1 kHz. MOSFET input op-amps have an even higher $1/f$ corner frequency, and cannot be considered for audio. Although this noise will be attenuated by the anode resistance of the valve, it will still contribute to the noise produced by our specially selected quiet triode.

For a power amplifier this noise will not be a problem, but for a sensitive pre-amplifier it needs to be considered. The best way of dealing with it in a pre-amplifier is to make the pre-amplifier insensitive to power supply noise. This implies low r_a and high R_L to obtain maximum attenuation of HT noise, and another ploy is to use differential pairs, which will further attenuate HT noise, particularly at low frequencies, which is precisely where we have our problem. These approaches will be explored in Chapter 6.

Regulator sound

Single ended amplifiers (whether pre-amplifiers or power amplifiers) supplied from a regulator or buffer force the error amplifier to track the musical waveform. This is because the amplifier draws a current proportional to the music, and the regulator or buffer strives to maintain a constant voltage in the face of this changing current. At high frequencies, the output shunt capacitor is a short circuit and maintains a low output impedance, but at low frequencies, it is the regulator that must do the work, and cope with the (musical) current waveform. The quality of the regulator is therefore inevitably audible.

A practical design

We now are now in a position to be able to design a complete HT and LT power supply. Since the requirements for a pre-amplifier are more critical, we will design a pre-amplifier power supply; we can then simply use whichever blocks are necessary for a less arduous application.

Specification of the power supply

Choice of HT voltage

Although the power supply should be designed to match the load, it makes good sense to consider how the power supply would have to be built to supply a given voltage, and thereby avoid specifying a power supply that would be expensive to make.

The world is full of computers. Computers are mass produced, and it is therefore economical to invest heavily in the design in order to shave production costs. Partly for this reason, computers invariably use switched mode power supplies. Switchers rectify the mains directly and then have a reservoir capacitor. In Europe, mains voltage is 220–240 V, resulting in $\approx 340\,V_{DC}$ when rectified. 385 V rated capacitors with low inductance are therefore readily available at reasonable cost. It makes sense to use them.

Old style HT transformers are once again becoming readily available, but isolating transformers, at any current rating we like, are available off the shelf. We now have a $240\,V_{AC}$ secondary and a $385\,V_{DC}$ electrolytic capacitor. We would need a very good reason for not combining the two with a silicon bridge rectifier and using this to supply our raw HT. Allowing for the voltage dropped by the regulator and isolating buffers between stages, we could achieve a final HT voltage at each valve stage of 285 V, and this is why most of the circuits in Chapter 2 were designed to operate from 285 V.

Occasionally, 'no-compromise' designs may force a higher HT voltage, and this has expensive repercussions which will be explored at the end of this chapter.

HT capacitors and voltage ratings

If we need a higher HT voltage, perhaps 430 V for a pair of EL34s, then a 450 V rated capacitor will be overstressed if mains voltage rises by 10% (as it is now allowed to do). There are two choices: we can either use a higher voltage capacitor, which will be a paper or plastic film capacitor and generally only available in quite low values, or we connect *equal value* electrolytic capacitors in series to obtain the required voltage rating.

Because the capacitors are connected in series, the same charging current passes through both capacitors and each capacitor receives an identical charge ($Q = It$). If the capacitances are equal, then the voltage across each capacitor must be the same ($Q = CV$).

Unfortunately, even if the capacitances are equal, the leakage currents in each individual electrolytic capacitor are unlikely to be equal, so the voltage across each capacitor will not be equal. In order to equalize the voltages, and prevent one capacitor from exceeding its rated voltage, each capacitor is bypassed by a resistor and the resulting potential divider chain forces the voltages to be equal. See Fig. 4.26.

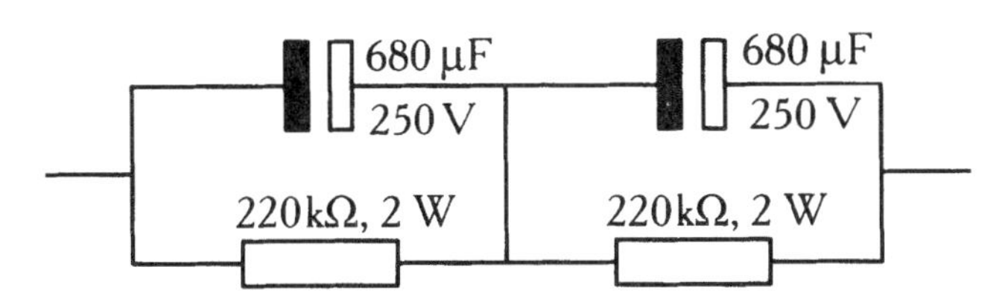

Fig. 4.26 *Bleeder resistors equalize capacitor potentials*

The divider chain should pass at least 10 times the expected leakage current of the capacitors to ensure correct operation; typically, a 220 kΩ 2 W resistor will suffice.

An even better method is to use separate HT windings and rectifier/smoothing circuits, and place the resulting floating DC outputs in series to

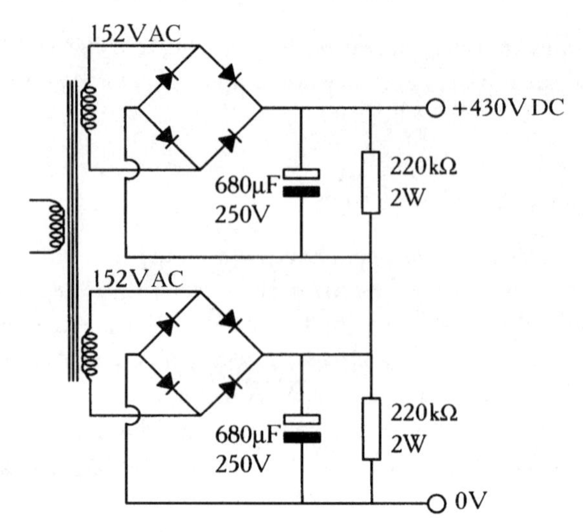

Fig. 4.27 *Achieving HT > $340V_{DC}$ with electrolytic capacitors*

obtain the required HT voltage. This ensures that each capacitor cannot exceed its rated voltage, but the mains transformer is now slightly more complex. See Fig. 4.27.

Can potentials and undischarged HT capacitors

Both of the previous schemes for producing a composite HT capacitor of high voltage rating resulted in one capacitor with its negative terminal at greater than ground potential. This is significant because the can potential of an electrolytic capacitor is very close to the potential of the negative terminal. Cans at an elevated voltage must not only be insulated from the chassis, but must also be properly insulated from the user to prevent shock hazard.

HT supplies represent a formidable shock hazard, and it is essential that provision is made for fully discharging the reservoir and smoothing capacitors when the equipment is switched off. The HT supply therefore needs a purely resistive discharge path to 0 V at some point, and the simplest way of providing this is to connect a 220 kΩ 2 W resistor across the reservoir electrolytic, which will not only discharge this capacitor, but (provided that there is a return path) will also discharge subsequent HT capacitors.

The switch-on surge

Because we have not used a valve rectifier, the HT will switch on instantly, and if this occurs whilst the valves are cold, it will damage the cathodes. Suddenly applying full voltage to electrolytics is not very desirable either (because of the high inrush current), so we should look to see if there is some way of avoiding these problems.

If we were to leave the heaters permanently powered, then we could apply HT instantly without causing cathode stripping. Keeping the valves warm will also reduce the time taken for the pre-amplifier to reach its peak performance, as opposed to merely operating. However, if the cathodes are kept at their full operating temperature without current being drawn they eventually become poisoned, causing increased noise. The solution is to operate the heaters at ≈63% of their normal operating voltage in standby mode, and then apply full voltage at full switch-on.

The power supply electrolytics still need to be protected. If we suddenly apply rectified AC to the reservoir capacitor, we may be unlucky enough to switch at the instant that the cycle is at its peak voltage. The instantaneous transition from 0 V to 340 V ($dV/dt \approx \infty$) applied to the capacitor causes a theoretically infinite current to flow:

$$I = C \cdot \frac{dV}{dt}$$

If, however, we always switch at the zero voltage point, then although dV/dt for a sine wave is at a maximum at this point, it is not infinite, and the inrush current is reduced.

Devices capable of performing this switching are known, predictably, as zero-voltage switching relays and are readily available. These relays require only low voltage DC to energize them, so we can remotely switch the power supply from the pre-amplifier using the permanently energized heater supply. If we are going to have a relay, why not also use it to switch mains to the rest of the equipment? The system is now becoming domestically acceptable, and requires only one switch to switch it all on.

The LT supplies

It is highly likely that we need at least two LT supplies, and perhaps three, because even the basic pre-amplifier has a valve with a cathode at an elevated voltage. LT supplies are not expensive to make if they are accounted for at the design stage. Adding them later is painful.

Note that *all* of the circuitry within the LT supplies superimposed on an elevated voltage is at this elevated voltage, and that it therefore represents a shock hazard if touched. Even though this circuitry contains components rated at a low voltage, this circuitry should be treated with as much caution as HT supplies.

RFI from external sources

RFI is Radio Frequency Interference. In its worst form it could be breakthrough from transmitters owned by taxi firms, but it could simply be a coarser treble than expected. Trying to cure RFI after it has been diagnosed is a nightmare. It is best to assume that it will be a problem, and take steps to deal with it at the design stage. Most of the precautions are constructional, and will be dealt with in Chapter 7, but one purely electrical precaution is a mains filter.

Although the power consumed by electronic equipment may not be particularly high, the ripple current (as we found earlier) could be much higher than the load current. A pair of large amplifiers with generous reservoir capacitors could easily draw 60 A pulses of current from the mains, even though power drawn from the mains might only be 600 W. (If 600 W seems an unlikely amount of power to be consumed by a pair of audio amplifiers, consider that each 'Beast' 35 W Class A push–pull triode amplifier consumes ≈160 W from the HT supply, the output valve heaters need another ≈75 W, and ≈65 W is consumed by the remaining heater supplies and regulators.)

Most of the commercial RFI filters are rated at 16 A or less; this is not enough for audio. If we want an RFI filter, we will need to make it ourselves.

We now know that our basic power supply will include an RFI filter, a pair of mains transformers, a semiconductor HT rectifier/regulator, at least two regulated LT supplies, and some mains switching. We are now able to draw a preliminary block diagram of the power supply. See Fig. 4.28.

There are many choices that can be made in power supply design, usually related to whether or not you have particular components, so this design is not rigid, and is loosely designed to power the basic pre-amplifier in Chapter 6.

We need to lay down some sort of specification, otherwise we cannot design the regulators. The following values allow for extra valves to be added later, if needed, without having to rebuild the supply. You could even treat this as a general purpose supply that will power any experimental, or permanent, pre-amplifier (this is what the author has done).

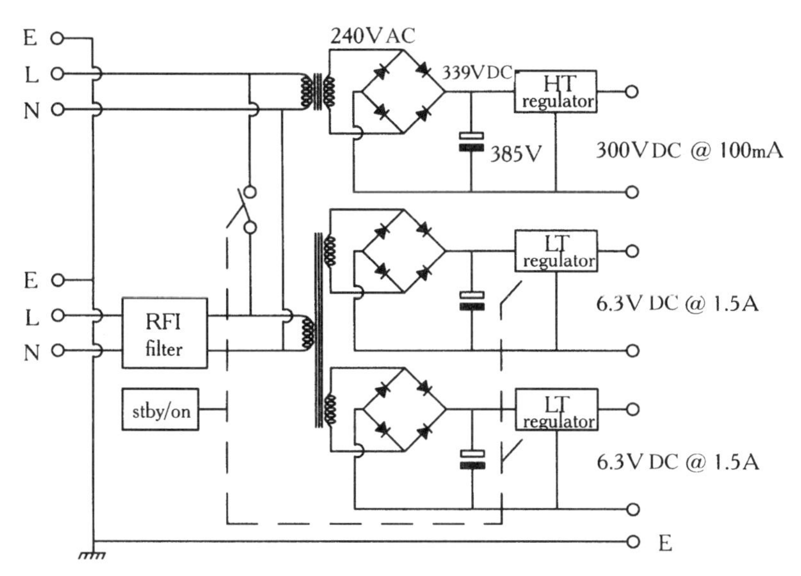

Fig. 4.28 *Preliminary block diagram of power supply*

HT supply: $300\,V_{DC}$ @ $100\,mA_{(max.)}$, $1\,mV_{pk-pk}$ ripple or less.

LT supplies: $2 \times 6.3\,V_{DC}$ @ $1.5\,A_{(max.)}$, minimal ripple. Standby mode of 4 V.

Design of the LT supplies

These are the easiest, so they make a good place to start. 1.5 A was deliberately chosen as being the maximum current that a 317 can supply. There are higher current regulators, such as the 3 A LT1085, the 5 A LM338, and the 5 A low drop-out LT1084, but high currents from a 6.3 V supply will waste almost as much power in the regulator as is dissipated in the load. The best solution is then to resort to a current regulated series heater chain, which will be described later in this chapter.

Power supply design begins with regulator drop-out voltage. A 317 needs at least 2.5 V across it to avoid dropping out of regulation at 1.5 A. We therefore need 8.8 V before the regulator.

Transformers tend to have standard secondary voltages, such as $6\,V_{RMS}$ or $9\,V_{RMS}$. Assuming that we use a capacitor input filter (a choke would probably need to be custom wound), $6 \times \sqrt{2} = 8.5\,V$ which is not enough, so we must use a 9 V secondary, which will provide 12.7 V. We will use a bridge rectifier, which always has two diodes in series, so we will drop 1.2 V across the rectifier, which brings the voltage down to 11.5 V.

If we have a rectified sine wave with a peak voltage of 11.5 V leaving the rectifier, then this is the maximum voltage to which a reservoir capacitor of infinite capacitance could charge. A capacitor of finite capacitance will charge to this voltage on the peaks, but will discharge to a minimum voltage determined by its ripple voltage. The absolute minimum voltage that we can allow is 8.8 V, so the maximum ripple voltage that we can tolerate is $2.7\,V_{pk-pk}$.

Using our earlier equation that related ripple voltage to current:

$$C = \frac{It}{V}$$

The equation requires 5600 μF, and we could perhaps obtain this value, but this would not allow for any tolerance on capacitor value, or mains voltage variation, so 10 000 μF would be a much safer choice, and would result in 1.5 V of ripple.

A 10 000 μF capacitor is capable of delivering a very high current into a low resistance, and a wedding ring placed across the terminals of this capacitor will discharge the capacitor perfectly. The explosive arc that will be struck at the instant of discharge will vaporize some metal, resulting in flying molten metal. *Do not* wear jewellery whilst working on low voltages.

1.5 V ripple is 13% of 11.5 V, and if this were the raw supply for an audio amplifier, this level of ripple would be unacceptable, but because a regulator is being used, we can allow a high ripple voltage, and benefit from the low ripple *current* that results.

The reduced ripple current eases rectifier current specification, and thermal considerations become more important. If the bridge rectifier drops 1.2 V, and passes an average current of 1.5 A, then it will dissipate ≈1.8 W (this is an approximation because $I_{average} \neq I_{RMS}$). This is a significant amount of heat to be lost from the typical 1.5 A bridge rectifier package, and they invariably become very hot. We should either use individual diodes, such as the 3 A 1N54xx series, or a 4 A bridge rectifier package.

Another possibility is to use Schottky diodes, perhaps the 31DQxx series, for the bridge rectifier. These have a lower forward voltage drop, reducing diode dissipation, but the main reason for their use is that they switch off cleanly, without the current overshoot exhibited by junction diodes. This overshoot is effectively a burst of high frequency noise that will be magnified by any resonances in the transformer/rectifier/reservoir capacitor system.

The output voltage from the rectifier is 11.5 V, so V_{RRM} need only be 23 V; 50 V is commonly the lowest rating available and allows for mains spikes, so this will be fine.

Each diode in the bridge should be bypassed with a film capacitor. 100 nF, 63 V_{DC} is a good choice, but almost anything will do, provided that the voltage rating is at least equal to twice the DC output voltage.

We should now consider how much power the 317 will be required to dissipate if the full 1.5 A load current is drawn. Ignoring ripple, the voltage applied to the regulator is 11.5 V, so the voltage across the regulator is 5.2 V, and the regulator therefore dissipates 7.8 W. Actually, the ripple will reduce dissipation slightly, but it is always best to be conservative in thermal calculations.

7.8 W is not an unreasonable amount of power for the (20 W) TO-220 package of a 317T to dissipate (with a suitable heatsink), so it seems to be worth continuing with the design. (The suffix T denotes TO-220 package, whereas K is the big TO-3 metal 'power transistor' package.)

The next step is to determine the value of resistors needed in the potential divider. Experience shows that the reference voltage tolerance on the 317 is actually very good, and that it is not usually necessary to include a variable resistor to tweak the output voltage. You may have a different view on this, but the author simply uses an upper resistor of 150 Ω, and a lower resistor of 600 Ω.

The Thévenin resistance of the 150 Ω, 600 Ω combination is 120 Ω. Theoretically, 13 μF should be used to bypass the ADJ pin to ground, but 10 μF or 15 μF will be perfectly acceptable. The manufacturer's application notes recommend that the output of the 317 be bypassed to ground with a 1 μF tantalum via a 2.7 Ω resistor.

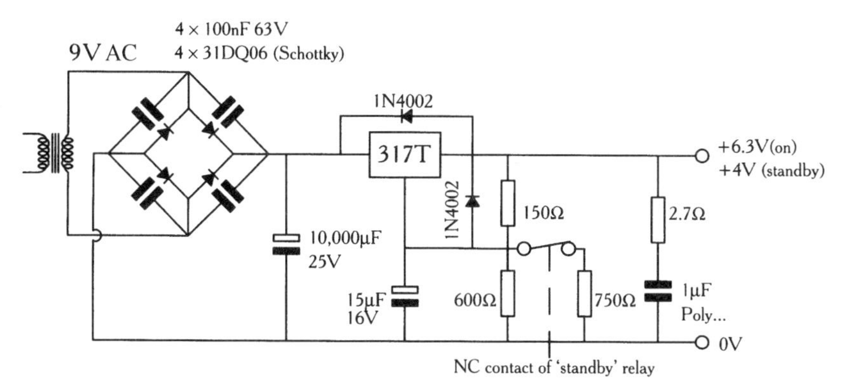

Fig. 4.29 *Practical 6.3 V regulated heater supply*

We have now designed a fully functioning 6.3 V regulator, but need a means of putting the pre-amplifier into 'standby' mode, and reducing the heater voltage to ≈4 V to avoid cathode poisoning. The simplest way to do this is to switch a resistor in parallel with the lower leg of the divider using the normally closed contact of a relay, a 750 Ω resistor will reduce the voltage to 4.04 V.

We can now draw a heater supply circuit diagram. There will be two of these, fed from the separate 9 V windings of the LT transformer, which should preferably be an E/I type with an electrostatic screen, but a dual chamber bobbin type is acceptable. See Fig. 4.29.

The HT rectifier/regulator

Unlike the LT supplies, the transformer voltage is already known (240 V), and so we will design the regulator to fit the raw HT voltage, rather than the other way round.

The bridge rectifier will charge the reservoir capacitor to 339 V. Although encapsulated bridge rectifiers are available that can cope with this voltage, it is safer to use individual diodes, as this allows greater spacing of the connections, with less risk of inadvertent short circuits.

If we are going to use separate diodes, it is well worth using ultra-fast soft recovery diodes such as RHRD4120 or STTA512D, which generate less current overshoot for less time than standard junction diodes, and therefore generate less noise. (Schottky diodes are not available in this voltage rating.) We only need a very low continuous current rating (100 mA), and should select the lowest current rating above this as higher current diodes are slower and noisier.

Our regulator will be designed to produce 300 V, so we can afford to drop 39 V by way of regulator drop, reservoir ripple, and diode drop. If we use our earlier 5% criterion for ripple voltage, then the ripple voltage will be ≈17 V. However, 17 V of ripple is a rather larger proportion of 39 V than we would ideally like, and it would be preferable to reduce this to 10 V or less, so a low ESR 220 μF reservoir capacitor would be ideal. It should be noted that such a capacitor, charged to 339 V, stores considerable energy, and *extreme* care should be taken when testing this circuit.

We can now consider the design of the regulator, starting with the potential divider chain. See Fig. 4.30.

If we pass 5 mA down the chain, the lower resistor will have ≈300 V across it, and a 60 kΩ resistor dissipating 1.5 W would be required. If we instead use a 220 kΩ 2 W resistor, this will dissipate only 0.4 W, which is perfectly

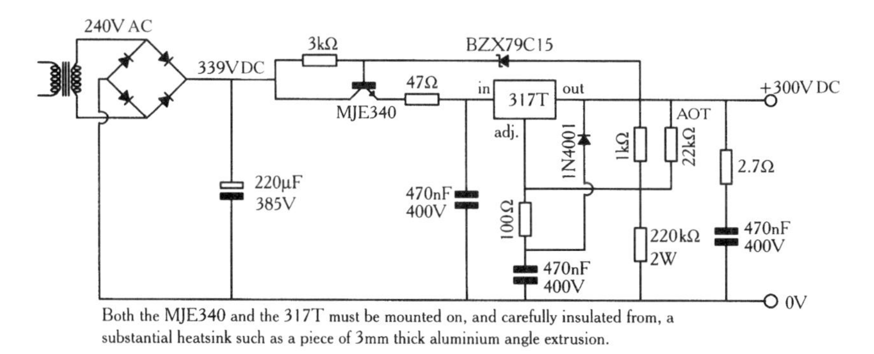

Fig. 4.30 *Practical 300 V regulated HT supply*

acceptable. This has the further advantage that because the upper leg of the divider has to rise in value, the Thevénin resistance rises, and a smaller capacitor is required to bypass the ADJ pin to ground. Because the bias chain does not draw 5 mA (the minimum load current required to allow the 317 to regulate correctly), unloading the output of the supply would cause the output voltage to rise; however, this situation is unlikely, so we need not worry about it.

The 220 kΩ lower resistor passes 1.358 mA, of which 50 μA is bias current from the ADJ pin of the 317. The upper resistor therefore passes 1.308 mA, and has 1.25 V across it, so a 955.7 Ω resistor is required. But the 317 reference voltage is 4% tolerance, and so there is little point in accurately fitting this calculated value of resistor. We could fit a variable resistor, but these are less reliable than fixed resistors, and component failure in high voltage silicon circuits is almost invariably catastrophic. The safest option is to fit a 1 kΩ resistor and leave space for it to be paralleled with an additional component that is Adjusted On Test (AOT).

Before constructing the circuit, we measure and record the exact value of the 220 kΩ 2 W resistor, which might be 221 kΩ. When the circuit is built, we measure the output voltage (which will be low), and this voltage might be 290 V. The voltage across the 220 kΩ resistor is therefore 288.75 V, so 1.307 mA of current is flowing. We subtract 50 μA of bias current to find the current in the upper resistor (1.257 mA), and multiply this by the 1 kΩ resistor to find the reference voltage (1.257 V).

We can now work forwards: 298.74 V divided by 221 kΩ is 1.352 mA. We now subtract the 50 μA bias current to give 1.302 mA, and divide this into the reference voltage of 1.257 V to find the required resistor (965.6Ω). A 27 k in parallel with the existing 1 k will now produce the correct 300 V

HT voltage. Although this is quite long winded, it is safer than adjusting a potentiometer, and the risk of subsequent failure is much lower.

The Thévenin resistance at the ADJ pin is ≈950 Ω, requiring a 1.5 μF bypass capacitor to ground. This is an expensive and bulky (400 V rated) component, so the value is commonly reduced, and 470 nF is a common choice.

In the application note that spawned this breed of regulators, a resistor was interposed between the emitter of the series transistor and the 317 to limit short circuit current. Other designs, such as those by J. J. Curcio, have retained this resistor for a different reason, although the value is often reduced to avoid excessive voltage drop. If a capacitor is connected to ground at the regulator end, we now have RF filtering, and stability of the regulator is improved. A slight disadvantage of this approach is that there is now no possibility of the regulator surviving a short circuit to ground.

References for elevated heater supplies

Several times we have mentioned that elevated LT supplies will be required for various circuits. Typically, we would need the superimposed voltages to be 100 V and 200 V. These voltages, although they will not supply any current, will need a low source resistance and adequately filtering. See Fig. 4.31.

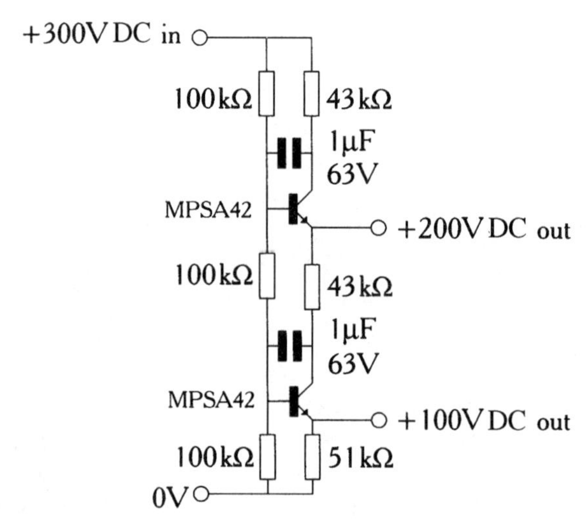

Fig. 4.31 *Superimposing smooth DC on heater supplies*

This circuit is connected across the output of the HT supply, and is a pair of emitter followers whose output voltage is set by the chain of 100 kΩ resistors. Collector current is set by the 51 kΩ resistor at 1.95 mA, resulting in 0.2 W being dissipated in this resistor. In order to reduce dissipation in the transistors, 43 kΩ resistors have been added as collector loads, and these set V_{ce} = 15 V. Output resistance is less than 2 kΩ, although supplementing each transistor with another, to form a Darlington, would lower this output resistance.

Filtering is achieved by placing the filter capacitor not from base to ground, which would require a high voltage component, but from base to collector. Gain to the collector $A_v = -1$, and Miller effect therefore multiplies this capacitor by a factor of 2, so that the effective value is 2 μF. Input resistance at the base of the transistors is purely that of the resistor chain to ground and the filter cut-off frequency is therefore 1.2 Hz. The lower emitter follower sees two cascaded 1.2 Hz filters and so noise is further rejected.

Putting the blocks together

We have LT supplies and an HT supply, and we now need to put these together with some relays and transformers. We will assume that a 240 V isolating transformer was used for the HT supply, and 2 × 9 V for the LT supplies. Depending on availability of mains transformers, your supply might be different. See Fig. 4.32.

An RFI filter has been fitted at the mains input, consisting of a 130 joule metal oxide varistor, a pair of RF chokes wound on the same ferrite core, and a pair of Class X2 capacitors. X2 capacitors are the only type of capacitors that may be legally connected from live to neutral on the mains, the reason for this is that they are specifically designed to fail safely. Many RFI filters also include a pair of Class Y capacitors from live to earth, and from neutral to earth. This is not appropriate for audio equipment, as it makes the earth noisy. If electrostatic screens are available on the transformers, these should be connected directly to the point where mains earth is bonded to the chassis.

The zero-voltage switching relay simultaneously switches mains to the HT transformer and ancillary equipment. The second, conventional, relay is used to bring the LT regulators out of standby mode. It is worth using a 4-pole relay even if you only have two LT supplies, then if you add another supply at a later date, the relay contacts are ready and waiting. The 'power/standby' switch operates by grounding the lower leg of each relay coil (although the zero-voltage switching relay is actually silicon, and does not actually possess

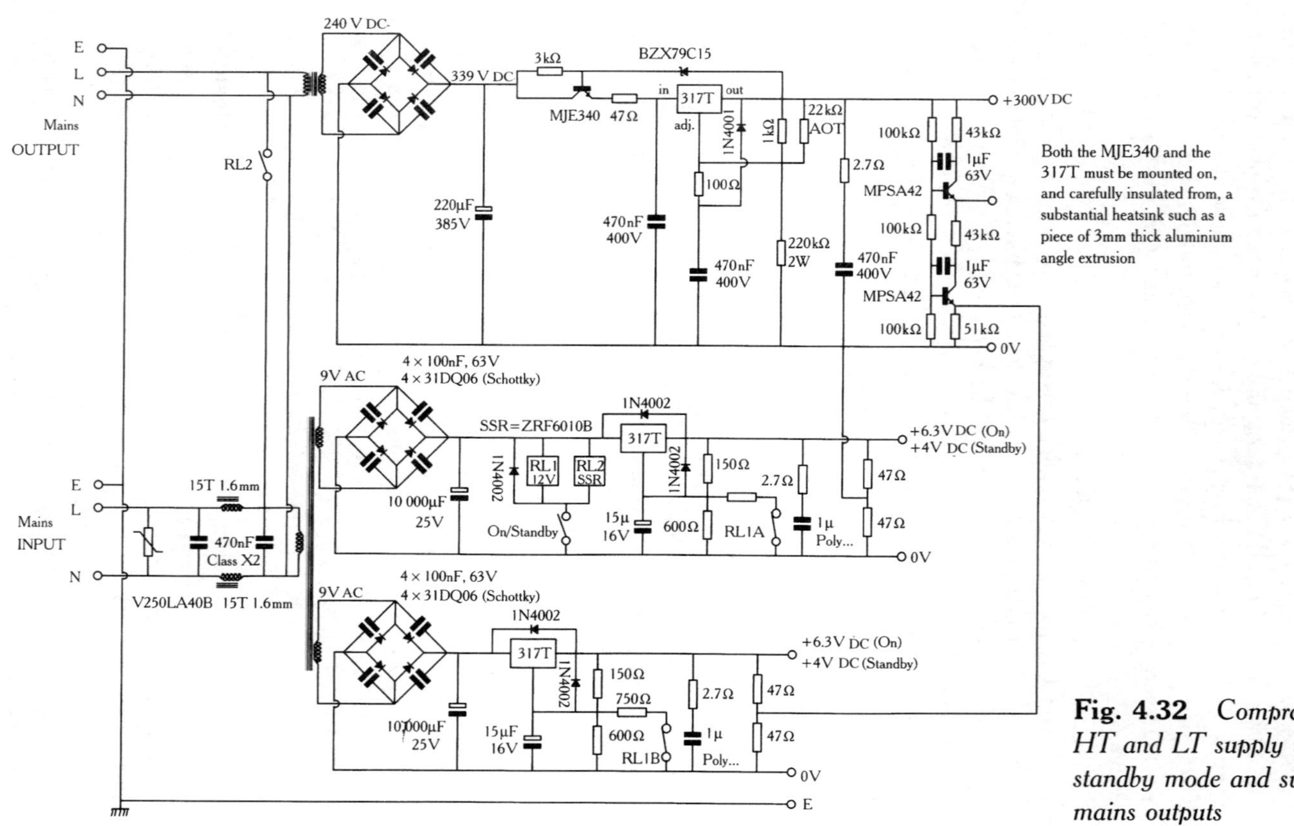

Fig. 4.32 *Comprehensive HT and LT supply with standby mode and switched mains outputs*

a coil), this means that the raw LT voltage is not present in the umbilical cable that links the pre-amplifier to its supply and hum is not induced.

A better power supply

The previous power supply was designed to a price, whereas the following supply was simply designed to be quiet enough to feed the 5842 RIAA stage in Chapter 6.

The foremost priority for a pre-amplifier power supply is low noise. Not only do we want to obtain good immunity to external sources of noise (such as mains RFI), but we also want to avoid generating noise ourselves.

The worst culprit for generating noise is the rectification and smoothing process, particularly if capacitor input, so choke input supplies are obligatory. Choke input HT supplies are reasonably conventional, but choke input LT supplies are unusual, and require some careful thought in conjunction with the pre-amplifier design. Power supplies should not be designed in isolation.

The LT supplies

Multiple LT supplies are a nuisance and cost money – and multiple choke input LT supplies are even worse. The 5842 RIAA stage was therefore specifically designed to avoid cathode followers, thus eliminating floating LT supplies. As mentioned earlier, high current 6.3 V_{DC} regulators are expensive to build; moreover, high current chokes are thin on the ground. These last two restrictions suggested that a constant current LT supply feeding a series heater chain would be ideal. After much thought, a dual mono valve line-up of 5842, 6J5 and 12B4-A was determined for the RIAA stage; each channel therefore requires 300 mA @ ≈25.2 V.

Series heater chains should not be connected in parallel, so we need to design a pair of 300 mA constant current regulators. The versatile 317 is ideal for the job, but instead of maintaining 1.25 V across part of a shunt potential divider chain, it now strives to maintain this voltage across a series current sensing resistor. See Fig. 4.33a.

As before, we also want to be able operate the heaters in standby mode, and because valve heaters do not obey Ohm's law (non-constant temperature), the required current cannot be directly calculated, even though we know that it would develop 63% of the operating voltage across the chain.

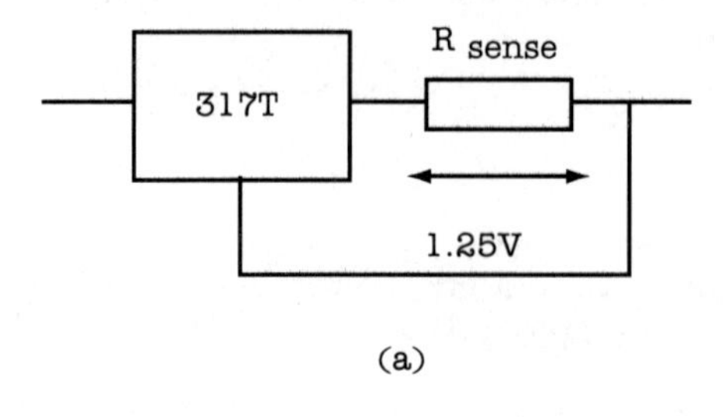

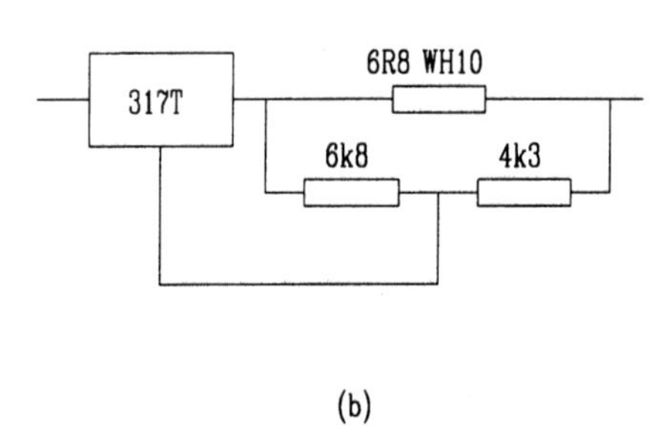

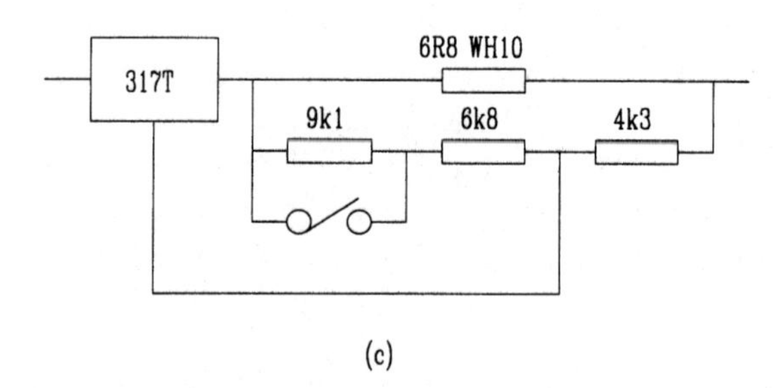

Fig. 4.33 *Using the 317 as an LT current regulator*

A series of experiments determined that operation at 78% current is equivalent to 63% voltage for indirectly heated valves.

The minimum allowable value for the current sensing resistor must take into account the standby mode, and is determined using:

$$R_{sense} = \frac{1.25\,\text{V}}{I_{required}} = \frac{1.25}{0.3 \times 0.78} = 5.342\,\Omega$$

Unsurprisingly, this is not a standard value, but (at the expense of wasting voltage) any higher value can be used, since it will simply develop >1.25 V at the required current, which can be attenuated back to 1.25 V using a potential divider in parallel with the sense resistor, so 6.8 Ω was chosen.

Paradoxically, it is easiest to determine the potential divider required to set 300 mA, and then modify this to set the standby current. Using Ohm's law, the voltage drop across the 6.8 Ω when 300 mA flows:

$$V = IR = 0.3 \times 6.8 = 2.04\,\text{V}$$

This can be attenuated to 1.25 V by a potential divider composed of a 6k8 and a 4k3 resistor. See Fig. 4.33b.

In standby mode, a reduced voltage is developed across the sense resistor:

$$V = IR = 0.234 \times 6.8 = 1.591\,\text{V}$$

The potential divider loss must therefore be reduced by increasing the value of the lower leg (the 6k8 resistor), and an additional 9k1 resistor restores 1.25 V. See Fig. 4.33c.

A normally open relay contact across the 9k1 resistor leaves the regulator in standby mode, energizing the relay shorts the 9k1 resistor, and full current is applied to the heaters.

Current regulated LT supplies for series heater chains have a number of advantages compared to conventional 6.3 V_{DC} regulated supplies for parallel heaters:

- The thermal shock to the valve heaters at cold switch-on is eliminated.
- They are far more efficient (in this example, halving regulator dissipation).
- Heater wiring resistance is now irrelevant.
- Individual heater resistance can be used as part of an *RC* RFI filter.
- Each valve heater is inherently at a slightly higher voltage than its cathode.
- They are inherently proof against accidental short circuits.

Specifying the LT mains transformer and choke

Working backwards from the load (≈25.2 V), we drop 2 V across the sense resistor, and need 3 V across the 317, so an absolute minimum of 30 V is required before the regulator. The chosen choke had an internal resistance of 1.5 Ω, and 0.6 A (two heater chains) causes a potential drop of 0.9 V, requiring a DC component of 31 V at the output of the rectifier. The bridge recifier adds two diode drops, although Schottky diodes will reduce this to 1 V total, so 32 V is required. Remembering that the DC component of a rectified sine wave is $0.9\,V_{in(RMS)}$, we need $36\,V_{RMS}$ from the transformer. However, this does not allow for mains voltage variation or valve heater voltage variation, so $40\,V_{RMS}$ would be a better choice.

It is easier to check a potential candidate for suitability than to specify a choke that can probably not be found. A swinging choke has the following specification:

$$L = 180\,\text{mH} \;@\; 200\,\text{mA}_{DC}$$

$$L = 90\,\text{mH} \;@\; 1.5\,\text{A}_{DC}$$

The lower value of inductance implies the highest AC current:

$$I_{AC(\text{positive peak})} = \frac{V_{in(RMS)}}{1155L} = \frac{40}{1155 \times 0.09} = 385\,\text{mA}$$

Allowing for a little current to be drawn by other parts of the LT supply:

$$i_{\text{total peak current}} = I_{DC} + i_{AC(\text{peak})} = \approx 600\,\text{mA} + 385\,\text{mA} \approx 1\,\text{A}$$

This is less than the 1.5 A maximum rating, and the proposed choke is therefore satisfactory.

Mains transformer current specifications refer to I_{RMS} (sine wave), so we may divide our peak current by a factor of √2 to give 0.71 A, so a 30 VA transformer would be just adequate, but the author chose a 50 VA type because it was virtually the same price, and derating reduces transformer core flux density and leakage flux, which in turn reduces induced hum into nearby circuits.

The HT supply

The 5842 RIAA stage requires 400 V @ ≈56 mA, and a 625–0–625 V transformer and 20 H choke salvaged from a Solartron Varipak were available. The Varipak was rated at 100 mA, so there is no need to check current ratings of the choke or transformer.

The HT choke and *Q*

Fortunately, there is plenty of HT voltage available, and we can afford to set the Q of the LC filter. Rearranging the standard equation, and assuming a 47 μF capacitor for the reservoir:

$$R = \frac{1}{Q}\sqrt{\frac{L}{C}} = \frac{1}{0.5}\sqrt{\frac{20}{47 \times 10^{-6}}} = 1304\,\Omega$$

Not only does the choke contribute to this resistance, but also the transformer winding resistance (secondary and reflected primary), plus any external resistance we may add. Therefore, to find the required additional resistance:

$$R_{ext.} = R_Q - \left[R_{choke} + R_{secondary} + \left(\frac{V_{secondary}}{V_{primary}}\right)^2 R_{primary}\right]$$

$$= 1304 - \left[265 + 129 + \left(\frac{625}{240}\right)^2 7.8\right] = 857\,\Omega$$

The nearest value is 820 Ω, and this will drop the voltage at the output of the choke to ≈493 V.

The HT regulator

Initially, a shunt regulator similar to that used by Allen Wright[5] for his FVP pre-amlifier was tried, but it was destroyed as a consequence of a valve heater failure, underlining the importance of considering the entire system when designing a single block.

Shunt regulators are inherently short-circuit proof, but they are vulnerable to open-circuits. If the load is disconnected, the shunt element has to pass all the expected load current in addition to its normal current, and it dissipates significant power. A single heater failure in a series heater chain removes heater power from all valves in that chain, switching off **all** the valves, thus unloading the HT supply. Shunt HT regulators and series heaters are an uncomfortable combination, particularly if HT currents >20 mA are required.

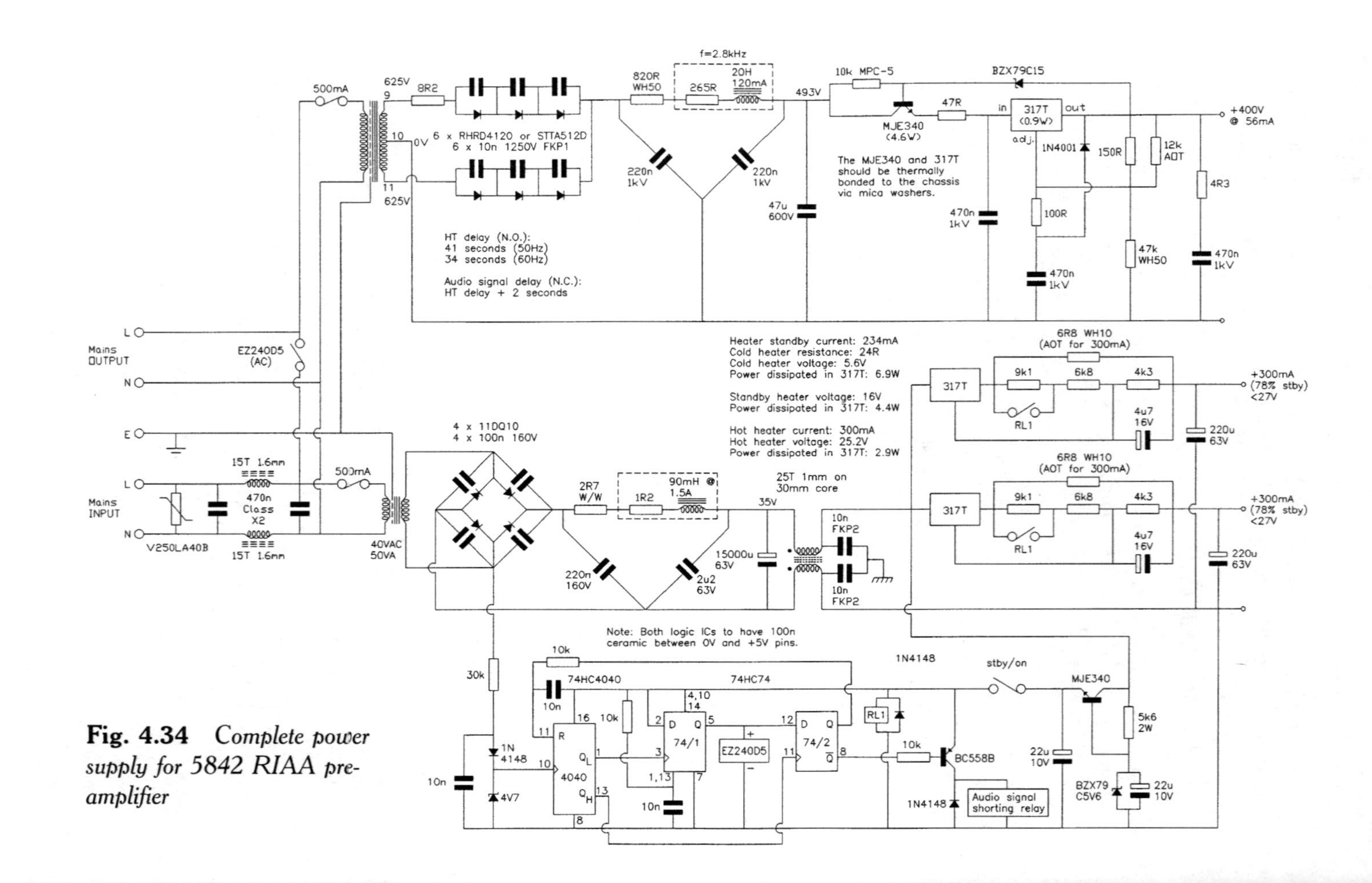

Fig. 4.34 *Complete power supply for 5842 RIAA pre-amplifier*

Valve regulators are extremely robust, and can be made to be fairly quiet, but they invariably require multiple heater supplies, and the error amplifier needs a regulated heater supply to reduce DC drift, further increasing complexity. Despite these problems, a valve shunt regulator might be worthwhile, if only because voltage reference valves such as the 85A2 are extremely cheap, and relative to their DC voltage, they are not unusably noisy. The author has a good stock of 85A2 and PL519, so experiments are continuing.

For the moment, the Maida design is the logical choice of regulator, although a minor improvement can be made in the potential divider chain. Previously, a 220 kΩ metal film resistor had been specified, since this value minimized wastage of HT current, and power dissipation in the resistor. Now that more current is available, a 47 kΩ wirewound resistor can be used, with the advantage that it does not produce excess noise.

With the new potential divider chain, optimum rejection of 100 Hz ripple would be obtained with a 10 μF bypass capacitor but this would slug the response of the regulator to changes in current, so a more practical value is actually beneficial. The chain is a 50 dB step network (400 V/1.25 V), and a 470 nF bypass capacitor sets the f_{-3dB} frequencies to ≈2.3 kHz and ≈7 Hz, with best ripple rejection at the higher frequency. Nevertheless, at 100 Hz, the bypass capacitor still improves rejection by ≈23 dB over the unbypassed case, and theoretically increases ripple rejection to 90 dB. Since the ripple at the output of the LC filter is ≈1 V, a further 90 dB reduction reduces it to an entirely negligible 32 μV, so the low value of the bypass capacitor is not a problem.

The HT rectifier/transformer

Like the Beast in Chapter 5, the choke input HT supply needs composite diodes and charge equalizing capacitors. The centre-tapped transformer economizes on diodes and capacitors, but adds a new problem.

Centre-tapped transformers invariably have their windings wound one on top of the other, so the average diameter of the outer winding is a little larger than the inner winding, resulting in a slightly higher internal resistance. Unless balanced by adding an external resistance to the inner winding, a ripple component at mains frequency will appear at the output of the rectifier, which will not be particularly well attenuated by the following smoothing. This is a very minor point, but the addition of a single cheap resistor eliminates it.

The HT delay circuit

At the beginning of this chapter, valve rectifiers were mooted as a solution for soft starting valve electronics. Valve rectifiers are noisy, expensive and wasteful of HT voltage, but semiconductor rectifiers normally apply the HT before we are ready for it.

As before, a zero-voltage switching solid state relay is used to gently energize the HT transformer, but this relay is delayed by ≈41 seconds to allow the cathodes to warm from their standby temperature to full operating temperature.

Additionally, an output is provided for driving a relay that is normally a short circuit across the output of the associated pre-amplifier. The relay energizes ≈2 seconds after the HT, but releases just before the HT begins to decay, thus eliminating thumps at switch-on and switch-off, which could damage transistor amplifiers.[6]

The delay circuit half-wave rectifies one output of the LT transformer via a 30 kΩ resistor (to reduce DC flowing through the transformer) and produces 50 Hz pulses which are clipped to ≈5 V by the 4.7 V Zener. The 10 nF capacitor removes RF noise that could otherwise falsely trigger the 4040 counter. Q_L of the 4040 counter changes from logic 0 (0 V) to logic 1 (5 V) after 2048 mains cycles, and the positive going edge triggers the following 74 D-type into loading the logic 1 on its D input onto its Q output, thus energizing the HT relay. Simultaneously, this Q output is applied to the D input of the next D-type, but is not loaded to its output until Q_H of the 4040 next changes from 0 to 1, which occurs 128 mains cycles later. The inverted Q output is used to switch on the BC558B transistor and energize the audio shorting relay, or relays. The relays are bypassed to prevent reverse EMF spikes from destroying the transistor.

Any significant loss of power to the LT supply discharges the LT reservoir capacitor, and resets the timer circuit, thus switching off HT and muting the audio. The standby/on switch merely applies power to the delay circuit and full heater voltage to the valves, so accidentally plugging the supply into the mains with the standby/on switch switched on still delays the HT.

References

1. Schade, O. H. (1943) Analysis of Rectifier Operation. *Proc. I.R.E.*, Vol. 31, No. 7, July, pp. 341–361.

2. 3-terminal Regulator is Adjustable. *Linear Applications Handbook*. National Semiconductor Application Note 181.
3. *1990 Linear Databook* and *1992 Linear Databook Supplement*. Linear Technology.
4. Maida, Michael. High Voltage Adjustable Power Supplies. *Linear Applications Handbook*. National Semiconductor Linear Brief 47.
5. Wright, Allen (1995) *The Tube Preamp CookBook*, 1st ed.
6. Self, Doug. (1996) Precision Pre-amplifier '96. *Electronics World*, September, pp. 708–716.

5

The power amplifier

The task of a power amplifier is to provide a fixed level of gain to a processed signal and deliver *power* into a load such as a loudspeaker. It should do this without introducing spurious signals, such as hum, noise, oscillation, or audible distortion, whilst driving a wide range of loads. Additionally, it should be tolerant of abuse, such as open or short circuits. It will be appreciated that this is not a trivial objective, and will therefore require careful design and execution if it is to be achieved.

The determining factor is the output stage. The solution adopted here will dictate the topology of the remainder of the amplifier, so we will begin by investigating the output stage.

The output stage

Typical audio valves are high impedance devices and can swing hundreds of volts, but deliver only tens of milliamps of current. By contrast, a loudspeaker of typically 4–8 Ω nominal impedance may require tens of volts and amps of current. The obvious solution to this problem is to employ an *output transformer* to match the loudspeaker load to the output valve or valves.

This is where the problems start. As was hinted earlier, transformers are rather less than perfect, and the ultimate quality of a valve amplifier is limited by the quality of its output transformer. Despite this, the transformer coupled output stage is a good engineering solution, and is used in most valve amplifiers (see later for Output Transformer-Less designs).

Valves designed specifically for audio use generally have optimized configurations that are detailed in the manufacturers' data sheets. Designing output stages for audio valves from first principles is reinventing the wheel, but an overview of the practicalities is most useful, therefore we will indulge in a brief analysis of a currently very fashionable topology.

The single ended class A output stage

A typical transformer coupled output stage is the familiar common cathode triode amplifier using cathode bias. See Fig. 5.1.

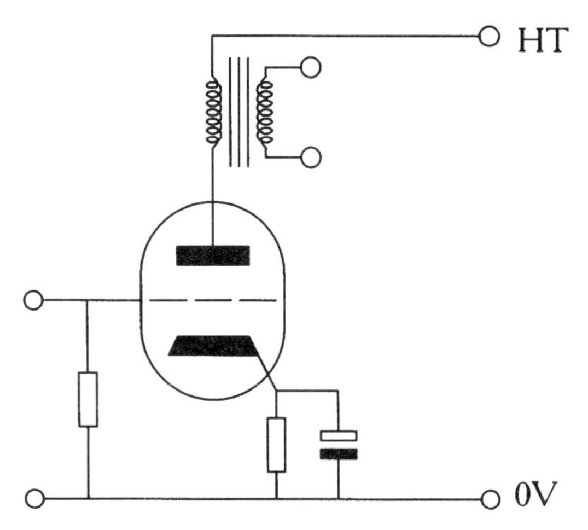

Fig. 5.1 *Single ended transformer coupled stage*

When we investigated voltage stages, we used a loadline to choose the value of anode load, and generally optimized for linearity, rather than voltage swing; this time, we need to maximize power. For this example we will use an E182CC double triode, which might be useful as a headphone amplifier. We would normally set the operating point at the intersection between maximum continuous anode voltage ($V_a = 300\,V$) and maximum anode dissipation ($P_a = 4.5\,W$). But as there is a grid curve intersecting at $V_a = 295\,V$, the operating point has been moved for convenience. For maximum output power, the optimum load for a triode is $2 \times r_a$. In our example r_a is $3.57\,k\Omega$, so $R_L = 2 \times r_a = 7.14\,k\Omega$, and we plot this loadline. See Fig. 5.2.

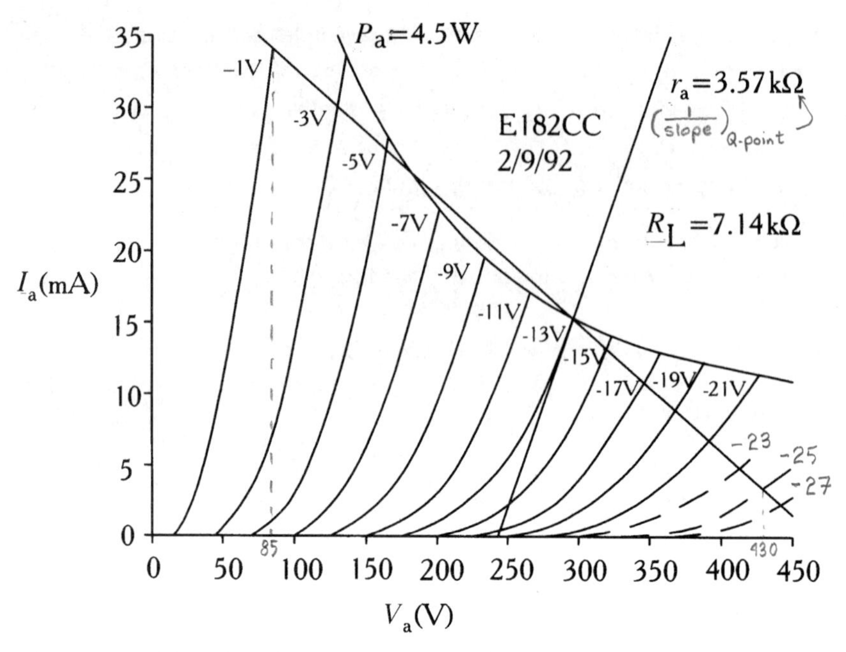

Fig. 5.2 *AC loadline for transformer coupled stage*

$V_{gk} = -1$ V is our positive limit from the bias point of $V_{gk} = -13$ V, therefore the negative limit will be $V_{gk} = -25$ V for a symmetric input voltage. This results in a peak to peak output voltage swing of 430 V – 85 V = 345 V, or 122 V_{RMS}, which equals 2.1 W dissipated in the load. Under these conditions, 4.5 W is dissipated in the valve, giving an efficiency of 32%.

We should now observe some important points about the operation of this stage:

- The loadline strays into the region where $P_a \geq 4.5$ W. Since the stage is driven only with AC (it must be, as we could not otherwise transformer couple to the load), this is not a problem. This is because although on one half cycle the anode dissipation is ≥4.5 W, on the other half cycle it is less, and the thermal inertia of the anode will average the dissipation out at ≤4.5 W.
- We set the operating point of the valve at 300 V. If the transformer is perfect, then there will be no DC voltage dropped across the primary winding, and so the HT voltage must be 300 V. Yet we have allowed V_a to rise to 430 V, which is considerably above HT. This is possible because

the transformer stores energy in the magnetic flux of its core. In theory, a perfect valve could swing V_a from 0 V to 2 × HT, which is a very useful feature in a power amplifier.

- We carefully set our anode load at 7.14 kΩ, but in doing so, we assumed that the loudspeaker was a resistor. Loudspeakers are not resistive, and the transformer is not perfect, so the actual load seen by the valve will not be a precise resistance, but a complex and variable impedance.

  The valve therefore sees an AC loadline that is an ellipse with its major axis roughly aligned with the theoretical resistive loadline. The gradient of the major axis is the resistive component, and the width of the minor axis indicates the relative size of the reactive component. This means that most of the calculations we can make for an output stage are informed guesses at best, and there is little point worrying about precise values.
- Because we wish to maximize the power in the load, we have to maximize the anode voltage swing, resulting in poor linearity. We could improve linearity by increasing the value of the anode load and plotting another loadline, but this will reduce available output power.

Although the linearity of single ended stages is not good, the distortion produced is mostly second harmonic, which, as we observed earlier, is relatively benign to the ear. We can estimate the percentage of second harmonic distortion from the following formula:

$$\%D_{\text{2nd harmonic}} \approx \frac{V_{\text{quiescent}} - \dfrac{V_{\text{max.}} + V_{\text{min.}}}{2}}{V_{\text{max.}} - V_{\text{min.}}} \times 100\%$$

In our example, $V_{\text{max.}} = 430\,\text{V}$, $V_{\text{min.}} = 85\,\text{V}$, and $V_{\text{quiescent}} = 295\,\text{V}$, resulting in 11% second harmonic distortion at full output. Clearly, $>10\%$ distortion is not hi-fi, but the attraction of single ended amplifiers is that their distortion is always directly proportional to level, and so at one-tenth output power, the distortion would be $\approx 1\%$, and so on. Since most of the time music requires very little power, it is often argued (oddly enough, by single ended enthusiasts) that it is the quality of the first watt that is important, not the remainder that are rarely used.

The measured distortion could be reduced using negative feedback, but this technique is almost universally shunned by the supporters of single ended amplifiers. Since the distortion is produced by unequal gain on positive and negative half cycles, a driver stage producing precisely the same imbalance of gain would tend to cancel distortion because of the inversion of the output stage. This is a very popular technique in single ended amplifiers, but although the theory is elegant, maintaining the cancellation in practice is rather more difficult.

Transformer imperfections

For this analysis we have treated the output transformer as being perfect, and we should now consider how its imperfections will affect the stage. Unfortunately, we are passing a constant magnetizing current ($I_{quiescent}$) through the primary of the output transformer. In order for the core not to saturate, which would cause odd harmonic distortion, we must have a very large core. Another method of avoiding core saturation is to reduce the number of primary windings, which reduces the magnetizing effect of the quiescent current, but also reduces primary inductance.

Usually both methods need to be used, which results in a physically large transformer of low primary inductance at the operating point. Because the transformer is so large, it has correspondingly large stray capacitances, and the high frequency performance is also compromised. Typically, the transformers used in this way are large, expensive and have poor LF and HF performance.

It might therefore be thought that the single ended transformer coupled power stage is a complete non-starter, but, curiously, this is not so. If we look at the hysteresis curve for transformer iron, this offers a clue to this topology's recent resurgence in popularity. See Fig. 5.3.

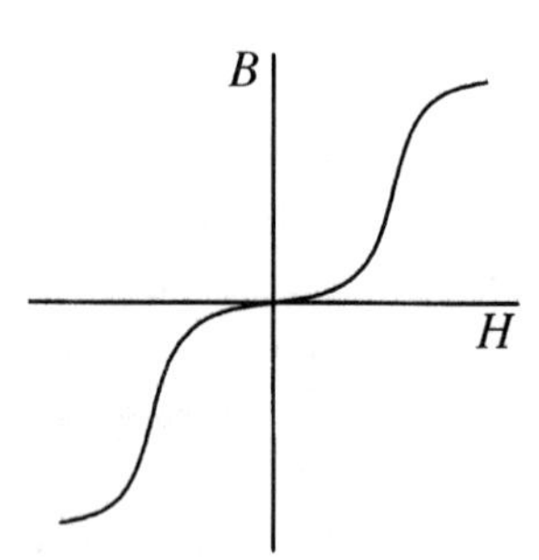

Fig. 5.3 *Exaggerated BH curve of iron*

When used as a transformer, the hysteresis curve may be treated as a transfer characteristic showing the relationship between V_{in} and V_{out}. If there were no DC current flowing through the transformer, then an AC signal would swing symmetrically about the origin. If we look at the small-signal performance, we see that there is a kink in the characteristic around the origin where the slope of the curve is reduced; this is equivalent to reducing gain. The cause of this kink is that the individual magnetic domains that make up the core have stiction in reversing the polarity of their magnetism. (The same effect is true in electrostatics, in that there is stiction in reversing electrostatic charges in polar materials such as polyester and polycarbonate.)

By passing the valve's quiescent current through the transformer, we avoid this crossover region, and the transfer characteristic is more linear, which perhaps explains the claims for excellent midrange detail in this breed of amplifiers.

Although the transformer has a low primary inductance, suggesting a poor bass response, the core does not saturate at low frequencies, since it had to be oversize to accommodate the quiescent current. Because of this, the inductance at full AC output power is substantially the same as that at zero AC output. This results in good subjective bass quality, since it does not change with level.

Unfortunately, we can make no excuses for the HF performance. The large, leaky output transformer produces significant losses at HF, although excellent construction will help matters.

In order to reduce distortion, single ended amplifiers ideally use true triodes with directly heated cathodes, such as 300B, 211 and 845, rather than beam tetrodes or pentodes connected as triodes. Unfortunately, directly heated cathodes are prone to hum when fed by AC from the mains transformer heater winding.

To sum up: single ended triode amplifiers have good low level performance, the bass, and especially midrange, are good, they tend to be low power (typically $\approx$10 W), and they are outrageously expensive due to the cost of the output transformer and the esoteric valves used. They are also surrounded by hyperbole.

The single ended pentode amplifier found in a 1960s Dansette record player shares none of these characteristics except low output power.

Classes of amplifiers

The 'class' of an amplifier refers to the proportion of quiescent anode current to signal current. Until now, we have only looked at Class A amplifiers, although the fact was not explicitly stated. If we relax that restriction, we will need some definitions.

Class A

Anode current is set at a level such that, even with maximum allowable input signal, anode current never falls to zero. In other words, the valve never switches off. (Maximum theoretical efficiency for sine wave output is 50%.)

Class B

There is zero quiescent anode current, and anode current only flows during the positive half cycle of the input waveform. The valve is therefore switched *off* for the negative half cycle of the input waveform, and considerable distortion of the signal occurs, since it has been half-wave rectified. Additional measures will need to be taken to deal with this problem. (Maximum theoretical efficiency for sine wave output is 78.5% for a push–pull Class B amplifier.)

Class C

Anode current flows for *less* than half a cycle of the input waveform. This method is only used in radio frequency amplifiers where resonant techniques can be used to restore the missing portion of the signal, and results in even greater efficiency and distortion than Class B.

Radio frequency engineers refer to the *conduction angle* to specify the proportion of time in which anode current flows. Using this description, we see that Class A amplifiers have a conduction angle of 360°, Class B is 180° and Class C <180°. The transition between Class A and pure Class B is quite broad, and therefore there is an intermediate class known as Class AB. See Fig. 5.4.

In Fig. 5.4, the transfer characteristic of the output device is assumed to be perfect, so the input sine wave is simply reflected through the diagonal transfer characteristic to produce the output. In Class B operation, the bias voltage cuts off the valve, and it is only on positive half cycles that the signal is able to switch the valve on. It will be noticed that the output waveform of the Class B stage is very similar to the power supply waveforms in Chapter 4, and for the same reason, half-wave rectification is taking place.

Note that as bias voltage is increased negatively, the conduction angle falls.

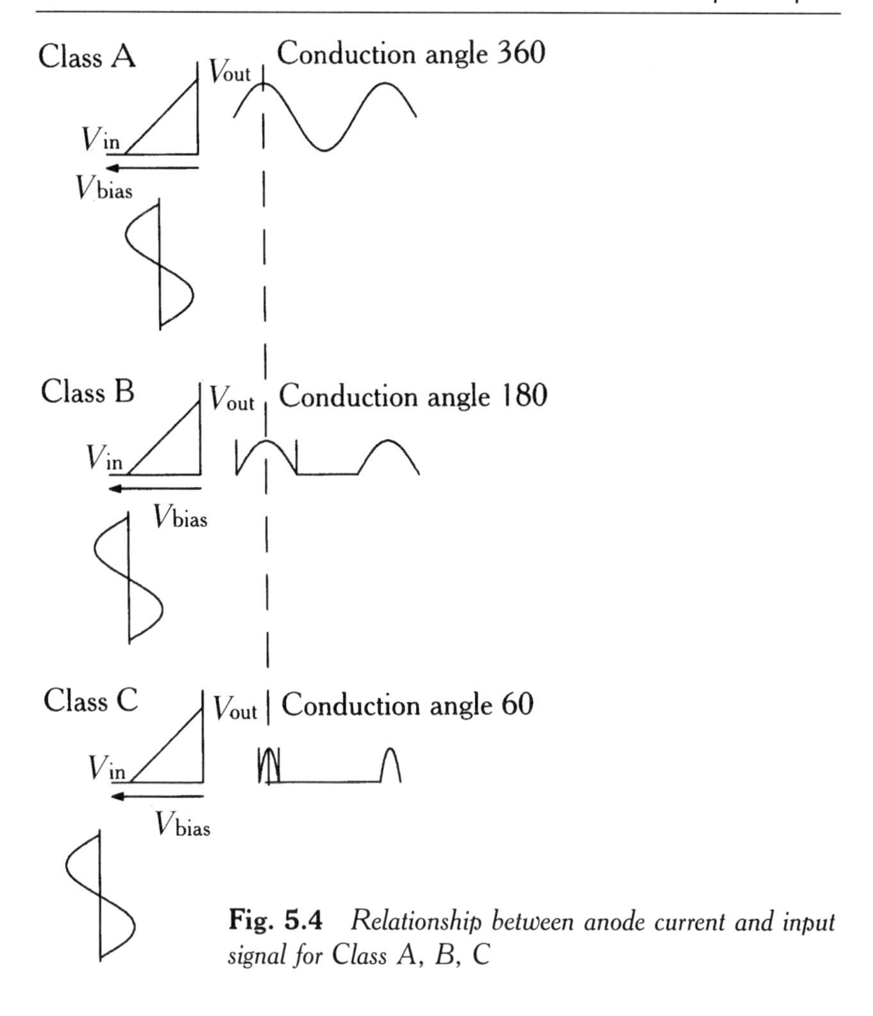

Fig. 5.4 *Relationship between anode current and input signal for Class A, B, C*

In audio, we normally refer to currents rather than conduction angles, and there are subdivisions of classes defined by the *grid* current of the valve. (RF engineers are unable to do this since they invariably operate with grid current to maximize efficiency at the expense of linearity.)

Class *1

Grid current is not allowed to flow. Many of the larger (≥50 W) classic amplifiers were push–pull Class AB1.

Class *2

The input signal is allowed to drive the grid positive with respect to the cathode, causing grid current to flow. This improves efficiency, since the anode voltage can now more closely approach zero, which is particularly relevant to triodes. At the onset of grid current, the input resistance of the output stage falls drastically (possibly approaching $1/gm$), and the driver stage needs a very low output resistance if it is to maintain an undistorted signal into this non-linear load without distortion. Some modern single ended amplifiers claim to be Class A2, whilst transmitters are commonly Class B2.

It will be apparent that efficiency can only be improved at the expense of linearity.

The push–pull output stage and the output transformer

We saw that the Class B stage introduced considerable distortion by half-wave rectifying the input signal. Clearly, this is a disadvantage for a hi-fi amplifier, since we require linearity.

Suppose, however, that we had two Class B valves, one fed directly with the input signal, and the other with an *inverted* signal. During time t_1 the upper valve conducts, whilst the lower is cut off, and during t_2 the situation is reversed. See Fig. 5.5.

So far all that we have achieved is to ensure that any *one* valve is switched on, no matter what the incoming phase. If we now invert one of the outputs and sum both outputs in the output transformer, we will recreate the shape of the original input waveform. The inversion is performed by reversing the connection of one winding, and is marked on the diagram with + and – symbols.

Whether achieved by a transformer or by a direct coupled series amplifier, such as a White cathode follower, this form of connection is known as *push–pull*, and is the only way of approximating linearity in a Class B amplifier.

Unsurprisingly, this dissection of the signal and its subsequent restitching is rather less than perfect, and pure Class B is very rarely used because of the distortion generated at the crossover region, where one valve takes over from the other. In practice, some quiescent current is allowed to flow

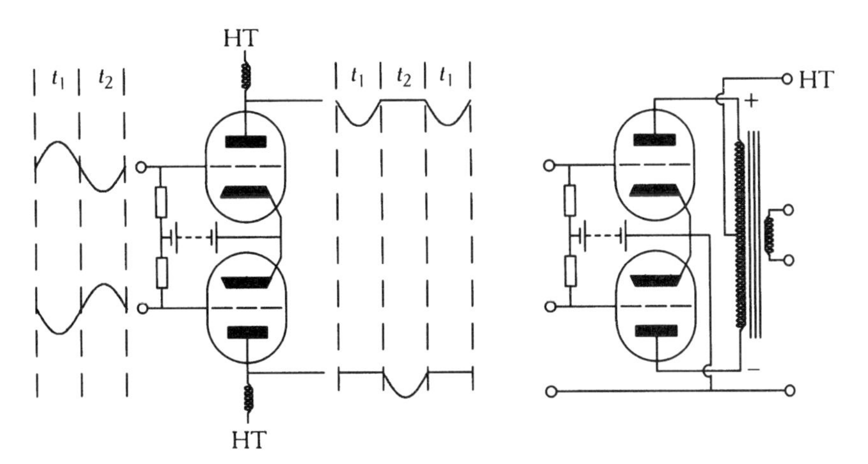

Fig. 5.5 *Summation of Class B signals in output transformer*

in an attempt to smooth this transition, resulting in Class AB operation. Although it is theoretically possible to determine an 'optimum' bias point for a Class AB amplifier, practical valves do not operate linearly down to cut-off and then suddenly switch off. Because of this, and because of individual differences between valves, the ideal point is ill-defined, and the low level distortion performance is poor.

Push–pull output stages can also be used for Class A amplifiers, giving additional advantages.

Because of the reversal of one winding, the magnetizing flux caused by the quiescent anode currents will cancel (provided that they are equal). This means that the transformer only has to handle signal current, so it can be far smaller for a given power – this is the *main* reason for using a push–pull output stage in a Class A amplifier.

Since the core is small, it is important that the quiescent anode current of each valve *is* identical, otherwise DC magnetization of the core will generate odd harmonic distortion. This can be done by having an adjustment for DC balance in the bias circuit, or by using pairs of valves with matched anode currents. See Fig. 5.6.

P2 sets total anode current, whilst P1 adjusts DC balance by biasing one grid more or less positively than the other.

If the core should become permanently magnetized (perhaps by failure of a valve), it will need to be degaussed, or it will generate additional (and

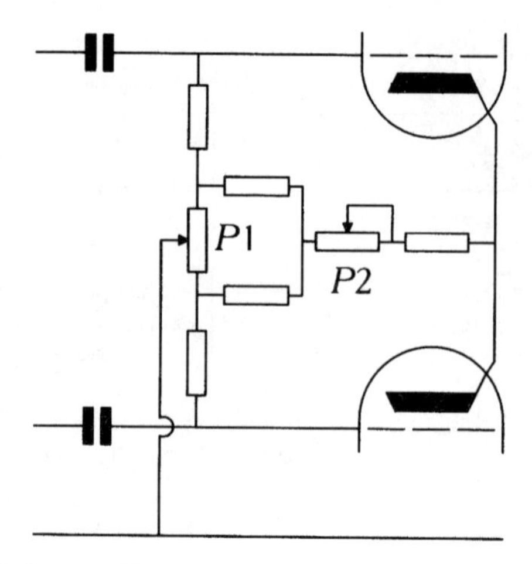

Fig. 5.6 *DC balance adjustment*

unnecessary) distortion. This can be done by applying a sufficiently large alternating magnetic field to the core to saturate it both positively and negatively, and then reducing the field to zero over a period of about 10 seconds.

A useful consequence of the reduced size of transformer is an improved HF response due to the reduction of stray capacitances.

Not only does quiescent anode current cancel in the transformer, but any power supply signals cancel, since they are also in phase in each winding, making the stage more tolerant of hum and noise on the HT line.

Additionally, even harmonic distortion, caused by unequal gain on positive and negative half cycles, is cancelled, whilst odd harmonic distortion is summed.

Since triodes generate primarily even harmonic distortion, this is useful, but pentodes generate primarily odd harmonic, and will therefore require considerable (>20 dB) negative feedback to reduce the distortion to acceptable levels. It is even arguable that cancellation of even harmonics in a push–pull pentode amplifier is undesirable, as the even harmonics may mask the audibility of odd harmonics.

This cancellation of even harmonic distortion will only be achieved if both windings are fed identical signal levels by their valves, so some amplifiers have adjustments for AC balance, whereas others specify gain matched pairs of valves. See Fig. 5.7.

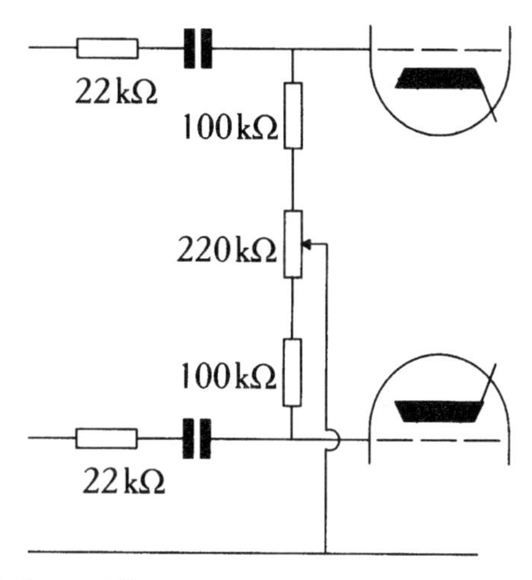

Fig. 5.7 *AC balance adjustment*

Modifying the connection of the output transformer

We have mainly considered triodes, and given pentodes scant regard because they produce copious amounts of odd harmonic distortion. But if we imagine the output transformer primary as a set of windings that could be tapped at any point, we see that for pentode operation g_2 would be connected to the centre tap (0%); whereas for a triode, it would be connected at the anode (100%). See Fig. 5.8.

What would happen if we were to tap at an intermediate point? This question was asked in the 1950s, and the 'ultra-linear' output stage was

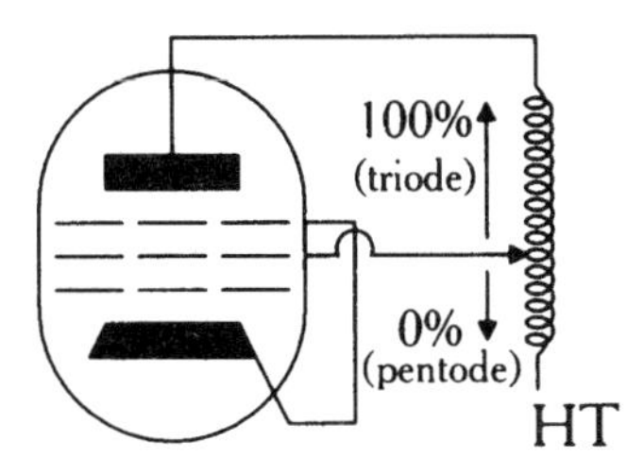

Fig. 5.8 *'Ultra-linear' output stage*

rediscovered. The popular tapping points are 43% (minimum distortion) and 20% (maximum power). This method became almost universal in the final days of valve supremacy, since it combined the efficiency and ease of driving the pentode with much of the improved linearity of the triode. It should be noted that:

$$I_a \propto V_{g2}^{3/2}$$

And as a consequence, negative feedback at g_2 is not as linear a process as one might wish; nevertheless, almost all power amplifiers using pentodes in the output stage use this scheme because it is far superior to pure pentodes.

Up until now we have placed the transformer in the anode circuit, but we could place the same transformer in the cathode circuit to form a cathode follower, resulting in an extremely low output resistance from the valve. As an example, a pair of EL34, connected as a triode, would each have an anode resistance of about 900 Ω, but used as cathode followers, the driving resistance would be a tenth of this at 90 Ω. Reflected through the transformer, the output resistance seen by a loudspeaker would be a fraction of an ohm even before global negative feedback.

Unfortunately, there are two crippling disadvantages to this topology. First, although the output stage is excellent, we have transferred its problems to the driver stage. Each output valve now swings ≈150 V_{RMS} on its cathode, and has a gain <1, requiring ≈500 V_{pk-pk} to drive it! This can be done, but it is not a trivial exercise to design the driver stage, since we must either use transformer interstage coupling, or a resistive anode load requiring a very high HT voltage.

Second, the high voltage on the cathode of the output valves severely strains the heater/cathode insulation, and can cause premature heater failure. Connecting the heater to the cathode solves this problem, but requires individual heater windings for each half of the output stage (to avoid shorting k_1 to k_2), and forces each valve to drive the interwinding capacitance (≈1 nF) of the mains transformer. A high frequency heater supply via a small transformer with individual windings would solve the problem, at the expense of possible RFI and increased complexity.

Alternatively, there are a few valves whose heater/cathode insulation can withstand 300 V, such as the 6080/6AS7G. Because this valve has such a low r_a, its optimum load resistance is fairly low, and its output voltage at full power is also quite low, reducing the strain on heater/cathode insulation. Unfortunately, the μ is also very low, so the gain of the output

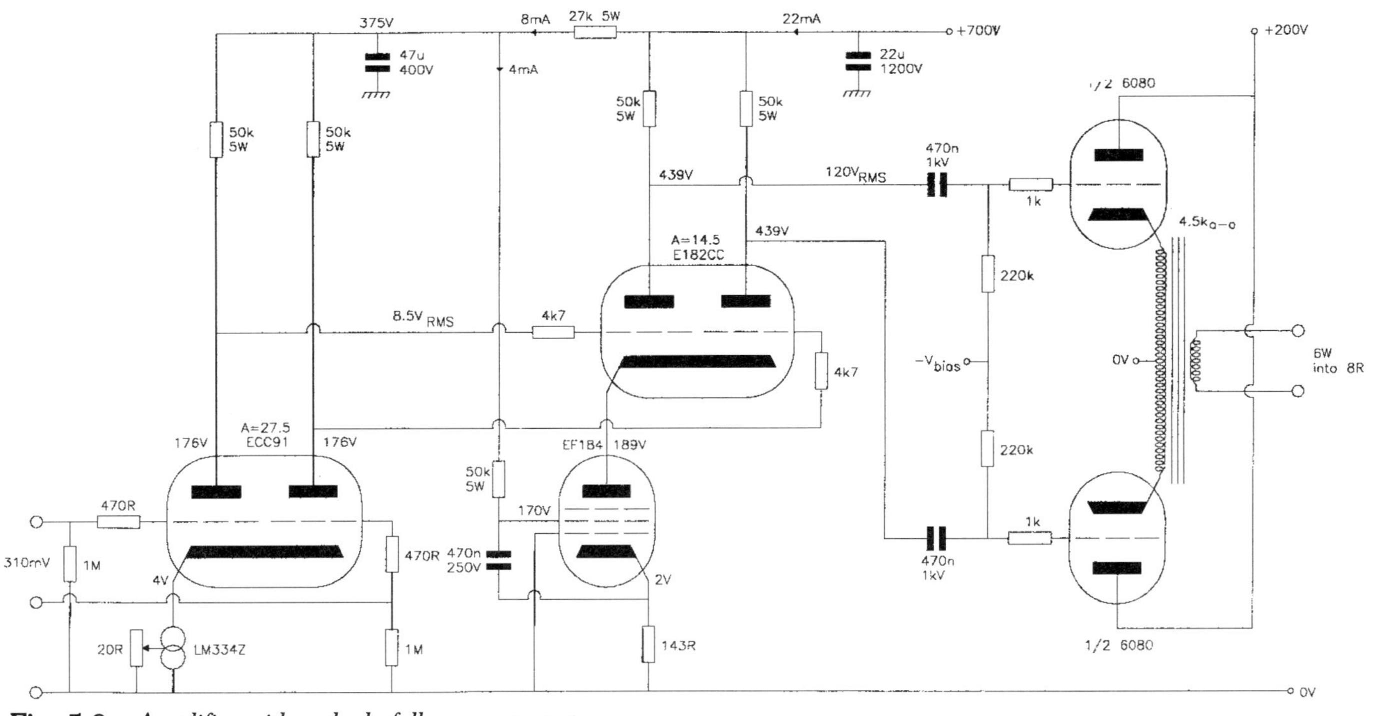

Fig. 5.9 *Amplifier with cathode follower output stage*

stage is substantially less than unity, and the driver stage has to be quite special. See Fig. 5.9.

As can be seen, a complex power supply would be required, simply to produce 6 W. Admittedly, the driver could cope with a number of 6080 in parallel, but the cure still seems worse than the complaint. The only reason that this design survived to the drawing stage is that the output stage should be quite tolerant of poor output transformers; conversely, a good output transformer would allow very good performance. The amplifier relies on balance rather than global feedback to reduce distortion, so a balanced input is possible, as shown.

Another possibility is to use a distributed load, where part of the load is in the cathode and part in the anode. This was used successfully by Quad in the UK and McIntosh in the USA, and relaxes the driving requirements whilst retaining some of the benefits of local feedback. See Fig. 5.10.

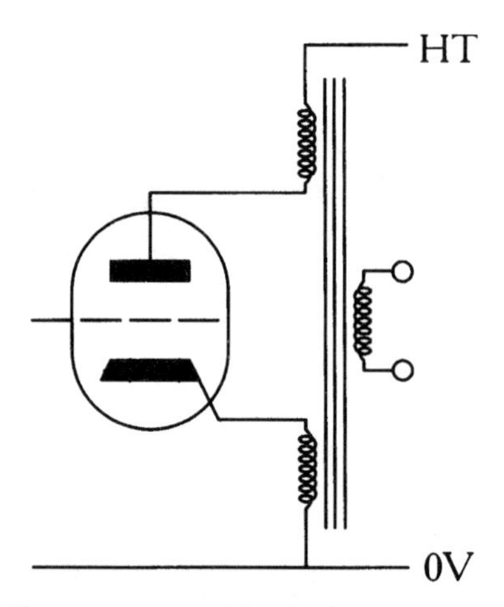

Fig. 5.10 *Quad II output stage (aka McIntosh configuration)*

An interesting version of the distributed load technique was used in the Michaelson & Austin TVA10 which had equal anode and cathode loads. Because the anode and cathode windings for each valve are wound in antiphase (the dot signifies the start of the winding), the quiescent anode current cancels, and there is no need to match valves for anode current.

Sadly, the drive requirements are almost as severe as for the cathode follower, and this technique has not been widely used. See Fig. 5.11.

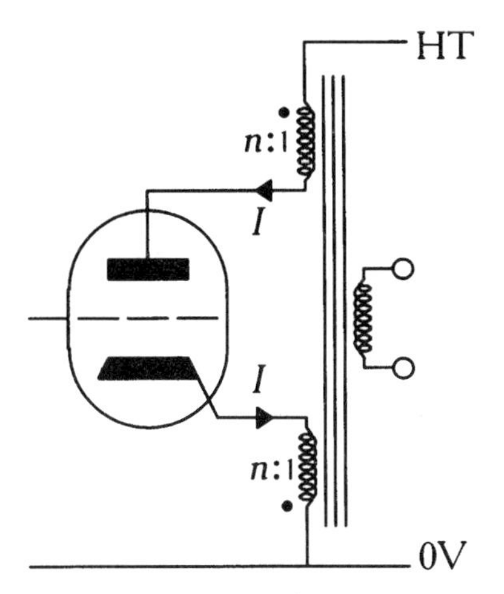

Fig. 5.11 *Complete cancellation of DC current in output transformer using only one valve*

Output transformer-Less (OTL) amplifiers

Almost all of the different output stage configurations were devised in an effort to reduce the adverse effect of the output transformer, so it is not surprising to find that there have been some designs that dispense with the output transformer. These are often known as Futterman[1] amplifiers (who first popularized the notion), or OTLs.

Driving low impedance loads directly is not natural for a valve, so radical approaches are needed. Special valves have to be used, which invariably were not designed for audio, and they therefore have extremely questionable linearity. Examples are the 6080/6AS7G double triode (r_a $\approx 300\,\Omega$, designed for use in power supplies), and television line scan output valves such as the PL504 and PL519 pentodes. Efficiency is generally on the low side of appalling. Output stages invariably use paralleled White cathode followers with plenty of global feedback to reduce the output resistance. See Fig. 5.12.

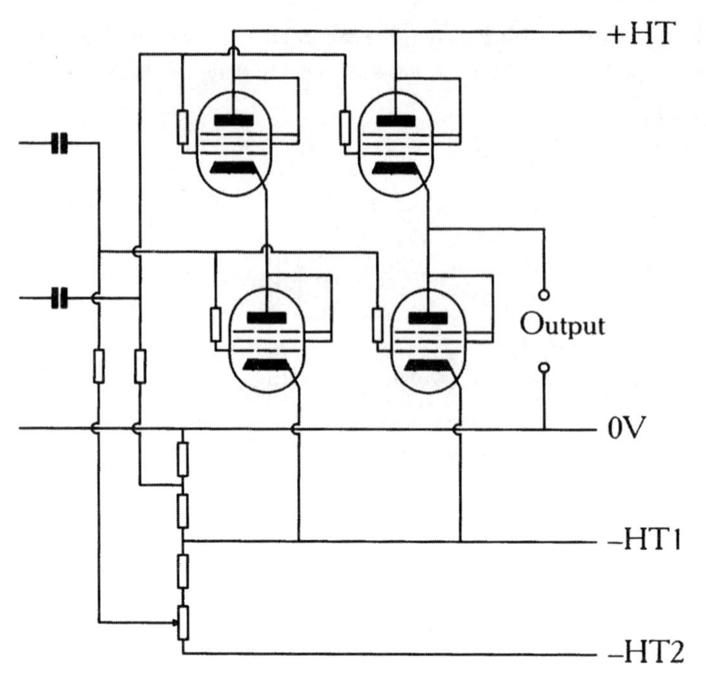

Fig. 5.12 *OTL output stage (paralleled White cathode followers)*

These amplifiers are quirky in the extreme, yet some designers think that the problems of output transformers are so severe that they persist in making successful OTL amplifiers.

The entire amplifier

Having looked at the problems of the output stage, we can now consider the support circuitry in detail. The output stage will not be sufficiently sensitive to be driven directly from a pre-amplifier, so it will need additional gain. If it is push–pull, it will need a phase splitter. Since linearity is unlikely to be ideal, we will probably need global negative feedback, which will further reduce gain, and this will need to be restored. A complete system might therefore comprise an input stage, a phase splitter, a driver stage and the output stage. See Fig. 5.13.

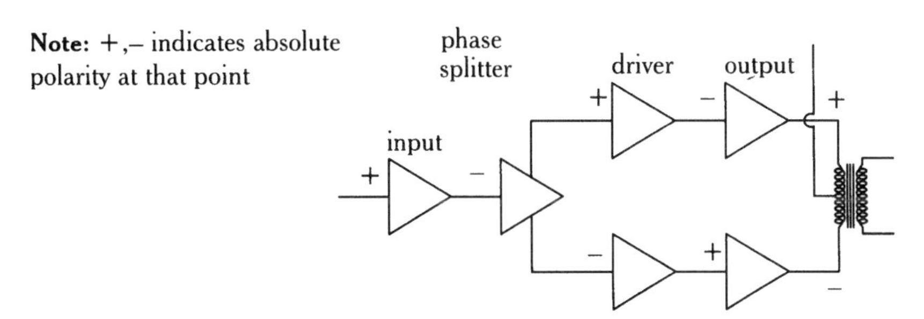

Fig. 5.13 *Block diagram of complete power amplifier*

Although a Class A output stage is a constant resistive load, a Class AB2 output stage heavily loads the driver stage when drawing grid current, and its driver stage must have very low output resistance to drive this load without distortion.

Unlike the output stage, and possibly the driver stage, the remainder of the stages in a power amplifier will be loaded by predictable resistive loads. It is therefore possible, and desirable, to design these stages with great care in order that they should not degrade the performance of the entire amplifier. We will consider push–pull amplifiers, since they make up the majority of designs, and the design principles may perfectly well be applied to single ended amplifiers.

The driver stage

Unless the amplifier is quite low power, it will require a dedicated driver stage. We need a stage with good linearity, low output resistance and good output voltage swing.

The differential triode pair is the ideal choice. The output stage probably requires about 25 V_{RMS} to each grid, and has an input capacitance of 40 pF or more. An output resistance of 10 kΩ coupled to an input capacitance of 40 pF gives a high frequency cut-off of 400 kHz, which is acceptable. Since $r_a \approx R_{out}$, the high μ valves, which tend to have a high r_a, are probably unsuitable.

In a properly designed power amplifier, the output stage should be the limiting factor, so we ought to design the driver stage to have at least 6 dB of overload margin. This requirement probably rules out our favourite valve, the E88CC. Very few commonly available valves satisfy our requirements.

Dual triodes suitable as drivers:

Type	r_a	Comments
6SN7	≤10 kΩ	Excellent reputation for clarity
ECC82	≤10 kΩ	Not as good as 6SN7
E182CC	≤5 kΩ	Good on paper, but can sound strident
6BX7	≤2 kΩ	Capable of driving 845. Robust

Single pentodes suitable as drivers (when triode connected):

Type	r_a	Comments
EF184	≤5 kΩ	Cheap and really plentiful. $\mu = 60$
A2134, N78, etc.	≤3 kΩ	Uncommon, but cheap
EL84	≤2 kΩ	NOS almost extinct, but current production OK

Even lower driving resistance can be provided by an additional direct coupled cathode follower stage. See Fig. 5.14.

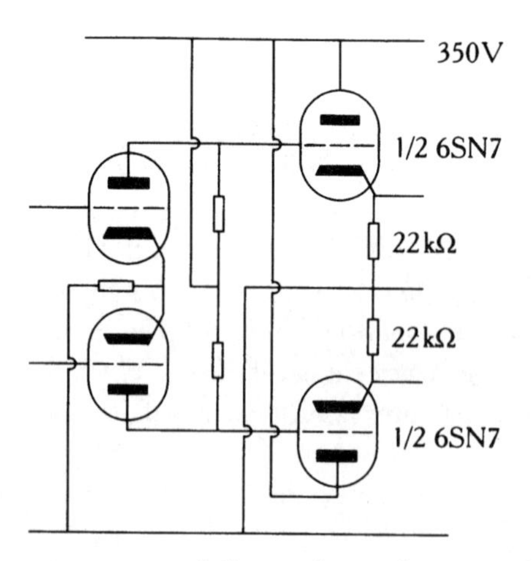

Fig. 5.14 *Driver stage using differential pair direct coupled to cathode followers*

The phase splitter

The driver stage is always preceded by the phase splitter, and traditionally the two stages have been combined – although as we shall see, this is not always a good idea. Design of the phase splitter is crucial to the success of a push–pull amplifier, so we will look at this in detail.

The phase splitter converts a single ended signal into two signals of equal amplitude, but with one of inverted polarity. There are three fundamental ways of achieving this goal. See Fig. 5.15.

- We use a centre-tapped transformer in the same way that we use an output transformer to provide inverted and non-inverted signals. All of the previous considerations about transformers apply with a vengeance because of the comparatively high impedances involved, and the technique is not widely used, even though balance is near perfect under all conditions. See Fig. 5.15a.
- We have two outputs, one is the original signal, and the other is simply the input passed through an inverter. See Fig. 5.15b.
- A single device controls the flow of current in two resistors, one of which is connected to ground, and the other to HT. Increased current causes the DC voltage dropped across each resistor to rise, so at any instant the absolute voltage at one output is falling, whilst the other is rising. See Fig. 5.15c.

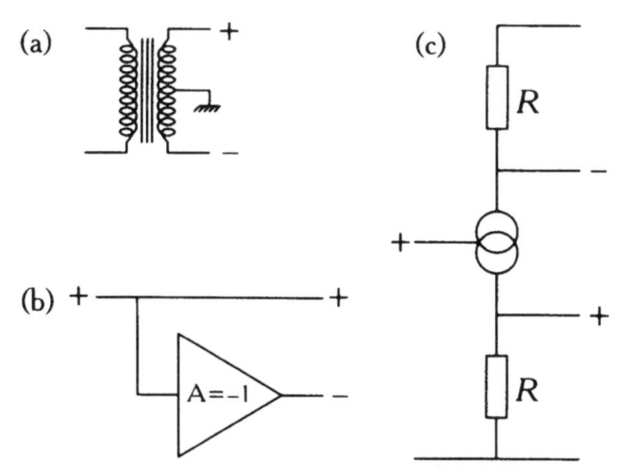

Fig. 5.15 *Fundamental basis of* all *phase splitters*

All of the phase splitters employing the second method are based on the differential pair, whilst the third method is the basis of the concertina phase splitter. Pure triode phase splitters with low resistance outputs are very sensitive to their loading, and produce different output resistances when each output is loaded equally, compared to only one output being loaded. Phase splitters with a high r_a have an output resistance dominated by R_L, so phase splitters using pentodes and cascodes are immune to loading symmetry problems.

Loading sensitivity means that pure triode phase splitters should only ever be loaded by a stage that can be guaranteed to be Class A1 at all times if the balance of the phase splitter is not to be upset. (Any transition into Class B temporarily eliminates Miller capacitance at the input of the switched-off valve, whilst Class A2 dramatically reduces input resistance of the switched-on valve once grid current flows.)

The differential, or long-tailed, pair and its derivatives

Most classic phase splitters are based on the differential pair, and much ingenuity was demonstrated in improving their performance.

A perfect differential pair comprises two devices (each with a load resistance) connected so as to allow a signal current to swing backwards and forwards between the load resistances without any loss whatsoever. Any loss of signal current from the cathode to ground impairs performance, so the tail resistance is crucial, and should ideally be infinite.

The $R_k \gg R_L$ solution

The differential pair can be optimized with a pentode or cascode constant current sink. An ECF80 pentode can achieve a tail resistance $>10\,\text{M}\Omega$, and even the larger pentodes such as the EL84 can manage $1\,\text{M}\Omega$ unaided. Performance at low frequencies could be improved by adding a transistor to make a hybrid cascode, but we will always be limited at high frequencies by C_{kh} from the differential pair's cathode, even if the sink is perfect. See Fig. 5.16.

The behaviour of the differential pair was discussed in Chapter 2. Provided that $R_k \approx \infty$, a balanced output must be achieved if R_{L1} =

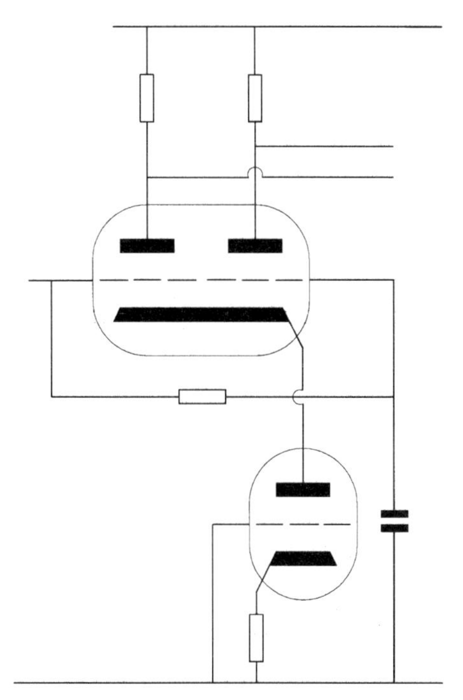

Fig. 5.16 *Differential pair with triode constant current sink as phase splitter*

R_{L2}. Output resistance must also be identical from both outputs, and $r_{out} = r_a \parallel R_L$, as before.

However, if only one output is loaded, then:

$$r_{out} \approx \frac{R_L(R_L + 2r_a)}{2(R_L + r_a)} \approx \frac{R_L}{2}$$

So the stage should only be loaded by a stage that never strays out of Class A1.

The $R_k \approx R_L$ compensated solution

We accept that we cannot easily achieve a high tail resistance, and do not even try. We use a resistor, typically from $22\,k\Omega$ to $82\,k\Omega$, as a tail, calculate what the errors will be, and try to correct them. This is known as the cathode coupled phase splitter. See Fig. 5.17.

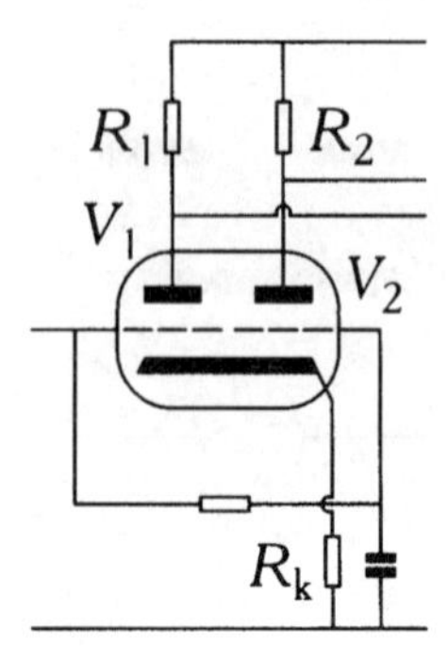

Fig. 5.17 *Cathode coupled phase splitter*

V_2 can be considered to be a grounded grid amplifier, fed from the cathode of V_1. It is this use of the first valve as a cathode follower to feed the second valve that results in the apparent loss of gain of the second valve, since for a cathode follower, $A_v < 1$. By inspection, we can see that $2V_{gk}$ is required to drive the stage, and so the gain of the compound stage to each output is half what we would expect from an individual valve.

If the output were in balance, then $V_1 = V_2$, so:

$$i_1 R_1 = i_2 R_2$$

The gain of V_2 is A_2, so:

$$v_{gk} = \frac{v_2}{A_2}$$

The signal current flowing in the cathode resistor will be the out-of-balance output signal current:

$$v_{gk} = (i_1 - i_2)R_k$$

The signal at the output of V_2 will be:

$$(i_1 - i_2)R_k \,.\, A_2 = i_2 R_2$$

Expanding and collecting I_2 terms:

$$i_1 \,.\, R_k \,.\, A_2 = i_2(R_2 + R_k \,.\, A_2)$$

Substituting $i_1 R_1 = i_2 R_2$, and simplifying:

$$\frac{R_2}{R_1} = \frac{R_2}{R_k \,.\, A_2} + 1$$

This shows that unless the gain or the tail resistance of the stage is infinite, the ratio of the anode loads should be adjusted to maintain balance. Note that A_2 is the individual, unloaded, gain of V_2, and not the gain of the entire stage.

As an example, the Leak TL12+ phase splitter/driver was investigated. This uses an ECC81 and gives a gain for V_2 of 42 (μ = 53, r_a = 26.5 kΩ); R_1 should therefore be 91 kΩ. See Fig. 5.18.

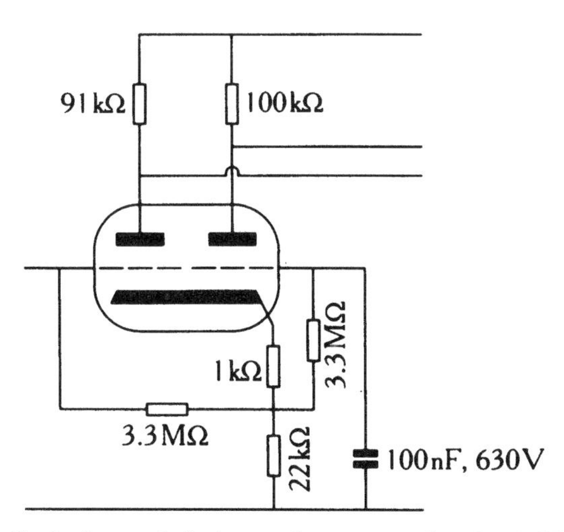

Fig. 5.18 *Cathode coupled phase splitter as used in Leak TL12+*

The output resistance of each half of the stage is slightly different because it is in parallel with a slightly different anode load, but curing this in order to preserve HF balance upsets the voltage balance at low frequencies. The nearest approximation that we can achieve is to include the grid leak resistances as part of the anode loads when calculating the necessary changes. The following grid leak resistors are 470 kΩ, so R_{L2}

= 100 kΩ || 470 kΩ = 82.46 kΩ, and the gain of V_2 falls to 40. The required total load for V_1 (including the 470 kΩ grid leak) is thus 75.7 kΩ, and R_{L2} = 90.2 kΩ.

LF balance is determined by the time constant of the grid decoupling capacitor and its series resistor, since it cannot hold the grid of V_2 to AC ground at very low frequencies.

Rather than tinker with resistor values whose calculation is critically dependent on valve parameters, the author would far rather add a constant current sink to the cathode to force the stage into balance.

The $R_k \ll R_L$ high feedback solution

We make no attempt at providing a large value of tail resistor, and rely on feedback to maintain balance, this circuit is known as the floating paraphase or see-saw phase splitter. Typically, this design uses a high μ valve, such as the ECC83, from whose data sheet this circuit was taken. See Fig. 5.19.

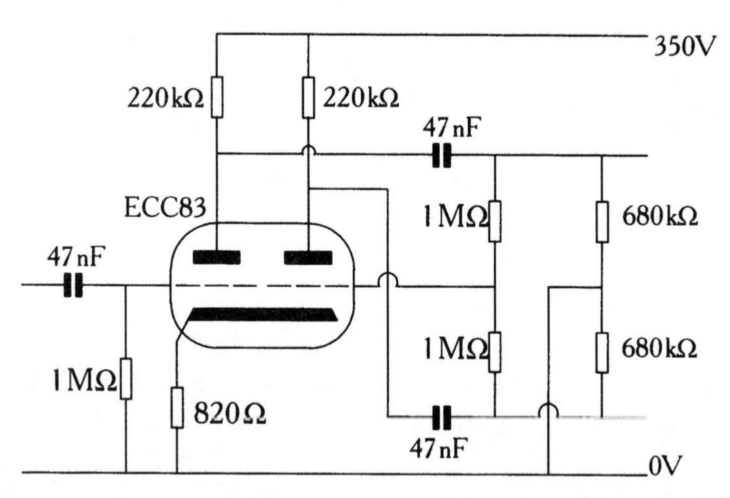

Fig. 5.19 *Floating paraphase or see-saw phase splitter (Reproduced by courtesy of Philips Components Ltd)*

If we redraw this circuit, we see that V_2 is simply a unity gain invertor, whose gain is defined by resistors R_1 and R_2. See Fig. 5.20.

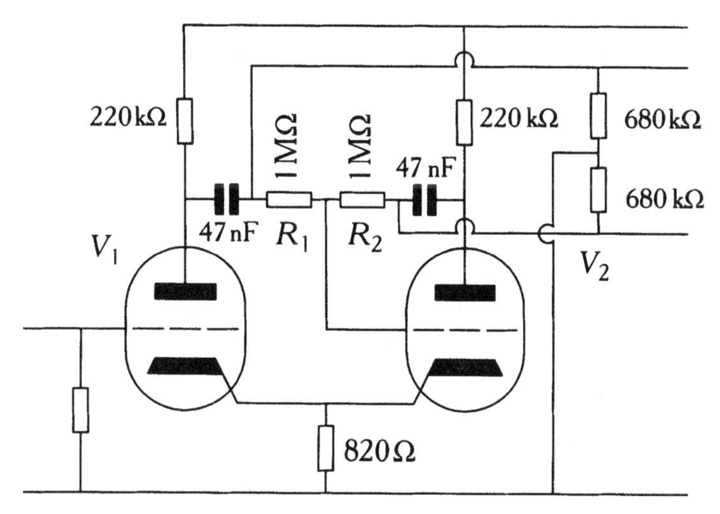

Fig. 5.20 *Floating paraphase phase splitter redrawn to reveal invertor*

Since the open loop gain of V_2 is not infinite, these values will need to be adjusted to give a gain of -1; this calculation will be further complicated by the fact that R_2 affects the loading and open loop gain of the stage. V_2 will also require a build-out resistor to equalize the output resistance, since its output resistance has been reduced by negative feedback. Once these corrections have been made, the balance of this phase splitter is good, since the operation of V_2 is stabilized by negative feedback.

The circuit is analysed first by drawing a DC loadline to correspond to the 220 kΩ anode load. The Mullard operating point is at $V_a = 163$ V. The 1 MΩ feedback resistor is in parallel with this at AC, so we draw an AC loadline through the operating point corresponding to 180 kΩ. From this we find that the AC gain of the valve is 67.

We need to find the value of β that will give final gain of 1, using:

$$\beta = \frac{1}{A} - \frac{1}{A_0}$$

We find that $\beta = 0.985$. The easiest way to achieve this is to increase the value of the feedback resistor:

$$R'_f = \frac{R_f}{\beta}$$

This gives a value of 1015 kΩ, so we would add 15 kΩ in series. So far, we have only discovered a 1.5% error, which is trivial, but if we consider the output resistances we find a much larger error. The output resistance of V_1 is r_a in parallel with R_L, and is ≈53 kΩ, but the output resistance of V_2 has been reduced by a factor of $(1 + \beta A_0)$, from 53 kΩ to ≈790 Ω. 52.2 kΩ of build-out resistance is therefore required, but the nearest standard value of 51 kΩ would be fine. These outputs are then each loaded by 680 kΩ, and if corrections are not made, the output from V_2 will be ≈6% high.

In practice, these corrections were never made, which perhaps accounts for the poor reported performance of the stage. It might be thought that the common connection of the cathodes would improve balance, but V_2 has such heavy feedback that it can easily overcome any self-balancing action generated at the cathode.

The concertina phase splitter

The phase splitters based on the differential pair were all able to provide overall gain, but this was obtained at the expense of an output balance that was partially dependent on the matching of μ between the valves.

Although the concertina phase splitter does not provide gain, its output balance is almost totally determined by passive components, and valve characteristics hardly enter the picture. Conceptually, operation is very simple. Modulation of grid voltage causes a signal current to flow in the valve, the anode and cathode loads are equal and they have the same current flowing through them, so the signals generated across them are equal, implying perfect balance. See Fig. 5.21.

Gain of the concertina

The gain of the concertina may be found using the standard triode gain equation, but noting that all undecoupled resistances down to ground via the anode resistance are multiplied by a factor of $(\mu + 1)$, so that:

$$r'_a = (\mu + 1)R_k + r_a$$

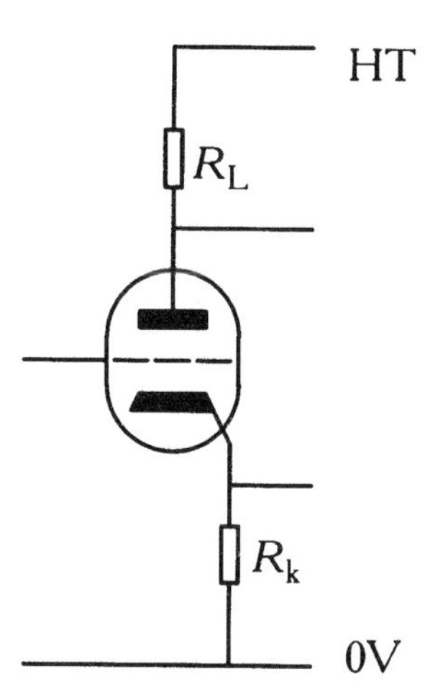

Fig. 5.21 *Concertina phase splitter*

$$A = \frac{\mu R_L}{r_a + (\mu + 1)R_k + R_L}$$

But for the concertina, $R_k = R_L$, so:

$$A = \frac{\mu R_L}{R_L(\mu + 2) + r_a} \approx 1$$

Because of this low value of gain to the anode, Miller capacitance is very low, and the stage has wide bandwidth.

Output resistance with both terminals equally loaded (Class A1 loading)

The concertina is a special case ($R_k = R_a$) of an unbypassed common cathode amplifier with outputs taken from both cathode and anode. The general feedback equation is:

$$A = \frac{A_0}{(1 + \beta A_0)}$$

The denominator of the feedback equation is the factor by which resistances are changed. Since we know the gain of the concertina and the gain of a simple triode amplifier, we can substitute them into the feedback equation to solve for the feedback factor:

$$\frac{\mu R_L}{R_L(\mu + 2) + r_a} = \frac{\dfrac{\mu R_L}{R_L + r_a}}{\text{feedback factor}}$$

Cross-multiplying to find the feedback factor:

$$\text{feedback factor} = \frac{R_L(\mu + 2) + r_a}{R_L + r_a}$$

The anode output resistance of a common cathode triode amplifier with no feedback is:

$$r_{out} = \frac{R_L r_a}{R_L + r_a}$$

The feedback works to reduce anode output resistance, so this value must be divided by the feedback factor:

$$r'_{out} = \frac{R_L r_a}{R_L + r_a} \cdot \frac{R_L + r_a}{R_L(\mu + 2) + r_a}$$

The $(R_L + r_a)$ terms cancel, leaving:

$$r'_{out} = \frac{R_L r_a}{R_L(\mu + 2) + r_a}$$

Initially, it seems most surprising that series feedback ($R_k = R_a$, after all) should *reduce* output resistance from the anode so that $r_{out} \approx 1/gm$, but this can be understood by considering an external capacitive load on each output. In the same way that $R_k = R_a$ defines a gain of 1 at low frequencies, so $X_{C(k)} = X_{C(a)}$ defines a gain of 1 at high frequencies, and changing this ratio of capacitances certainly would change the gain, or frequency response at high frequencies, since it would change the feedback ratio β.

Because $Z_k = Z_a$, the frequency response at each output is forced to be the same, so the output resistances must also be equal, and $r_{out(k)} = r_{out(a)}$.

Concertina output resistance, only one output loaded (Class B loading)

Looking into the cathode, we see R_k down to ground, in parallel with r_k the anode path to ground:

$$r_k = \frac{R_a + r_a}{\mu + 1}$$

Substituting:

$$r_{out(cathode)} = \frac{R_k \dfrac{R_a + r_a}{\mu + 1}}{R_k + \dfrac{R_a + r_a}{\mu + 1}}$$

Simplifying, and noting that $R_a = R_k = R_L$:

$$r_{out(cathode)} = \frac{R_L(R_L + r_a)}{R_L(\mu + 2) + r_a}$$

The cathode output resistance has to be calculated fully, but generally gives a value of about 1 kΩ.

Looking into the anode, we see R_L to HT, which is AC ground, in parallel with the cathode path to ground:

$$r_a' = (\mu + 1)R_L + r_a$$

Substituting:

$$r_{out(anode)} = \frac{R_L((\mu + 1)R_L + r_a)}{(\mu + 1)R_L + R_L + r_a}$$

Simplifying:

$$r_{out(anode)} = \frac{R_L^2(\mu + 1) + R_L \cdot r_a}{R_L\,(\mu + 2) + r_a}$$

If we inspect this equation closely, we see that the terms involving μ are the only significant terms, and that if μ is reasonably large, then $(\mu + 1) \approx (\mu + 2)$, so that $r_{out} \approx R_L$.

High μ valves, such as the ECC83, are not really suitable for the concertina, ECC88 and E182CC are good, but 12BX7 and 12B4-A are better. It is usual to direct couple to the anode of the input stage, and let that determine the DC conditions of the concertina, resulting in the saving of a coupling capacitor and a low frequency time constant. Although the concertina is often criticized for its lack of gain, when directly coupled to the input stage, the two valves together give considerably more gain than the same two valves used as a phase splitter based on the differential pair.

The input stage

The input stage is where global negative feedback is applied, so it must provide an inverting and a non-inverting input, both with low noise. The triode differential pair is an obvious candidate for this stage, but the common cathode triode or pentode can also be used, in which case global feedback is applied at the cathode. See Fig. 5.22.

Design of the input stage is fairly trivial, but can be slightly complicated by the usual practice of direct coupling to the phase splitter, which restricts choice of anode operating conditions.

Stability

When we looked at *RC* networks in Chapter 1, we saw that a single *RC* network was asymptotic to 90° of phase shift. To make an oscillator, we need 180° of phase shift, so a single stage amplifier with one *RC* network causing an LF or HF cut-off *cannot* oscillate. If we cascade two such stages, we can approach 180° of phase shift, and if we feed this back into

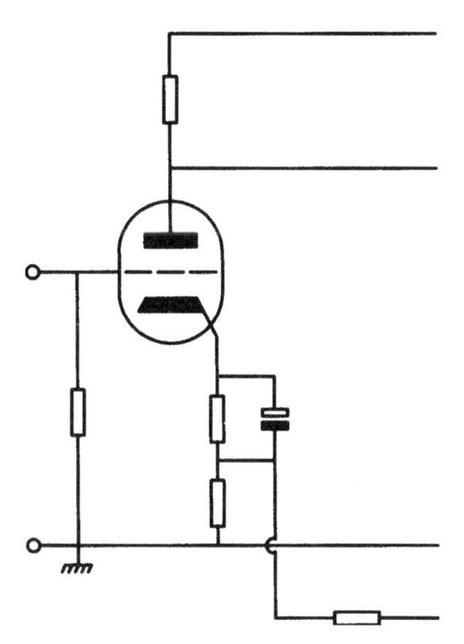

Fig. 5.22 *Application of global feedback at the input stage*

the input, it will ring, but not oscillate. If we have three such stages, it is a racing certainty that we can make the cascade oscillate when we apply feedback, and this is the basis of the phase shift oscillator.

To achieve oscillation, we need more than just phase shift. We also need sufficient *loop gain*. When we apply our feedback signal, even if its phase has been shifted by 180°, we may not necessarily generate oscillation. The basis of oscillation is that it is self-sustaining; the gain of the amplifier must be sufficiently high to overcome the losses in the feedback loop before oscillation will occur. Loop gain is defined as the gain of the amplifier multiplied by the loss of the feedback loop.

If we have a phase shift of 180°, *and* loop gain ≥ 1, the circuit *will* oscillate.

Now that we have this definition, we can see how to avoid designing oscillators. We have two weapons at our disposal:

- We can reduce the number of stages, such that phase shift never reaches 180°. We rarely achieve this ideal, because the output transformer plus output stage plus driver stage harbours so many phase shifts, but the principle of minimizing the number of stages within a feedback loop is still valid.
- We attack the second condition of the oscillator statement, and reduce loop gain to ≤ 1 at the troublesome frequencies. This is the basis of all

the methods that you will see for stabilizing amplifiers, and it is a powerful weapon capable of oppressing anything. Whether the resulting amplifier is of any use may be more debatable.

Slugging the dominant pole

This rather vibrant and mystifying description is actually very simple.

A pole in electronic jargon is simply another way of saying 'HF cut-off', just as a zero is an LF cut-off. What we aim to do is to make the amplifier *look* as if it only has one HF cut-off, which, with a maximum phase shift of 90°, is unconditionally stable. We look for the *RC* network with the lowest HF cut-off frequency, i.e. the dominant one, and we 'slug' it with more capacitance to make it even lower.

Suppose that as a worst case, we cascaded four identical amplifiers, each with an HF cut-off frequency of 300 kHz, and a gain of 10. At 300 kHz, each amplifier will contribute a phase shift of 45°, making a total shift of 180°. The gain of each amplifier will be 3 dB down at 300 kHz, so the gain of each amplifier will be $10/\sqrt{2} = 7.071$, and the gain of the total amplifier will be:

$$A_{total} = \left(\frac{10}{\sqrt{2}}\right)^4$$

$$= 2500$$

In a typical amplifier we might want to reduce this gain from 2500 to 125, which would be 26 dB of negative feedback, and would reduce distortion to a twentieth of its original value. In order to do this, the feedback loop would have a loss of 0.0076. If we now check for stability, $0.0076 \times 2500 = 19$. The amplifier has a loop gain ≥ 1, *and* a phase shift of 180°, so it will oscillate.

We need to reduce the open loop gain at 300 kHz by a factor of 19, or 25.5 dB, to achieve stability (this has little effect on the frequency response of the final amplifier). Remembering that 6 dB/octave is equivalent to 20 dB/decade, reducing *one* cut-off from 300 kHz to 30 kHz will give us 20 dB reduction, and halving from 30 kHz to 15 kHz will give us another 6 dB, making 26 dB in total. This procedure may be formalized by the following statement:

The loop gain may be as large as the ratio of the two most dominant time constants.

To apply this rule, we simply choose how much feedback we want, calculate the loop gain, and adjust the dominant time constant until the ratio between it and the adjacent time constant is equal to the loop gain.

The amplifier is now stable, but only just, and it *will* ring. We should distance the dominant time constant still further to increase stability and remove ringing, certainly by a factor of 2 and preferably a little more. It should be realized that overzealous stability compensation will also reduce the feedback, and therefore distortion reduction, at these frequencies.

Many practical amplifiers, having exhausted the two possible methods of achieving stability described, resort to manoeuvring the amplitude response independently of phase response using *step networks*. Traditionally, these were adjusted on test, but if we are forced to use this method, computer aided AC analysis is an excellent tool for observing the effects of changes – there are many analysis programs available.

Realistically, we ought to consider that the means of achieving stability, in priority order, are:

- Reduce the number of time constants or stages within the loop.
- Slug the dominant pole (or zero).
- Fudge the phase/amplitude response using step networks.

There are some stability problems that are peculiar to valve amplifiers, and they have well-known symptoms and cures.

LF instability, or motorboating

This is an oscillation at about 1 or 2 Hz, and is invariably caused by unintentional feedback travelling through the power supply circuits, due to the rising impedance of filter capacitors at low frequencies. The traditional cure was to insert an LF step network, or to reduce the value of the coupling capacitors, *in the signal path*, so as to reduce the available gain. This solution molests the second condition of the stability statement, but only treats the symptoms.

The *real* solution is to attack the first condition by *removing* the filter capacitors and their associated RC time constants by fitting HT regulators. This generally kills the problem stone dead. It is this improvement in stability that is the reason for the superior bass in designs using regulated supplies, since it removes previously unidentified LF ringing. It should be noted that this problem does not necessarily need a global feedback loop to make itself felt, and 'zero feedback' pre-amplifiers are not immune.

It is not unusual to discover that not only is the amplifier motorboating, but that it also has bursts of HF oscillation, known as *squegging*. If possible, it is best to cure the squegging first, since it indicates marginal stability when the amplifier is under maximum stress and may be concealed once the LF instability has been cured.

Parasitic oscillation of the output stage

The solution is almost encompassed by the description. Parasitics are the unwanted stray capacitances and inductances that result from the practical attempt to build an amplifier.

Miller capacitance in the valve will combine with series inductance in the grid wiring to form a resonant circuit, and valves with a high *gm* are prone to oscillation because of this resonant circuit. The cure is to damp the resonant circuit by fitting *grid stopper* resistors in series with the grid *as close as possible* to the grid pin of the valve base. The physical positioning of the resistor reduces the grid inductance whilst the resistance of the resistor makes the resonant circuit more lossy:

$$Q = \frac{1}{R}\sqrt{\frac{L}{C}}$$

For small-signal valves that are prone to this problem (E88CC) a surface mount resistor actually touching the pin is ideal. Typical values range from 100 Ω to 10 kΩ, and are usually found by experiment since individual layout is critical.

'Ultra-linear' amplifiers with poor output transformers or paralleled pairs of output valves sometimes need a series RC network between anode and g_2. This is because this section of the winding has resistance in series with leakage inductance – the additional network attempts to return this impedance to a pure resistance. For 43% taps, the impedance between the anode and g_2 tap is ≈9% of the total anode to anode impedance, so this is a good starting point for the value of resistor, but both have to be determined empirically, and are often around 1 nF and 1 kΩ.

Stability of valve amplifiers is often described by the amount of additional feedback that would be required to cause oscillation. This is simply the ratio by which the dominant time constants were further distanced, over and above the necessary minimum required for stability. The Mullard 5–20 is proud to say that 10 dB more feedback would be required to cause instability, whereas the Williamson is questionable at LF even without additional feedback.

From this it should be realized that keeping a valve amplifier stable after applying global negative feedback is not a trivial task.

Classic power amplifiers

Now that we can recognize and analyse individual stages, we can investigate the design of some classic amplifiers such as the Williamson, the Mullard 5–20 and the Quad II.

The Williamson

This amplifier was published in *Wireless World*[2] in 1947, and set a standard of performance that was years ahead of its time.

The input stage is the standard common cathode triode with 20 dB of global negative feedback applied from the loudspeaker output to the cathode. The phase splitter is a concertina circuit direct coupled from the input stage, and feeds a differential pair using both halves of a 6SN7. See Fig. 5.23.

The output stage is a push–pull pair of KT66 beam tetrodes operated as triodes that provide 15 W output in Class AB1, operating mostly in Class A. $RV1$ adjusts the DC balance of the output valves in order to minimize distortion due to the transformer core, whilst $RV2$ sets the quiescent current to 125 mA for the entire stage.

The linearity and headroom of each stage is excellent due to the careful choice of operating points, but because this amplifier has four stages enclosed by the feedback loop, stability needs to be taken very seriously.

The input stage initially has an output resistance of ≈7.5 kΩ, but this is raised by the feedback to ≈47 kΩ. In combination with 12 pF of input capacitance from the concertina, this gives a high frequency cut-off of ≈280 kHz. However, this has been modified by adding the step compensation components $R2$, C1 to the anode circuit of $V1$. This circuit puts a step in the amplitude response which begins to fall at ≈130 kHz, but the phase response remains virtually unchanged until 280 kHz.

The concertina drives a driver stage with an input capacitance of 60 pF, and because the output transformer for the Williamson was very carefully specified, it seems unlikely that losses in the output transformer would cause the global feedback loop to force the driver stage out of Class A operation. The concertina thus faces a balanced load, and has an output resistance ≈350 Ω, resulting in $f_{-3\,\text{dB}} = 7.5\,\text{MHz}$, which is sufficiently high to be insignificant.

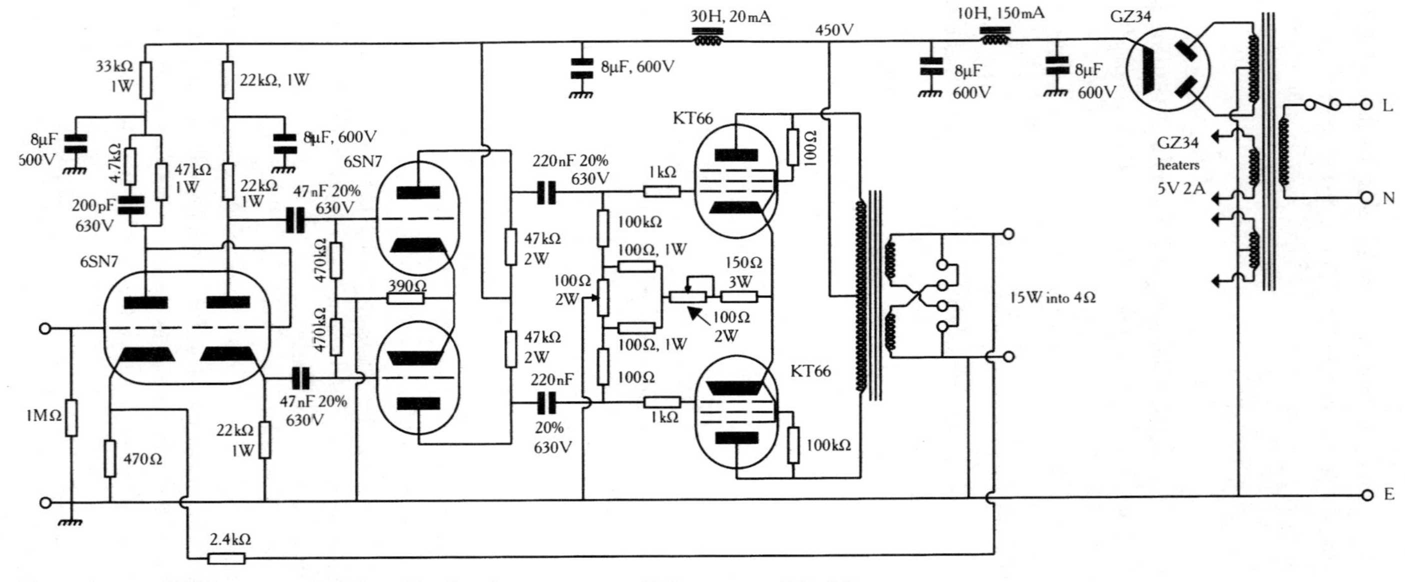

Fig. 5.23 *Williamson amplifier (By kind permission of* Electronics World)

The driver stage has an output resistance of ≈8.7 kΩ, together with 55 pF of input capacitance from the output stage, the cut-off is ≈330 kHz, and the output transformer is specified to have a cut-off of 60 kHz.

The number of HF cut-offs within the feedback loop has not been minimized, and the dominant HF cut-off (the output transformer) is rather close to the pair which are next most dominant. Thus the only remaining way to achieve stability at HF was to adjust the phase response independently of amplitude response by means of a step network.

At low frequencies it is more useful to consider time constants than –3 dB points. The input stage is direct coupled to the concertina, so we can ignore this. The concertina feeds the driver stage with a *CR* of ≈22 ms, as does the driver to output stage, and the output transformer is set to 48 ms. In view of this, it is not surprising that LF stability is questionable, as was conceded in the original *Wireless World* article. A little of this may also be due to the fact that the input stage is not operated from the same HT as the concertina. In the author's experience, this fact alone can sometimes induce motorboating.

It should be remembered that in 1947 circuits were designed using long multiplication or tables of logarithms, and if speed was needed – slide rules. Computer aided AC analysis was not an option! Most amplifiers were designed as carefully as possible, and then adjusted on test for best response.

The Mullard 5–20

This is a 20 W 'ultra-linear' design[3] introduced by Mullard to sell the EL34 pentode. There is a great deal of similarity between this design, the Mullard 5–10 (10 W using EL84), and some Leak amplifiers. See Fig. 5.24.

The input stage is an EF86 pentode, which is responsible for the high sensitivity, but poor noise performance, of these amplifiers. Most of the cathode bias resistor is bypassed, since it would otherwise reduce the gain from around 120 to 33, which would be a waste of open loop gain that could be used to correct distortion produced by the output stage.

Unadorned, the pentode has an output resistance of 100 kΩ, and drives ≈50 pF of input capacitance from the phase splitter, which would give a cut-off of 32 kHz, but this is modified by the usual compensation components across its anode load.

A slightly unusual feature is that the g_2 decoupling capacitor is connected between g_2 and cathode, rather than g_2 and ground. In most circuits, the cathode is at (AC) ground, and so there is no reason why the g_2 decoupling

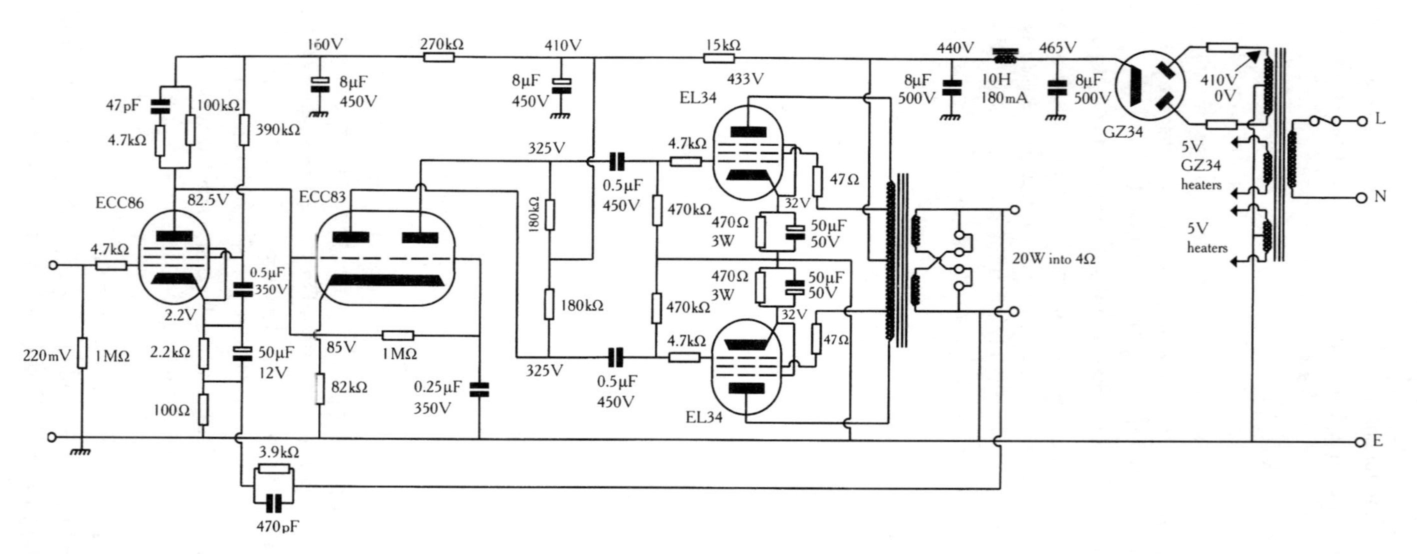

Fig. 5.24 *Mullard 5–20 (Mullard Ltd, originator of this design is now included in Philips Components Ltd)*

capacitor should not go to ground. In this circuit, there is appreciable negative feedback to the cathode, and so the g_2 capacitor must be connected to the cathode in order to hold g_2–k (AC) volts at zero, otherwise there would be *positive* feedback to g_2.

The cathode coupled phase splitter is combined with the driver circuit using an ECC83; when loaded by the output stage, for V_2, A_v = 54, but gain to one output is half this at 27.

The anode load resistors have not been modified to give perfect balance. With the 470 kΩ grid leak resistors of the output stage in parallel with the 180 kΩ anode loads, the *effective* anode load is 130 kΩ. Using the formula derived earlier, this means that V_{2b} should have an AC anode load 3% higher, and R_L for V_{2b} would then be 187 kΩ. Mullard did actually state this, but probably assumed that most constructors would not have access to sufficiently high precision resistors to use the information.

The output stage has an input capacitance of ≈30 pF, and combined with the 53 kΩ output resistance of the driver stage, this gives a cut-off at ≈100 kHz, which is quite poor.

Looking at the stage as a driver stage, we should investigate whether it is capable of driving the output stage. 85 V will be wasted across the 82 kΩ tail resistor, but with 410 V of HT, this still leaves us with 325 V. With the component values given, this puts the operating point at 240 V on the 180 kΩ DC loadline. Drawing the AC 130 kΩ loadline through this point shows that the stage would generate ≈4% second harmonic distortion at full drive (V_{out} = 18 V_{RMS}), if it were not operated as a differential pair. Mullard claimed 0.4% distortion for the entire driver circuitry.

Although distortion appears satisfactory, the driver stage has only 10 dB of overload capability. When output stage gain begins to fall, for whatever reason – cathode feedback, input capacitance loading the driver, or insufficient primary inductance in the output transformer – the global feedback loop will try to correct this by supplying greater drive to the output stage, and the 10 dB margin will quickly be eroded, raising distortion.

The driver circuitry was designed to produce an amplifier of high sensitivity even after 30 dB of feedback had been applied, and this has forced other factors to be compromised. Whereas the Williamson sacrificed stability for linearity, the Mullard 5–20 achieves stability at the expense of linearity.

The output stage is a pair of EL34 in 'ultra-linear' configuration, with 43% taps for minimum distortion. Unlike the Williamson, there is no provision for adjusting or balancing bias, and this might seem to be a backward step.

Bias adjustment implies connecting the cathodes together and using a proportion of grid bias to provide the balance adjustment. Because the biasing is firmly set by the potentiometers, there is no self-regulation of bias current, and as the valves age, balance will need to be reset. In short, providing this adjustment ensures that it has to be used regularly.

By contrast, the Mullard 5–20 has separate cathode bias resistors and relies on automatic bias to hold the anode currents at their correct, and therefore equal, levels. In practice, this works quite well, although it does not quite achieve the low transformer core distortion of a freshly balanced adjustable system.

This system does have a disadvantage in that the individual cathode bias resistors apply series negative feedback to the output valves, raising their output resistance. The output transformer could be redesigned to maintain the match to the load, but this is undesirable as it would require a higher primary to secondary turns ratio, which makes a high quality design more difficult to achieve. Because of this, the cathode bias resistors must be bypassed by capacitors, and this is where the problems really begin.

The capacitor is a short circuit to AC, and so prevents feedback, but its reactance rises at very low frequencies, so it is no longer a short circuit, and allows feedback. Because the output stage is load matched, this feedback causes an immediate rise in distortion and reduction of output power due to the mismatch. The obvious solution to this is to fit a large enough capacitor to ensure that the LF cut-off for this combination is below all frequencies of interest, perhaps 1 Hz. Remembering that the resistance that the capacitor sees is R_k in parallel with r_k, we can easily calculate the value required.

For a pentode, $r_k = 1/gm$; a typical output pentode has $gm = 10\,\text{mA/V}$ at its working point, so $r_k \approx 100\,\Omega$, which is in parallel with a bias resistor of $\approx 300\,\Omega$, giving a total resistance of $75\,\Omega$. For 1 Hz, we therefore need $2000\,\mu\text{F}$ of capacitance.

$2000\,\mu\text{F}$ 50 V capacitors were simply not available at the time, and they weren't fitted. They are readily available now, but there are two reasons why you might wish to use a smaller value:

- A $2000\,\mu\text{F}$ capacitor has considerable inductance compared to its capacitance, and therefore allows feedback at HF. However, we can avoid this problem by using a low inductance electrolytic designed for use in switch-mode power supplies and bypassing it with a smaller capacitor.
- This reason is rather more subtle. If the output stage is driven into Class B by overload, each cathode then tries to move more positively than negatively, it can't turn off any further, but it can certainly turn *on* harder.

The cathode capacitor smooths these changes into a gently rising DC bias voltage, which biases the valve further into Class B, and the problem continues. The effect of this is that a momentary overload can cause distortion of following signals, even though they would normally have been within the capabilities of the amplifier. As the cathode bias capacitor becomes larger, this recovery time from overload lengthens. Theoretically, we never overload amplifiers, and this is not a problem, but occasional overload is inevitable, and the effects should be considered.

The ideal way to deal with all of these problems is to reduce the cathode bias resistor to $\leq 1\ \Omega$, so that it no longer causes noticeable feedback, and measure the current through it using an operational amplifier. This then feeds an asymmetric clipper so that when the valve strays into Class B and clips one half cycle, the clipper removes an equal amount from the other half cycle before feeding the processed signal to an integrator. The integrator can have an RC time constant of almost any value we choose, and 10 s is not unusual. The output of the integrator is a smoothed DC voltage proportional to anode current, which can be compared to a fixed reference, and the difference between the two levels drives an amplifier whose output sets the negative *grid* bias for the output valve.

If the anode current of one valve is set as a reference, then the other valve, or valves, can share this reference, which then forces anode currents into balance. The increased complexity of this scheme is (partly) offset by its improved performance and reduction in HT voltage required, since the cathode bias scheme wastes HT. See Fig. 5.25.

This circuit was designed to sense a 40 mA anode current by developing 40 mV across the 1 Ω resistor; the rest of the circuit is based on this 40 mV signal, so if a different current is to be sensed, the sense resistor should be changed to suit. The 5534 has a gain of 100, and amplifies the mean DC level to 4 V, with AC peaks rising to 8 V. Any peak above 8 V is clipped by the diode/transistor clamp, since the other half cycle will already have been clipped by the valve. The clipped signal is integrated by the 2.2 MΩ resistor in combination with the 470 nF capacitor, giving $\tau = 6.5$ s. The 071 compares this smoothed DC with a reference derived from the potential divider chain, and uses this to control the bias transistor. The reference and clamp voltage are made adjustable by the 2 kΩ variable resistor in order to allow for fine adjustment of anode current. Although this circuit was designed to provide −11 V bias, this can easily be changed by returning the bias transistor's collector load to a more negative supply as necessary; no other changes are required.

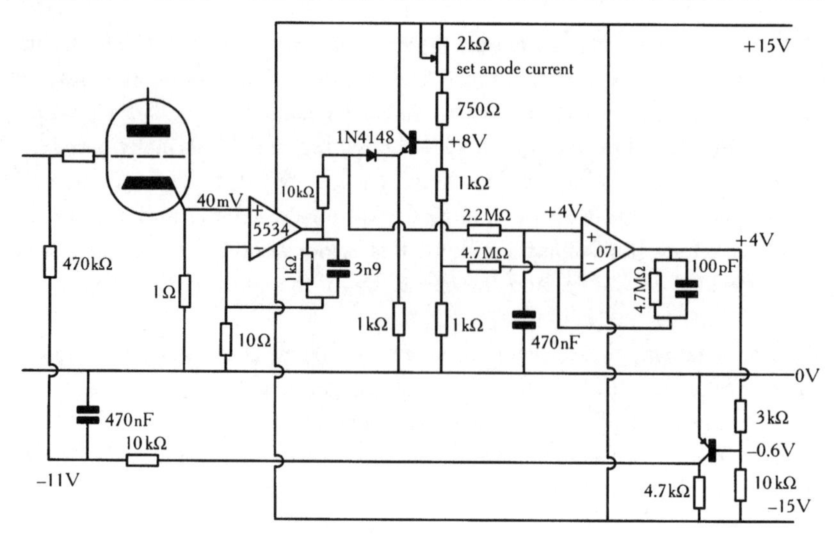

Fig. 5.25 *Principle of output bias servo*

The Quad II

The Quad II is an unusual design, which at first sight does not look too promising, but works because the design is synergetic.

In this design, not only has the phase splitter been combined with the driver stage, but it has also been combined with the input stage. In order to achieve the necessary gain, pentodes have been used. Output resistance is therefore high, as is input noise. To make matters worse, a variant of the see-saw phase splitter has been used. The output stage has local feedback, requiring increased drive voltage. See Fig. 5.26.

The output stage is a pair of KT66 beam tetrodes with anode and cathode loads split in the ratio 9.375:1. The cathode therefore provides little drive to the loudspeaker, and this may be considered to be series feedback from the output transformer. However, the cathode current in the output transformer is the sum of the anode and g_2 currents, and it was found that this summation reduced third harmonic distortion by a further 8 dB over that due to the negative feedback.

The effect of this feedback on output resistance is the opposite to what might be expected.[4] If we simply leave a cathode resistor unbypassed, then this will generate series feedback which will increase r_a, whereas the transformer coupled feedback *reduces* r_a. This can easily be explained if we apply a short circuit as a load. Clearly, the output stage will be unable to drive

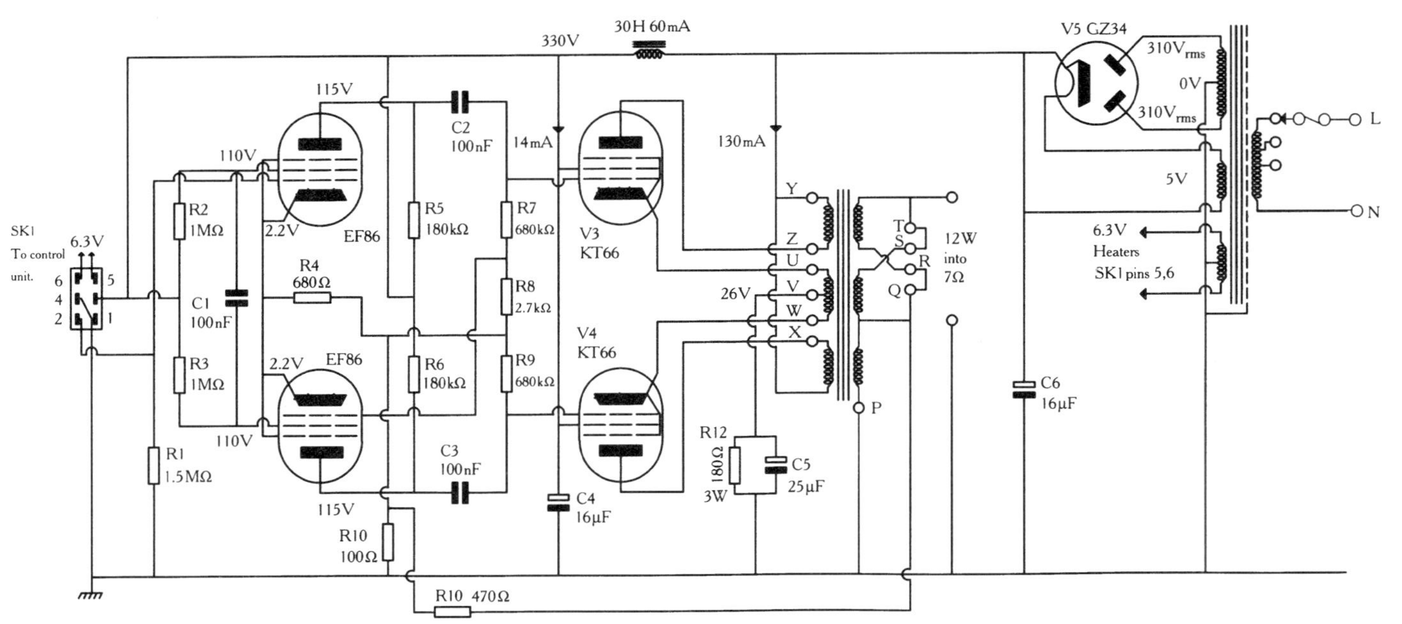

Fig. 5.26 *Quad II (By kind permission of Quad Electroacoustics Ltd)*

any voltage into this load, but conversely, there will be no feedback signal applied to the cathodes. The grids will then be driven by the full input signal, rather than the input signal minus the feedback, and the output stage will be driven harder, as it attempts to maintain its voltage into a short circuit. This action is directly equivalent to reducing output resistance, and the new value of output resistance can be found using the normal feedback equation.

The transformer primaries are equivalent to 3 kΩ anode to anode. With tetrodes, this low value of anode load results in a reduction of third harmonic distortion, and an increase in second harmonic, which is then cancelled by push–pull action in the output transformer.

There is no provision for balancing anode current, and the automatic bias is shared, so we can expect an increase in distortion at low frequencies due to saturation of the transformer core. Curiously, the cathode resistor was only rated at 3 W, yet it dissipates 3.8 W. If your Quad II distorts, a burnt-out cathode bias resistor may well be the cause.

Even with pentodes, there is not a great deal of gain from the driver circuitry, and input sensitivity is low; 1.4 V for full output. This is an excellent choice of input sensitivity for a power amplifier, as not only does it guarantee impeccable noise performance (even from a pentode), but it means that the input is far less susceptible to hum and noise from input cables or heater circuitry. The Quad II was only beaten in signal to noise performance by the Williamson, which was quieter because it had a triode input stage.

Despite being a variant of the see-saw phase splitter, the phase splitter/input stage does not rely on feedback for balance, and its operation is quite elegant. The output valves *must* each have a grid leak resistor, so instead of applying additional loading to the driver valves, a tapping is taken from one of these to provide the input for V_2. In theory, if this tapping has an attenuation equal to the gain of V_2, then the output of the phase splitter is balanced. Because of component variation, this will not always be true, so the cathodes of the two valves are tied together to improve balance.

Pentode stages have output resistance $\approx R_L$. Since R_L for the Quad input/phase splitter/driver is 180 kΩ, this would appear to be very poor at driving the ≈30 pF input capacitance of the output stage, resulting in a cut-off of ≈30 kHz. However, apart from the output transformer, this is the only HF cut-off in the circuit, and it is therefore not a problem. Each output valve requires a swing of $\approx 80\,V_{pk-pk}$, which is easily provided, because pentodes can approach 0 V more closely than triodes, and also because *LC* filtering, rather than *RC* filtering, was used on the HT line, thus increasing the available HT. This *LC* filtered supply is also used to feed g_2 of the output valves, which has the valuable advantage of reducing hum, since the anode current of a tetrode or pentode is more dependent on g_2 voltage than anode voltage.

Pentodes need to have g_2 decoupled to ground. Instead of each EF86 having a capacitor to ground, a single capacitor is connected between g_2 of the two valves. This has three advantages:

- If we had two individual capacitors, they would effectively be in series with a centre tap to ground. Since each valve is connected to an equal but opposite signal, the centre tap would be at ground potential even if it were disconnected from ground. If we disconnect the centre tap from ground, we have two capacitors in series that can be replaced by a single capacitor of half the value.
- Since this one capacitor is connected between two points of equal potential, it need not necessarily have the full voltage rating to ground. However, it is as well to consider the effect of fault conditions when determining the voltage rating, so this is not a great advantage.
- Connecting g_2 of each valve together at AC helps maintain balance in the same way as commoning the cathodes.

Although substituting one stage that combines the functions of input, phase splitter and driver does not achieve the linearity of purpose designed stages, it achieves better linearity than the Mullard circuit because less gain is demanded from it.

With only a simple driver circuit and output stage within the feedback loop, the elegant Quad II has no stability problems.

A practical design

We have investigated individual stages, we have looked at functional blocks, and we have seen how classic designs were configured. Rather than merely observing, it is now time to put that knowledge to use and design an amplifier. We need to be realistic about this and concede that we are not going to design a world beating amplifier overnight, and that by restricting our ambitions we stand a much better chance of making something that actually works.

Accordingly, we will design a 10 W Class AB1 push–pull 'ultra-linear' amplifier using EL84 output valves.

There are a number of reasons for this choice:

- It is cheap. If we need a 340 V HT supply, this can be smoothed by 385 V capacitors intended for switch-mode power supplies, and the HT can be provided by an 240 V isolating transformer with a silicon bridge rectifier. If any mistakes are going to be made, then it is best to make them with reasonably inexpensive components rather than expensive ones.

- Even now, there are plenty of second-hand amplifiers, such as a tatty Leak Stereo 20 or a couple of tatty Leak TL12+, or lesser known amplifiers, that can be cannibalized for transformers and chassis. They are reasonably cheap, or you might manage a swap for them.
- Powerful amplifiers require considerably more skill in layout and construction, and generate bigger bangs – they are not really recommended for the novice constructor.

The Bevois Valley amplifier

This design acquired its name because the prototype was built from a pair of mono amplifiers bought for £15 (including pre-amplifier) in Bevois Valley. Sadly, the junk shop that sold these amplifiers closed (permanently) shortly after their sale in 1982.

As was noted earlier, once the output valves have been chosen, the transformer configuration is limited, and therefore the entire output stage is fixed. We will use a transformer with a primary impedance of ≈8 kΩ anode to anode, and ideally with 43% taps for minimum distortion. This might have been scavenged from a Leak (20% taps), or it might even have been bought new. Either way, we will need an HT of 320 V, and each valve will require 8 V_{RMS} for full output.

Our task is to design superior driving circuitry. To do this, we will need to define what we need, so we will write a specification. This will enable clear thinking, and the electronic design will follow on naturally:

- *Low noise*. With the low noise obtainable from CD or a good LP, we really can't afford to have *any* noticeable noise in the power amplifier. 100 dB S/N, relative to full output power, is not an unreasonable figure to aim for, so this rules out pentodes and high sensitivity.
- *No hum*. This is the new millennium, not the 1920s! Hum is not acceptable under any conditions. This implies superb standards of construction, and/or DC heaters for the input stage.
- *Stability*. This might seem obvious, but some commercial designs that the author has endured were a little relaxed about this. Good stability requires an absolute minimum of stages.

- *Distortion.* This is a tricky topic. If you want distortion measured in ppm (parts per million), then you had better buy a decent transistor amplifier. If you think that hearing is everything and measurement is nothing, then sell the house and buy a single ended triode amplifier. We have to be honest about this. Valve amplifiers do not always measure well, but they *do* sound good. Presumably, we listen to music to enjoy it, so this quality is important.
- *Simplicity.* Valve designs should be simple. Simple systems tend to have simple shortcomings. Additionally, they are repairable. Complex systems are built on silicon, have lots of legs, are repeatable and disposable.

Put together, these criteria demand that we use a concertina phase splitter direct coupled from the input stage without a driver stage, and we can instantly draw a circuit diagram. That this circuit is quite similar to the GEC 912+ demonstrates that there is little new under the sun. See Fig. 5.27.

Since we are driving the output valves directly from the phase splitter, linearity of the phase splitter is paramount. Our chosen phase splitter only has a gain of ≈ 1, so the input stage will also need excellent linearity. Only three valves are really suitable for a concertina stage with limited HT current: 6SN7, ECC82, E88CC. We will use the E88CC.

Optimization of DC conditions

Because the two stages are DC coupled, their design is interactive. As before, the way to deal with an awkward problem like this is to garner as many facts as possible, label the drawing, and see if anything useful appears. Having chosen a concertina stage, we can start by labelling the anode and cathode loads as 22 kΩ. (This traditional value is used because $R_{out} \approx R_L$ if loaded by a stage entering Class B, and whilst we need to minimize output resistance, a significantly lower value would result in excessive power dissipation.) See Fig. 5.28.

Previously, when we used the E88CC, we observed that linearity was best when $V_{gk} = -2.5$ V, this tends to happen at $V_a = 80$ to 90 V. Although the concertina operates under heavy feedback, it would be preferable if it were linear *before* feedback. So we want to juggle conditions such that *both* valves are biased with $V_{gk} = -2.5$ V.

The only way of doing this is by an iterative process. We know that both stages will have an HT of ≤ 300 V since we have to drop volts from the output stage. We also know that each stage will have $V_a = 80$ to 90 V, at $V_{gk} = -2.5$ V. We draw the loadline for the concertina first and find V_a for

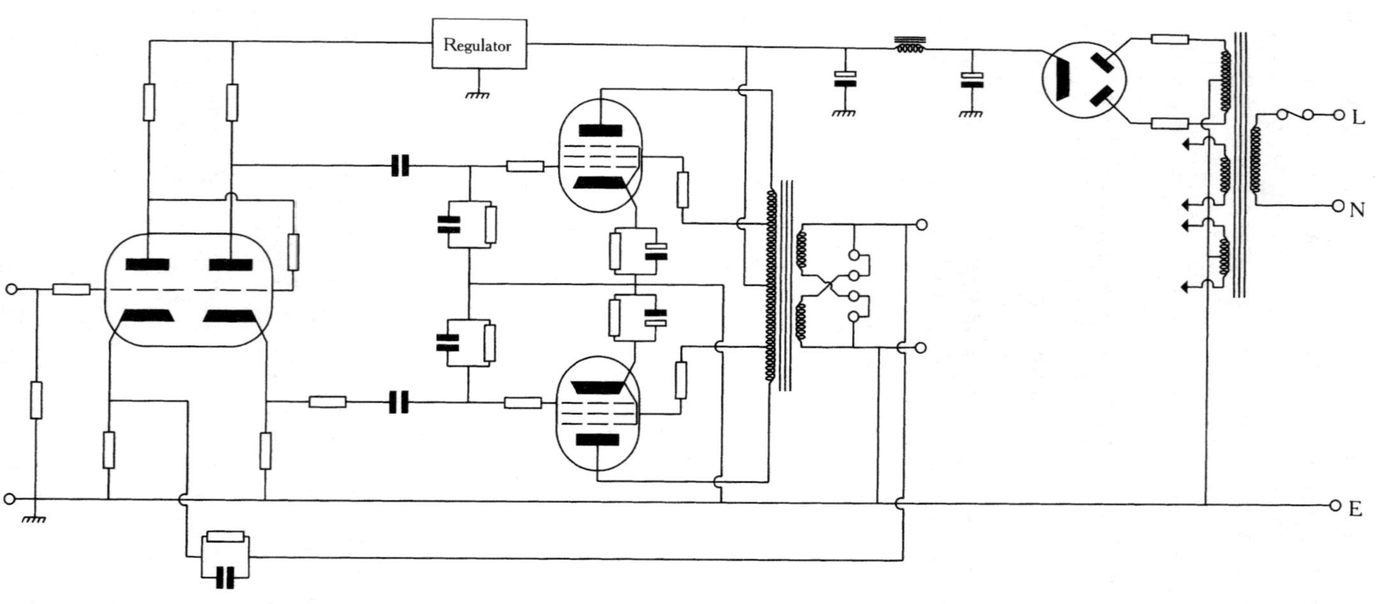

Fig. 5.27 *Power amplifier using concertina phase splitter*

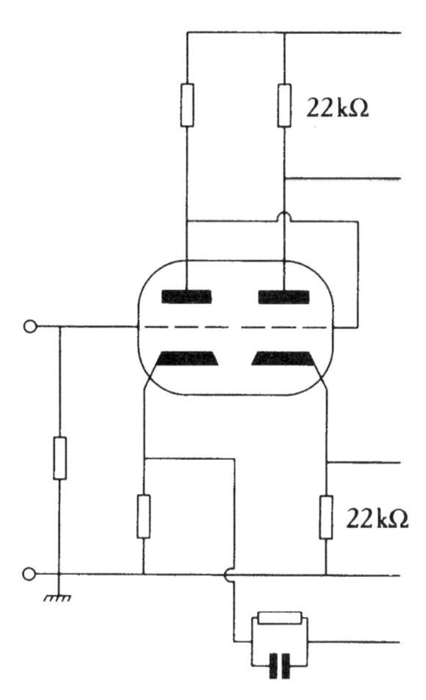

Fig. 5.28 *Determining the operating conditions of the driver/input stage*

$V_{gk} = -2.5$ V. This value is then subtracted from the HT voltage to give the voltage across R_k and R_a, and divided by 2 to give the voltage across R_k. The voltage on the grid will be 2.5 V lower than this, and this will be equal to the anode voltage of V_1. We now draw a loadline for V_1 and see if the optimum anode voltage corresponds with the voltage we have just derived. If it doesn't (and it almost certainly won't) our only variable is HT voltage. Fortunately, a few iterations found that an HT voltage of 285 V met the requirements, and this will be provided by a regulator.

Ideally, we would also juggle the AC current in the input stage to be equal and opposite to that in the concertina, requiring:

$$R_{L(\text{input stage})} = \frac{(\mu + 2) + r_a}{\mu} \cdot R_{L(\text{concertina})}$$

The previous equation is simply $R_{L(\text{concertina})}$ divided by the gain of the concertina to one output, thus ensuring equal currents in the load resistances of each stage. A compromise value of $R_L = 33\,\text{k}\Omega$ would allow this, but results in a high I_a for the input stage (minimizing noise, due to high *gm*) but the uncomfortably high P_a would reduce valve life. In this respect, the amplifier is flawed and should be adjusted, but the author enjoys listening to music through it and doesn't feel inclined to fiddle.

Now that we know the HT voltage for the two stages, we can calculate all the AC parameters.

Calculation of the cathode bias resistor and feedback resistor

This is easily the most complex calculation in the design of a power amplifier with negative feedback applied to the cathode of the input stage. However, if we stay calm and draw lots of neat diagrams, *fully labelled* with every piece of information that we have, then the problem is reduced to manageable proportions.

Attempting to solve this problem on the back of an envelope is doomed to failure.

We have four main factors at work:

- We need to set our cathode bias voltage correctly. This would normally be a trivial application of Ohm's law, but in this case the bias current flows through the cathode resistor *and* the feedback resistor.
- The input valve itself will generate a feedback current through the cathode resistor in addition to any current sourced from the output of the amplifier.
- We need to set the ratio of the two resistors so as to obtain the desired amount of negative feedback.
- As far as AC is concerned, the cathode resistor is shunted by r_k of the valve.

Now that we have stated our restrictions, we should be able to label a diagram and derive some equations.

Since we want 2.5 V bias on the cathode, and anode current is 190 V/47 kΩ, the total resistance to ground from the cathode must be 618.4 Ω.

The anode signal swing for full output is $8.636\,V_{RMS}$. This means that the anode signal *current* must be $8.636\,V/47\,k\Omega = 0.1837\,mA_{RMS}$. This current also flows in the cathode circuit and will develop a feedback voltage across any unbypassed cathode resistor.

We wish to make the input sensitivity of the amplifier $2\,V_{RMS}$, and we know that the unmodified sensitivity is $298\,mV_{RMS}$, so the feedback voltage required at the cathode will be $2 - 0.298 = 1.702\,V_{RMS}$. We know that for the full output of 10 W, the signal at the output of the amplifier will be

$8.944\,V_{RMS}$. Therefore there will be $7.242\,V_{RMS}$ across the feedback resistor.

Since r_k will shunt the cathode resistor at AC, we need to find r_k:

$$r_k = \frac{R_L + r_a}{\mu + 1}$$

Using this equation, we find that $r_k = 1.559\,k\Omega$.

We will assume that the output of the amplifier is a true Thévenin source driving the network through the feedback resistor 'y'. We will represent the valve's own feedback current as a Norton current source, and we will shunt the cathode resistor 'x' with r_k. We will label our currents and include *every* piece of information available. Note that this is an AC diagram. See Fig. 5.29.

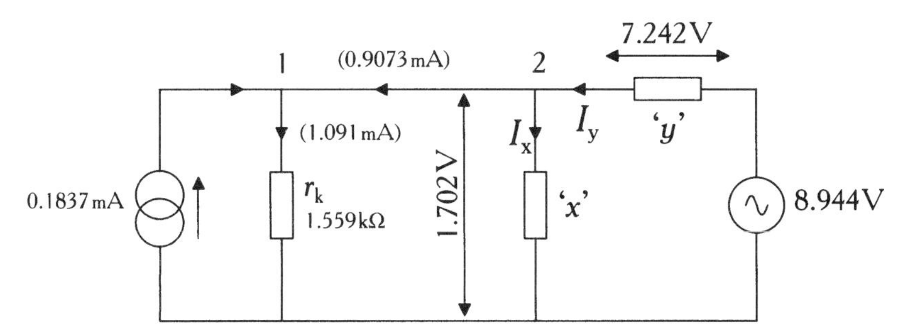

Fig. 5.29 *Equivalent circuit of AC conditions at the input stage*

From now on the solution is easy, if a little tedious. Our first observation is that there is a resistor of known value (r_k) with a known voltage (1.702 V) across it, so we ought to calculate the current through it. This current is 1.091 mA, and we can label it on the diagram.

Having labelled this current, we can now see that node 1 has two known currents flowing through it, so we can find the third, using Kirchhoff. If there is 0.1837 mA flowing into the node, but 1.091 mA leaving it, then 0.9073 mA must be supplied by the other node.

Moving to node 2, we can see that any current coming into the node must be supplied by I_y, and that this splits through the resistor 'x', and to node 1. Formalizing this:

$$I_x + 0.9073 = I_y$$

We can use Ohm's law to make statements about the currents in resistors 'x' and 'y':

$$I_y = \frac{7.242}{y}$$

$$I_x = \frac{1.702}{x}$$

The final restriction is the DC restriction, which says that x and y in parallel = 0.6184 kΩ.

$$0.6184 = \frac{xy}{x + y}$$

Now that we have some equations, we can solve them. This is easy, generating the equations was the hard bit! The way to do this is to substitute the second and third equations into the first:

$$\frac{1.702}{x} + 0.9073 = \frac{7.242}{y}$$

Rearranging and simplifying:

$$7.982x - 1.876y = xy$$

We can now substitute this into the fourth equation and solve it to give the ratio $y = 2.953x$. Substituting this ratio back into the equation yields $x =$ 828 Ω, and using the ratio, we find that $y =$ 2.44 kΩ.

Using standard values, we therefore need 1.2 kΩ in parallel with 2.7 kΩ for the cathode resistor, and 4.7 kΩ in parallel with 5.1 kΩ for the feedback resistor.

Some of V_1's cathode current is now flowing through the output transformer, and it might be thought that this would cause distortion. Assuming that the DC resistance of the transformer secondary winding is negligible, the current flow will be 2.5 V/2.44 kΩ ≈1 mA. Now the current turns ratio of the transformer is 31.6:1 (secondary to primary), so 1 mA of DC flowing in the secondary is equivalent to 31 μA out-of-balance DC flowing in the primary. Compared to 40 mA each side, this is entirely negligible, since output valve balance is highly unlikely to be as good as this.

We now know all the component values for the driving circuitry, so we need to determine values for the output stage.

The EL84 is allowed a maximum grid leak resistor of 300 kΩ with grid bias, but as we are using cathode bias, we can cheerfully increase this to 470 kΩ. We then need a 0.1 μF coupling capacitor, which should be polycarbonate, or preferably polypropylene, with a rating of $\geq 400\,V_{DC}$. 4.7 kΩ is the traditional value for grid stopper resistors on the EL84; it is possible that we may not actually need them, but it seems sensible to fit them just in case.

47 Ω in series with g_2 is alleged to reduce distortion whilst reducing peak power; the author has not tested this, so fitting them is a matter of personal choice. The Mullard circuits did fit them, whilst the Leaks did not.

The cathode bias resistor should be 270 Ω (from the data sheet), and this will dissipate 0.45 W. 2 W resistors are commonly used here, but a 15 W chassis mounting aluminium-clad type with tabs is a much better choice. This is because we will place an electrolytic capacitor very close to this resistor, and we want to keep it cool. Additionally, the resistor provides convenient tags for anchoring the capacitor.

The cathode bypass capacitor should be 2200 μF for a 1 Hz cut-off, but as discussed earlier, this value would cause additional problems, so we will compromise on 470 μF 63 V. 63 V may seem excessive, since it will only see ≈11 V, but the higher voltage component has a lower ESR, which becomes significant when we are trying to bypass the 67.5 Ω seen at the valve (r_k = 90 Ω, R_k = 270 Ω).

Because we have only one RC network plus the output transformer in the entire amplifier, LF stability will not be a problem. HF stability is not assured, and so this should be investigated.

The input stage has its basic sensitivity reduced from 298 mV to 2 V, which corresponds to a gain reduction of 6.71. From this, we can calculate the new r_a for the stage:

$$\frac{\mu R_L}{R_L + r_a} = 6.71\,\frac{\mu R_L}{R_L + r'_a}$$

Solving this, and using r_a = 5 kΩ, gives r'_a = 302 kΩ, in parallel with R_L = 47 kΩ; this gives Z_{out} = 41 kΩ. It will be found that applying global negative feedback invariably causes $Z_{out} \approx R_L$ for the input stage.

The concertina has 3.2 pF of Miller capacitance, and if we allow for strays, 5 pF is a reasonable total value. In combination with 41 kΩ, this gives a cut-off of about 780 kHz.

The output stage has input capacitance that loads the output resistance of the concertina. Although the EL84 is a pentode, it will still have Miller capacitance, albeit greatly reduced, so this should be included in the calculation.

We can find the anode gain of the output stage by calculating the voltage across the 8 kΩ transformer primary for 11 W, and knowing that we need 16 V_{RMS} from grid to grid to drive the stage. This gives a gain to the anode of 18.54. Since $C_{ag} = 0.5$ pF, this would result in a Miller capacitance of 9.8 pF. Unfortunately, this value of C_{ag} is for the pure pentode connection, whereas we will be using the 'ultra-linear' connection, where g_2 does not stay at a constant potential. We should therefore make an allowance for Miller effect from C_{g2}; unfortunately, the Mullard data sheet does not give a value for this, but it is probably wise to allow another 10 pF. If we add these to $C_{in} = 10.8$ pF, we have a total input capacitance of ≈35 pF including strays. Provided that the output stage never strays out of Class A1, this combines with concertina output resistance (≈190 Ω) to give a cut-off frequency of 24 MHz.

We can easily afford to slug this slightly since any additional capacitance will swamp the variations in capacitance between valves, and will therefore improve HF balance. Shunt capacitors of 68 pF across the EL84 grid leak resistors slug this pole to 8 MHz, and tend to preserve equal loading on the concertina if the output stage should stray into Class B.

We can now draw a full circuit diagram of the amplifier with component values. See Fig. 5.30.

The author's prototypes

The author has found this design of EL84 amplifier to be very satisfactory, and has built four variations on this circuit. A pair of scrapped Leak TL12+ contributed the output transformers, whilst a pair of surplus mains transformers and chokes are used for the HT, hence the GZ34 rectifiers. See Fig. 5.31.

The photograph shows the project in the middle stage of construction, before the driver PCB had been fitted. For the last few years, the front cover amplifier has powered a pair of Rogers LS3/5a and is driven directly by a Marantz CD63 mkII KI. A later version drives tweeters in the author's main system.

Over the past few years, the author has built three versions of the Bevois Valley amplifier, each with output transformers from Leak Stereo 20 or TL12+, but experimented with component types:

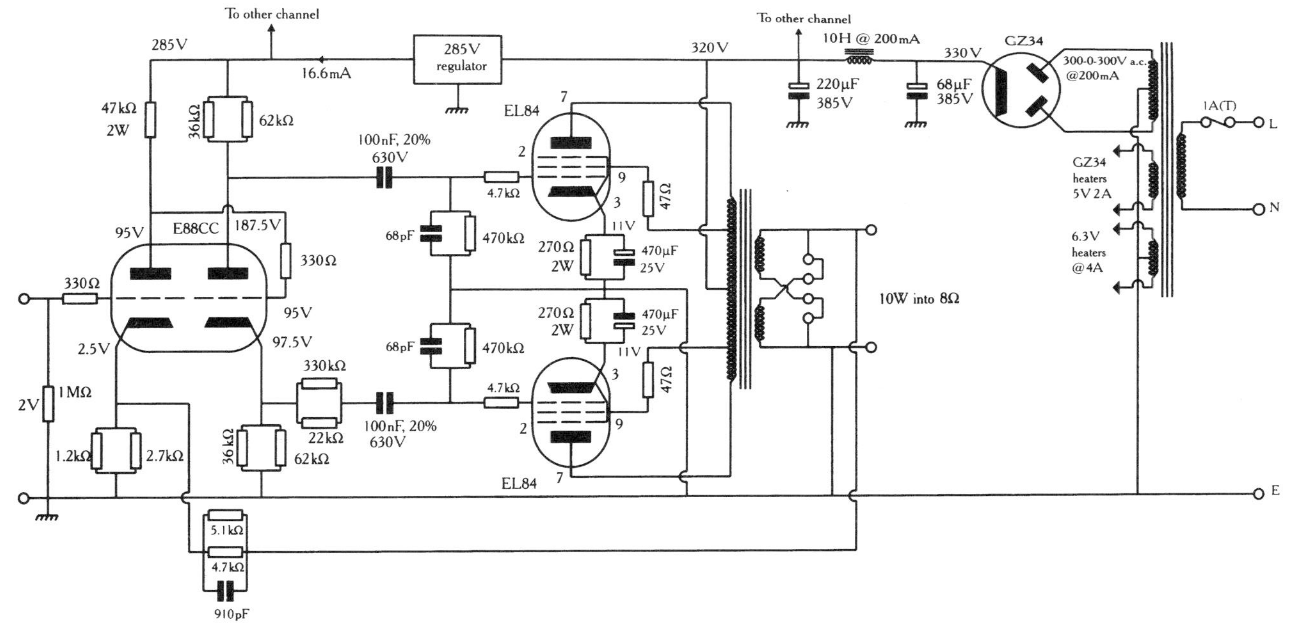

Fig. 5.30 *Practical 10W EL84 power amplifier*

Fig. 5.31 *The author's prototype*

- Ideal for LS3/5a, and produces a traditional rounded valve sound: film load resistors, polycarbonate coupling capacitors and polypropylene HT capacitors.
- A more modern, analytical sound: aluminium cased wirewound load resistors, polypropylene coupling capacitors and polypropylene HT capacitors.
- A beautifully sweet 5 W: film load resistors, polycarbonate coupling capacitors, electrolytic HT capacitors. Output valves strapped as triodes (R_k = 560 Ω).

Higher powered amplifiers

Whilst the previous amplifier was a good design, it has to be admitted that 10 W is not a great deal of power, and we may require rather more. The traditional way of doing this was to use a more powerful valve such as the EL34, or even the KT88. Another method is to use Class AB; using these techniques, we can obtain 50 W from a pair of EL34 or KT66, and 100 W from a pair of KT88. After this, we resort to transmitter valves at enormous cost.

This traditional method has a number of disadvantages:

- Transmitter powered valves are invariably disproportionately more expensive.

- They use high HT voltages, which makes the smoothing capacitors expensive, and the HT supply a major safety hazard.
- They tend to need higher impedance anode loads, which make the design of a good transformer difficult.
- They have savage drive requirements, and often need a power valve as a driver.

However, there are ways of avoiding these problems.

Sex, lies and output power

In the late 1960s and early 1970s, some quite unpleasant audio amplifiers were made using *transistors*. Compared to the valve behemoths, these transistor amplifiers were very small and light, but they didn't actually sound any better (in fact, most sounded a lot worse), so something was needed to make them sell. The one thing that early transistor amplifiers could do was to provide plenty of power, and thus the power rating war started . . .

To make a truly powerful amplifier, a large power supply is needed, but this is expensive. Now (classical) music generally has peaks of only a short duration, and nobody listened to anything else (or at least, nobody whose opinions were taken seriously), so amplifiers were designed that could manage higher output powers, but only for a very short time. This allowed power ratings to be increased further, and the 'music power' rating was born. We measure the maximum output power @ 10% distortion, or the onset of *clipping* (the point at which a sine wave begins to have its peaks clipped off), with bursts of 1 kHz driving one channel *only* into a resistive load. By this means, it is perfectly possible to convert a 20 W amplifier with a poor power supply into a 50 W model, and if we now double the output to account for two channels, we have a 100 W amplifier.

At least four fallacies were used in the previous argument!

Loudspeaker efficiency and power compression

We *can* build more efficient loudspeakers. This is an excellent solution, since inefficient loudspeakers frequently suffer from power compression; an effect whereby the resistance of the voice coil rises due to temperature, and reduces sensitivity until the coil has cooled down.

Active crossovers and Zobel networks

We can drive the loudspeakers more effectively. If the drive units are driven by dedicated amplifiers preceded by an active crossover, many benefits result.[5] For the purposes of this discussion, it is sufficient to say that a two-way loudspeaker system, driven actively by 10 W amplifiers, will go surprisingly loud, very cleanly. (Details of active crossover design are given in the Appendix.)

However, there can occasionally be a problem. Modern moving-coil tweeters tend to be Ferrofluid® damped, resulting in a very nearly resistive electrical impedance, but bass drivers cannot often use this technique. Their voice coils have more turns, and self-inductance is significant, so bass amplifiers see a rising impedance due to voice coil inductance which can compromise HF stability. Fortunately, this is easily corrected by adding a Zobel network directly across the loudspeaker terminals. See Fig. 5.32.

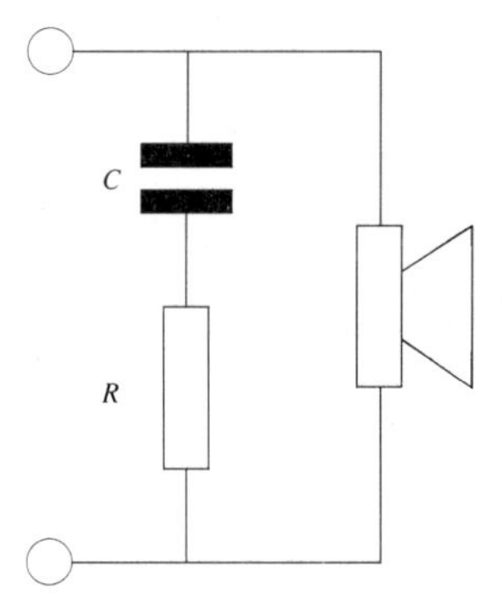

Fig. 5.32 *Zobel network for cancelling voice coil inductance*

The required additional resistor is equal to the DC resistance of the loudspeaker, and the capacitor value is found using:

$$C_{\text{Zobel}} = \frac{L_{\text{voice coil}}}{R_{\text{DC}}^2}$$

Because the loudspeaker can be considered to be a transformer with its voice coil loosely coupled to the shorted turn of the pole pieces which have hysteresis losses, the simple model of pure inductance in series with resistance is somewhat inaccurate, but sufficiently good for our present purpose.

Parallel output valves and transformer design

This is a cracking solution, and gives many advantages. If we use multiple pairs of paralleled output valves, we can keep the HT voltage within safe bounds, perhaps even at 320 V, if we are prepared to use many pairs of EL84. With each additional pair of valves, the transformer primary impedance falls, as does the turns ratio, making it easier to design a good quality component. Statistically, total anode current per side will be better balanced as we increase the number of valves, and deliberate selection will improve this still further.

Driving higher power output stages

Whether they are composed of paralleled devices or not, higher powered output stages always require more of the driver circuitry. When we investigated the Williamson, we found that it had a dedicated driver stage, but that the large number of stages made stability a problem. Clearly, a better approach is needed.

As before, listing the requirements will help solve the problem:

- We need low output resistance to drive the increased input capacitance of the output valves, and a cathode follower may be needed.
- We need to provide a large output voltage with low distortion, this invariably demands some form of a differential pair.
- Wide bandwidth and high gain are also desirable, because we would like to have only one set of coupling capacitors to ensure LF stability, and the cascode would be ideal, although a carefully designed cascade of DC coupled differential pairs can be even better.

We will first investigate a cascode differential pair with direct coupled cathode followers, sometimes known as the Hedge[6] circuit after its designer (although the original Hedge circuit did not include cathode followers). See Fig. 5.33.

Design of the individual parts of this circuit was covered in Chapter 2, so we need not go into great detail on this circuit other than to make a few observations.

A single differential pair is not the ideal phase splitter, so we must take extra care over this to obtain a good result. The anode load resistors must be matched, aged and generously rated to avoid drift. The constant current sink

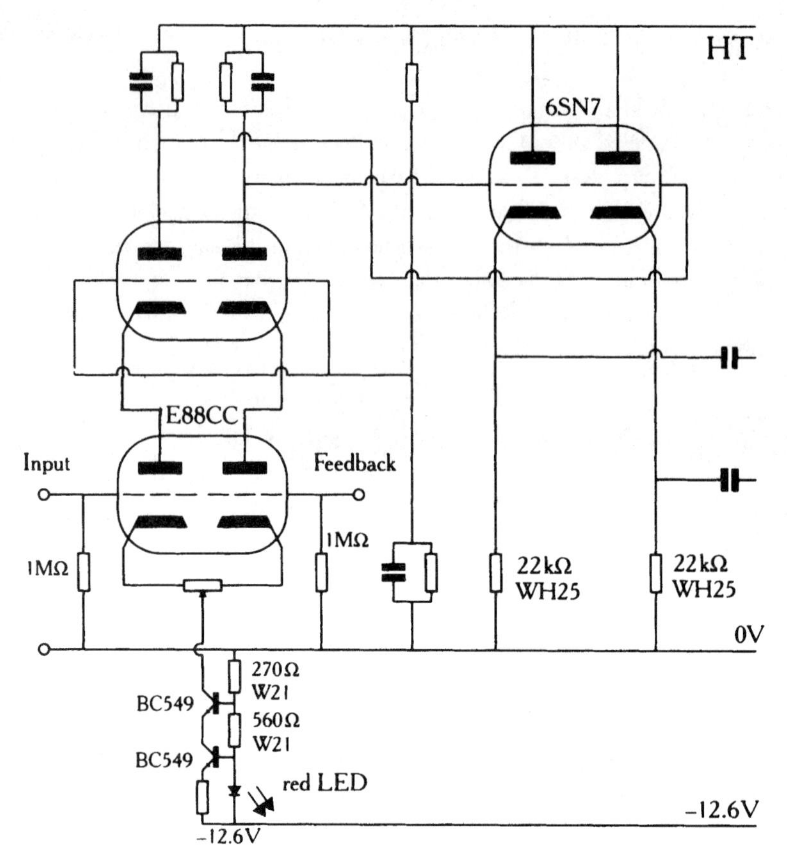

Fig. 5.33 *Cascode differential pair with direct coupled cathode followers*

should be made to have as high an output resistance as possible, and stray capacitance to ground from the cathode should be minimized to maintain a high impedance at HF. Matching the valves would be useful if possible.

Each pair of valves requires a separate heater supply. Sad, but true. The cathode followers need ≈200 V superimposed on their heaters, the upper pair of the cascode need ≈100 V, and the lower pair 0 V. Flirting with this rule will generate problems related to heater cathode insulation breakdown/ leakage, and emission from the heater to the cathode will be summed with the intentional cathode current. You have been warned!

As was mentioned before, the only really satisfactory valve for use as the lower valve in a cascode is the E88CC, any other type will waste HT. The cathode voltage on the lower valves is usually quite low, ≈2.5 V, and this is a little too low to allow a constant current sink to operate reliably, so the tail of the sink is usually connected to a subsidiary negative supply.

Feedback from the output can be applied to a grid, which makes the calculations of feedback network much easier, or the stage can accept a balanced input.

Power amplifiers, phase splitters and balanced alternatives

Phase splitters based on a single differential pair have previously been criticized, but the performance of a cascade of DC coupled differential pairs is almost beyond reproach, and has the advantage of allowing a balanced input if required . . .

Most of the amplifier designs discussed so far accepted an unbalanced input signal, which was then presented to the phase splitter, and generated a balanced signal to drive the push–pull output stage. If we had a pre-amplifier whose output was already balanced, we would not need a phase splitter, and the transmission of the signal from the pre-amplifier to the power amplifier in balanced form would give a great advantage in rejection of induced noise.

The only possible contender for a balanced input stage is the differential pair, but now that the input signal is applied to both grids, we need a means of implementing global negative feedback. The solution is to add a small resistor in series with each cathode and the constant current source, and inject a balanced feedback signal to each cathode. See Fig. 5.34.

The cathode resistors reduce the common-mode performance of the differential pair because the cathodes are no longer tied tightly together, and so their value should be minimized; 47 Ω is probably the lowest practical value.

Having set the value of R_k, R_F can be calculated. Fortunately, this calculation is not nearly as traumatic as that for the 10 W amplifier because the constant current source takes care of the DC conditions, and each R_k causes such a small amount of local feedback that it can be neglected.

If valve balance were perfect, there would be no signal voltage at the top of the constant current source, and we could therefore treat this as being at ground for the purposes of the following calculation. The output of the amplifier is no longer firmly ground referenced, and balances itself about the notional ground of the constant current source. Each leg of the output can

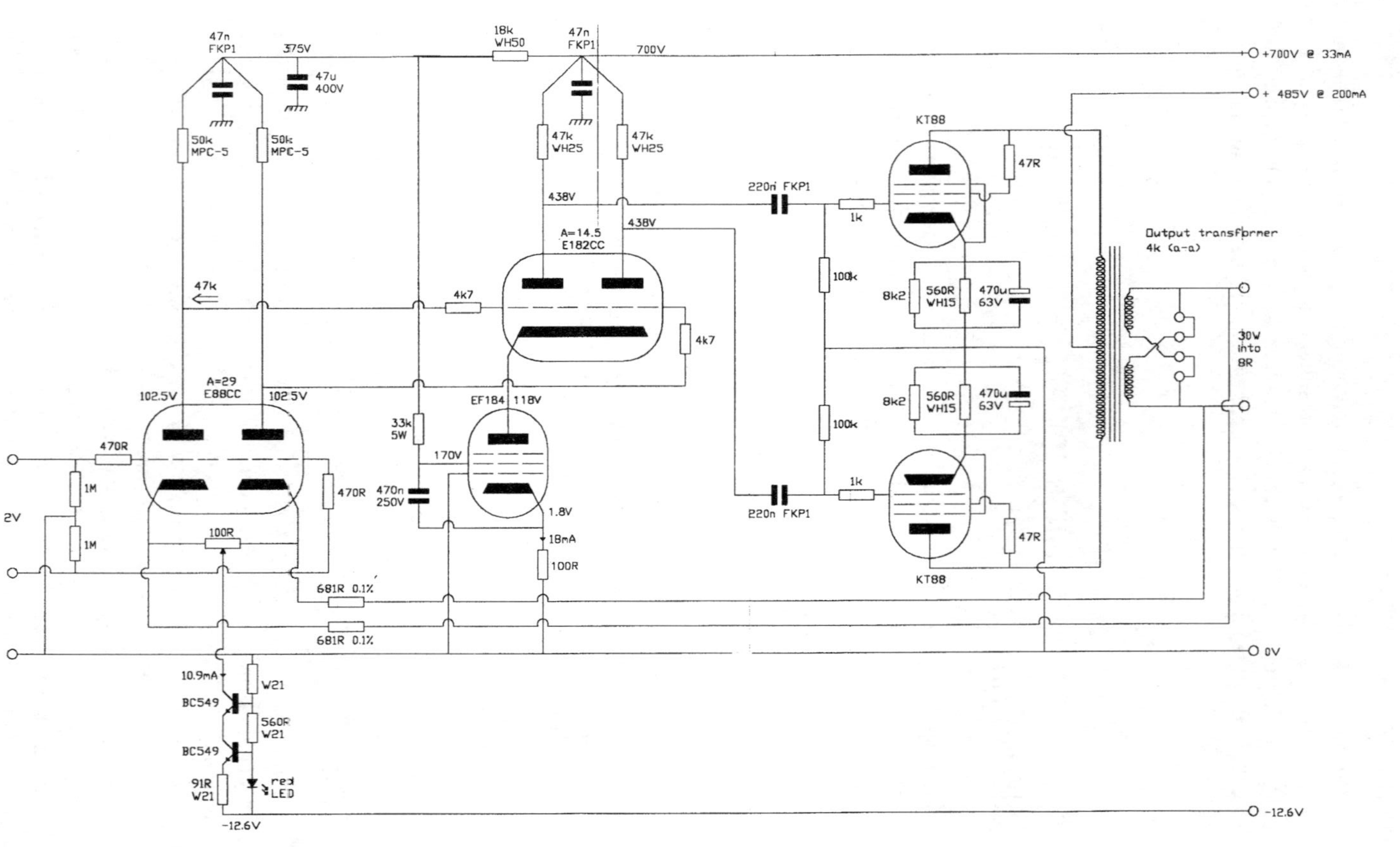

Fig. 5.34 *Balanced input amplifier using differential pair and balanced feedback*

therefore be treated as a signal, referred to this notional ground, of half the full output voltage.

The feedback R_k, R_F combination is a simple potential divider where the lower arm is loaded by the valve's r_k, and the input signal to the divider is half the full output voltage of the amplifier. The required value of R_F is now easily found using the normal feedback equation and the potential divider equation, without having to invoke Kirchhoff. However, it is essential that these resistors are accurately matched to avoid unbalancing the circuit, and 0.1% tolerance resistors are recommended.

The disadvantage of this arrangement is that the output of the amplifier now has the cathode voltage of the input stage superimposed on both its output terminals. This will not matter to a loudspeaker, even if the cathodes are at slightly different potentials, because this small voltage will be heavily attenuated by the potential divider formed by R_F and the loudspeaker voice coil, and practical values of R_F would reduce the likely offset to 10 mV or less. However, it does mean that neither output of the amplifier may be connected to ground, because this would upset the bias of the input stage.

The electrostatic Beast

This design was provoked by a request for a pure Class A, triode, amplifier that could drive home-made electrostatic loudspeaker panels directly. Electrostatic loudspeakers need very high signal voltages, so transmitter valves operating from 1 kV HT supplies are an advantage rather than a disadvantage. Electrostatic loudspeakers are inherently push–pull, so the search for a push–pull pair of suitable output valves began, culminating in the choice of a pair of 845 triodes.

The 845 is a directly heated triode designed in the early 1930s (the RCA anode characteristics are dated November 1933), yet the valve displays outstanding linearity. The downside is that the 845 is extraordinarily difficult to drive.

Driving circuitry for the Beast

The 845 is a low μ triode ($\mu = 3$), which means that although it is reasonably low input capacitance, each valve requires $\approx 300\,V_{pk-pk}$ to drive it to full output voltage. Developing $300\,V_{pk-pk}$ is not trivial, and it is no wonder that most designers resort to using power valves as drivers – with the 300 B being a popular valve. The author took one look at the price of a

300 B and decided that an alternative had to be found – using a driver valve that cost as much as the *pair* of output valves seemed bizarre.

A cascade of DC coupled differential pairs with a pentode constant current sink for the second pair seemed ideal. Because the 6BX7 has a low μ, it is very linear and capable of swinging many volts on its anode, thus making it the ideal choice for the second pair.

The input stage is also the phase splitter, and to ensure best balance should use high μ valves. The traditional choice would be the dual triode ECC83 ($\mu = 100$), but the little known EC91 is a high μ ($\mu = 100$) single triode with a much lower r_a than the ECC83. (The electromagnetic Beast uses a pair of EC91.)

Despite this, the electrostatic Beast uses the ECC91 ($\mu = 38$) as its input stage, for the following reasons:

- The slightly lower μ means that $V_{gk} = 4\,V$ at the operating point, and this is high enough to allow the use of a 334Z constant current source without a subsidiary negative supply – simplifying the design.
- The ECC91 has a true common cathode, with planar grids either side. Initial balance should therefore be maintained as the valve ages.
- A high μ input valve is not quite so important in a cascade of differential pairs since the second pair will correct any slight imbalance created by the first pair.

All of the design techniques needed for the design of the driver circuitry have been covered previously and need not be repeated. However, some resistors have high power ratings, not necessarily because of the power dissipated within them, or the voltage across them, but because of the voltage rating across their insulation resistance to chassis. See Fig. 5.35.

The Beast's HT power supply

Providing an HT of ≈1 kV with a conventional capacitor input supply implies pulse switching of the diodes between the mains transformer output and the smoothing capacitor when the transformer reaches its peak voltage. It is hard to imagine any better way of causing noise than deliberately producing large pulses of current in the HT supply.

The HT supply uses a choke input supply, to ensure that a constant (non-switching) current is drawn from the mains transformer rather than pulses. In practice, the rectifier diodes must switch off when the voltage across each of them falls to 0.7 V, meaning that the transformer secondary current is briefly

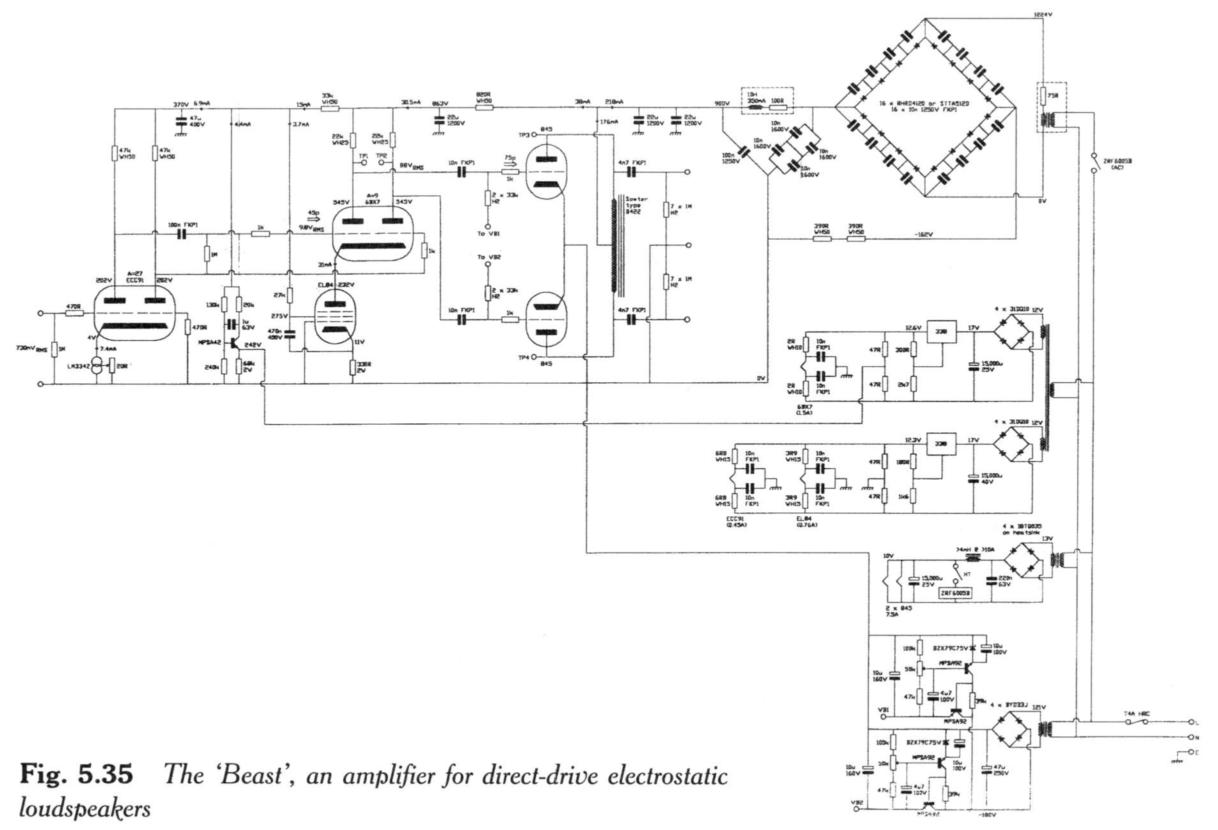

Fig. 5.35 *The 'Beast', an amplifier for direct-drive electrostatic loudspeakers*

interrupted at a 100 Hz rate. In order to reduce the disturbance caused by this interruption, the rectifier diodes are fast switching, soft recovery types, and are bypassed by capacitors.

As a further measure, the choke (and its stray capacitances) have been tuned to match the leakage inductance of the chosen mains transformer to ensure optimum filtering (see Chapter 4).

It should be noted that the tuning capacitor has to withstand a maximum voltage of:

$$V_{max.} = \sqrt{2} \, . \, V_{RMS(transformer)}$$

Thus the tuning capacitor shown in the diagram is made up of 1600 V capacitors in series to give a total voltage rating of 3200 V.

The low frequency resonance (in this case, a potential 9 dB peak at 8 Hz) has been critically damped by series resistance; fortuitously, part of the necessary resistance can be made up of the unavoidable transformer and choke series resistance. Because of the high voltages involved, the additional resistance needed to achieve $Q = 0.5$ has been added in the 0 V side of the supply, easing insulation problems.

The Beast's 845 LT supplies

Although choke input supplies are normally associated with HT supplies in valve equipment, they can be even better in low voltage supplies. This is because of the relationship between minimum current that must be drawn and input supply voltage that is associated with choke input power supplies:

$$I_{min(mA)} = \frac{V_{RMS}}{L(H)}$$

This relationship means that power supplies operating with a low input voltage only need a low value choke. Low value chokes have very much smaller stray capacitances, so their self-resonant frequency is higher, making them a more nearly perfect component.

It makes a great deal of sense to use a choke input filter for a low voltage supply, not just because the resonant frequency of a practical choke will be higher, but also because it reduces the peak current demanded from the mains transformer.

If we demand a large peak current from a mains transformer, then we stand a very good chance of saturating that transformer, and generating noise due to leakage flux. A choke input supply draws a constant current from the transformer equal to the DC current, whereas a capacitor input supply typically demands pulse currents of $> 10I_{DC}$ – *ten* times the DC current!

If large pulses are demanded, they must flow somewhere. In a capacitor input supply, they flow through the reservoir capacitor, so this capacitor has to be able to withstand large pulses of current, known as *ripple* current. Ripple current causes heating within the capacitor because the capacitor has internal resistance, and current through resistance dissipates heat ($P = I^2R$). Internal heat within the capacitor helps the (liquid) electrolyte to evaporate, and so pressure builds up within the capacitor, which is either released by a safety vent (thus allowing electrolyte to escape and reducing the remaining electrolyte), or by the capacitor exploding destructively. Either way, electrolytic capacitors should be kept as cool as possible to ensure longevity.

Summarizing, it is expensive to construct an electrolytic capacitor that can cope with high ripple currents, and it would be far better to reduce the current, particularly if $4\,A_{DC}$ (for an 845 heater) is required, which might conceivably cause a ripple current of 40 A.

The Beast's driver LT supplies

Because each of the stages of the Beast is a differential pair and does not have the cathode bypassed down to ground, r_k becomes important. It is an unfortunate, but inevitable, consequence of increased immunity to HT noise that r_k rises, and the stage becomes more susceptible to noise capacitively induced from the heater. Admittedly, heater to cathode capacitance is quite small, but once r_k begins to rise, C_{hk} becomes significant, especially if we consider that individual mains transformer windings are *not* screened from one another. Typical transformer adjacent winding capacitance is 1 nF, so if we were especially unfortunate, we might have 1 nF between the 1200 V HT winding and the 6.3 V winding destined for the input stage. There are three ways of avoiding the problem:

- Wind a custom transformer with electrostatic screens between each winding (very expensive).
- Use a separate heater transformer for each valve (possibly expensive and definitely bulky).
- The noise signal is common mode (on both wires), so simply regulating heater supplies does not help. We need common-mode RF filtering on the supplies, and this can be achieved by matched series resistance or inductance in each wire combined with capacitance down to the chassis as

close as possible to the valve base. The series resistance also has the minor advantage of reducing heater current surge at switch-on.

The electrostatic Beast thus has a separate heater transformer and common-mode filtering. However, the combination of series resistance and voltage regulation is not ideal, so see Chapter 4 for even better LT filtering strategies.

Testing

The Beast is quite the most frightening amplifier that the author has ever tested. The combination of a choke input 900 V HT *and* high current (≈250 mA) inspired terror and the purchase of a new 1 kV rated DVM (with leads also rated at 1 kV). In retrospect, it is perhaps unsurprising that the teething problems were not in the death volts department, but in the heater supply to the output valves.

Together, the output valves draw 7.5 A, making internal resistance of the LT supply crucial. Schottky diodes selected for low forward drop are essential not only to maintain correct LT voltage, but also to reduce self self-heating – but the diodes still needed to be thermally bonded to the chassis to keep them cool. A 12 V_{AC} transformer should have been ideal, but resulted in a low heater voltage, and 13 V_{AC} was needed. The 4.8 mH 10 A LT choke turned out to be of quite poor quality, and buzzed when operated so close to saturation, so it will be replaced.

A completely unforeseen problem involved oscilloscope probes. The Beast's predicted high output voltage (≈1.5 kV_{pk-pk} at each output) meant that ×10 probes would be inadequate for testing, so a ×100 probe was bought in the logical expectation that it would have an increased voltage rating. It didn't. The Beast's full output voltage could not be measured with this probe because its insulation broke down, causing spurious spikes on the oscilloscope.

'Daughter of Beast'

Neat, sweet and petite, this direct drive amplifier for electrostatic headphones is essentially the driver circuitry from the electrostatic Beast with a few minor changes, hence its name. The amplifier is designed to produce a maximum undistorted output of 200 V_{RMS} at each terminal, plus 550 V bias from 10 MΩ, in order to match the principal electrical characteristics of the Sennheiser HEV70 it is designed to replace. See Fig. 5.36.

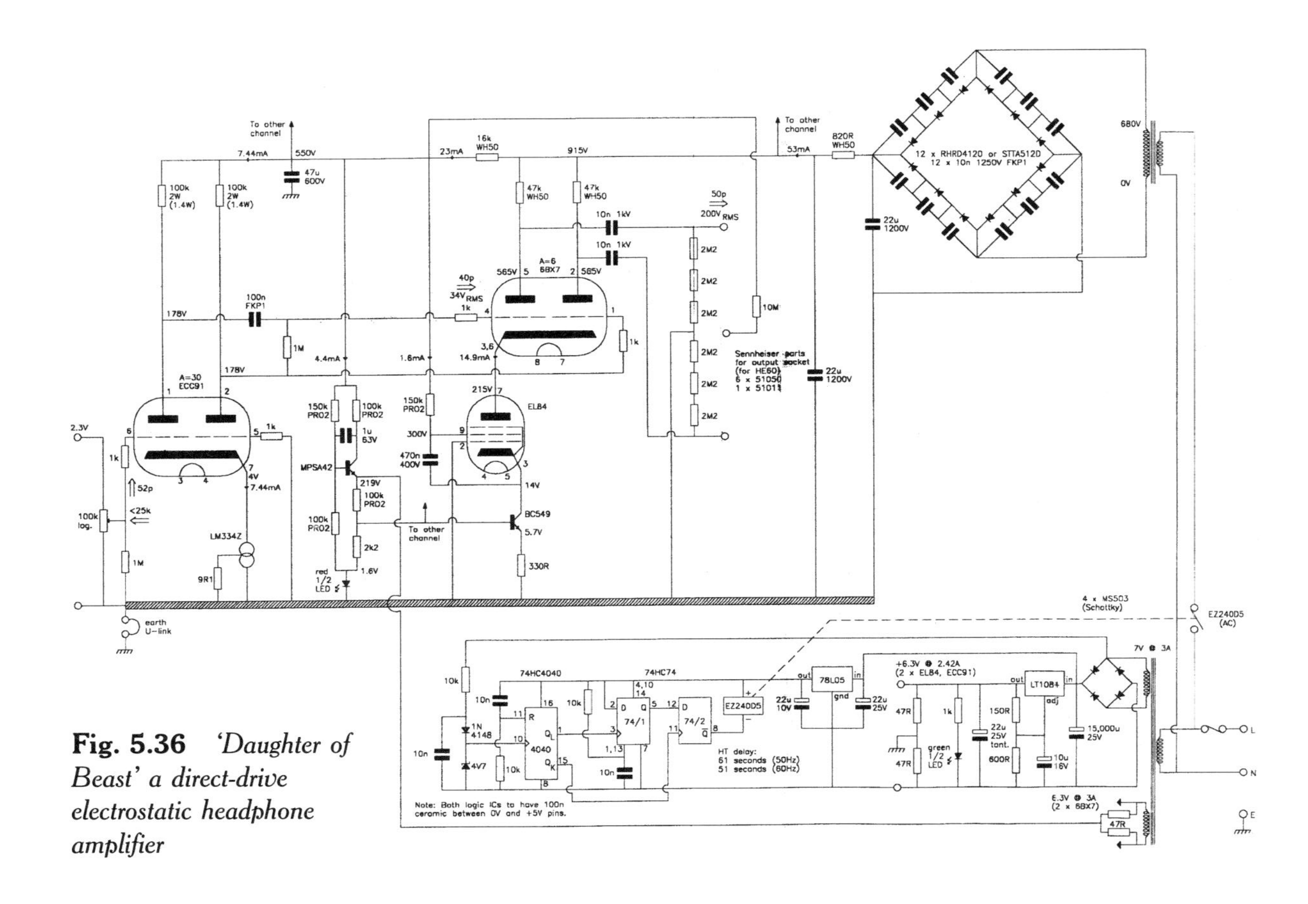

Fig. 5.36 *'Daughter of Beast' a direct-drive electrostatic headphone amplifier*

The amplifier is ultimately intended to be driven directly from a balanced 2 V_{RMS} DAC (modern 1 bit DACs actually produce a differential output which is normally converted to single ended by external circuitry, which seems a waste).

Ideally, the amplifier would have 3–4 dB gain in hand in order to be fully modulated by low level recordings. However, each time it was tried, substituting higher μ valves for the input differential pair invariably raised V_a for the first pair, raising V_k for the 6BX7, reducing V_{ak} and therefore reducing maximum possible output voltage swing. Additionally, the author dearly wanted to retain the ECC91 as the input valve, because it has a true common cathode, with planar g_1 and g_2 mounted either side. In any valve, the cathode deteriorates throughout its life, but as the same physical cathode is in use for both sections in the ECC91, we can reasonably expect initial balance to be maintained – a worthwhile advantage. Sadly, other electrically 'common cathode' valves, such as the E90CC and E92CC, actually have individual cathodes connected internally. . .

The operating point of the 6BX7 is critical to maximize output voltage, so a transistor was added to the cathode to make a hybrid BJT/pentode cascode, because this allowed I_k to be determined purely by semiconductor parameters, enabling valve replacement without having to reset I_k.

In a desperate effort to reduce the amplifier's size and weight, the power supplies have been greatly simplified compared to the Beast, and a conventional capacitor input HT supply has been used, so we need to check that this simplification is permissible.

Calculating output hum due to the HT supply

The principles of the following argument can be applied to other areas, so the topic is covered in some detail. Additionally, an interesting result emerges, which is well worthy of further thought.

The ripple voltage across the 22 μF reservoir capacitor is:

$$V = \frac{It}{C} = \frac{0.053 \times 0.01}{22 \times 10^{-6}} \approx 2V_{pk-pk}$$

This is then attenuated by the following 820 Ω and 22 μF network. If we consider 100 Hz (double mains frequency) to be the only significant component of the ripple sawtooth waveform, then to a very close approximation, the loss is:

$$\text{loss} \approx \frac{X_{C(100\,\text{Hz})}}{R} \approx \frac{72}{820} \approx 0.088$$

The ripple voltage fed to the output stage is therefore ≈175 mV_{pk-pk}.

As mentioned in Chapter 2, the attenuation of HT noise at the anode is quite low because for a differential pair, r'_a looking into one terminal is quite high:

$$r'_a = R_L + 2r_a$$

Substituting this into the potential divider equation, the loss of the potential divider formed by r'_a and R_L is:

$$\text{loss} = \frac{R_L + 2r_a}{R_L + R_L + 2r_a} = \frac{R_L + 2r_a}{2(R_L + r_a)}$$

Provided that $R_L \gg r_a$, this will tend to ½, or –6 dB. We would therefore expect to see ≈90 mV_{pk-pk} of ripple on each anode of the differential pair.

At this point, we must be very careful to compare like with like, so when we compare the ripple to the signal voltage, the signal voltage must be expressed as V_{pk-pk}. We must also ensure that we are comparing voltages between the same terminals. Because the output is taken from between the two anodes (which are inverted with respect to one another), we see double the voltage seen to ground from one anode. The maximum peak to peak voltage between one stator plate and the other of the headphones is therefore:

$$V_{pk-pk(max.)} = 200 \times \sqrt{2} \times 2 \times 2 \approx 1100\,V_{pk-pk}$$

The power supply ripple voltage may be related to maximum output voltage:

$$\frac{\text{ripple voltage}}{\text{maximum output voltage}} = \frac{0.09}{1100} \approx -82\,\text{dB}$$

Ideally, the headphones only respond to the *difference* in voltage between one anode and the other, and would therefore completely reject any common-mode ripple. The practical headphones' rejection of common-mode signals is determined by the equality of the insulating spacer thickness separating the diaphragm from either stator. Typically, the spacers might be made of polystyrene sheet, which can vary in thickness across the sheet, so two spacers stamped from a sheet could be of differing thicknesses, perhaps as much as 3% in error, expressed in dB:

$$\text{dB} = 20\log(0.03) \approx -30\,\text{dB}$$

It is therefore reasonable to suppose that the headphones are only likely to reject common-mode ripple by 30 dB, although they could very well be much better. We can now see that the audible power supply hum is likely to be ≤–112 dB below maximum output level, which is probably satisfactory.

However, we should not be complacent about low levels of hum, because hum is at a fixed frequency, and the eye, or ear, brain combination is ≈15–20 dB more sensitive to fixed pattern noise compared to random noise. A practical example of this difference in sensitivity is that when analogue music circuits were commonly provided by British Telecom, they would allow the customer to reject a circuit with random noise worse than –43 dBu, but single tone interference could be rejected at –60 dBu (17 dB lower in level).

Using the previous argument, the power supply hum could be considered to be equivalent to random noise ≈17 dB louder, which would be 95 dB below maximum level.

A digital meander

A correctly dithered '*n*' bit digital system (perhaps CD) through A–D and D–A has a theoretical (unweighted) Signal to Noise (S/N) ratio referred to maximum output of:

$$\text{S/N} = 6n\,(\text{dB}) - 3\,\text{dB} = 6 \times 16 - 3 = 93\,\text{dB}$$

The 3 dB factor is the unavoidable dither (noise) that should be added to the A–D's quantizer to linearize quantization and avoid noise modulation of music signals, although carefully chosen digitally synthesized signals may not require full dithering, allowing an improvement in S/N ratio of 4.76 dB to be achieved.

To summarize the previous digital arguments, the maximum unweighted S/N ratio of a digital system is defined by the channel, rather than the converters, so a 24 bit D–A on a 16 bit channel (such as CD) simply allows a closer approximation to the theoretical S/N ratio. However, it is possible to gain a purely subjective improvement by manipulating the frequency distribution of the quantizing noise using a technique known as 'noise weighting'. Currently, this allows a usable improvement of ≈18 dB, measured via an A-weighting filter, which should also result in a subjective improvement of ≈18 dB, so a noise weighted 16 bit channel could have an S/N of 111 dB(A).

Practical significance of the digital meander

The full output of a balanced output CD player should just drive the amplifier to its maximum output level, so the random noise from a standard CD should just mask the power supply hum, but a true 20 bit signal (perhaps DVD), or a noise weighted signal (such as Sony's Super Bit Mapped system) applied through a superior converter would be compromised. The current design would therefore ideally like to see headphones with rather better mechanical construction than postulated in order to ensure acceptable hum levels – unusually, a normally electronically determined characteristic is now critically dependent on the mechanical construction of a transducer.

Construction

As can been seen from the front cover photograph, rules were broken in the mechanical construction of this amplifier:

- The chassis is too small. However, a larger chassis would not have been tolerated – this is a headphone amplifier, after all!
- The output valves are far too close together. Fortunately, they are operated at only 44% of the maximum power rating for the envelope, but cooling is still an issue.

A small chassis is permissible *if* the cooling problems can be solved. The valves are mounted on perforated aluminium sheet, whilst the bottom cover of the amplifier is entirely perforated, and the feet are quite tall, allowing a free flow of cooling air past the valves and the components immediately beneath. The 6BX7 anode load resistors and LT Schottky diodes are thermally bonded to the chassis in order to minimize air temperature within the chassis.

References

1. Futterman, Julius (1956) A Practical Commercial Output Transformer-less Amplifier. *Journal of the Audio Engineering Society*, October.

2. The Williamson Amplifier. A collection of articles reprinted from *Wireless World*, April 1947, May 1947, August 1949, October and November 1949, December 1949, January 1950, May 1952.
3. Mullard (1959) Tube Circuits for Audio Amplifiers. Now reprinted by Audio Amateur Press, Peterborough, New Hampshire, USA (1993).
4. Williamson, D. T. N. and Walker, P. J. (1954) Amplifiers and Superlatives. *Journal of the Audio Engineering Society*, April, Vol. 2, No. 2.
5. Colloms, Martin (1997) *High Performance Loudspeakers*, 5th ed. Wiley.
6. Hedge, L. B. (1956) Cascade AF Amplifier. *Wireless World*, June.
7. Sinclair, Ian R (ed). (1998) *Audio & Hi-Fi Handbook*, 3rd ed, pp. 180–184. Newnes.

Further reading

Borwick, John (ed.) (1994) *Loudspeaker and Headphone Handbook*, 2nd ed. Focal.

Linsley Hood, John (1997) *Valve & Transistor Audio Amplifiers*. Newnes.

Manley, David (1994) *The Vacuum Tube Logic Book*, 2nd ed. Vacuum Tube Logic.

6

The pre-amplifier

The pre-amplifier receives a variety of signals at different levels, processes them, and passes them to the power amplifier. It should do this without adding noticeable noise or distortion and should be operationally convenient to use.

The block diagram of a typical pre-amplifier is shown in Fig. 6.1.

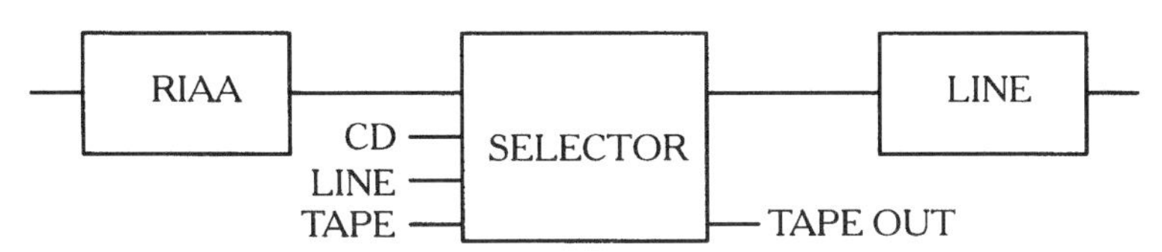

Fig. 6.1 *Block diagram of pre-amplifier*

Working back from the output, we have a line stage which will provide a limited amount of gain, it may be designed to drive long cables, and it may also include tone controls. This is preceded by the volume control and input selector; associated with this will probably be some form of switching to provide facilities for tape machines, which may or may not be buffered. One of the sources to the selector switch will be an RIAA equalized disc stage, although many modern pre-amplifiers neglect this stage because it is so difficult to design. The excuse for this is that the LP is obsolete; this may be so, but there are still many treasured collections of LPs that need to be played, so a proper pre-amplifier should include a disc stage.

The line stage

Determination of requirements

The pre-amplifier only has to provide a very limited output voltage to the power amplifier; even the most insensitive power amplifiers will not require more than 4 or 5 V_{RMS} to drive them into clipping. This means that we are only concerned with linearity and not headroom in this stage. The stage will have to drive the capacitance of the cable without loss of high frequencies, and it may need to be able to drive power amplifiers of lower input impedance, such as transistor amplifiers. Since the stage will be preceded by the volume control, the power amplifier will amplify *all* of its self-generated input noise, so we will need to ensure that this noise is minimized.

These requirements dictate a low output impedance, coupled with low gain which ought to be quantified.

Low capacitance screened cable has a capacitance of ≈100 pF per metre. To avoid inducing hum from the mains transformers into the pre-amplifier, we probably need to separate them by one metre. By the time we allow for the routing of the cable between the line stage and the input plugs of the power amplifier, we will probably use about 1.5 m of cable, which is equivalent to 150 pF. The power amplifier itself will have input capacitance, and we ought to allow ≈20 pF for a valve amplifier, and ≈200 pF for a transistor amplifier. This means that as an absolute minimum, we should be able to drive 170 pF, but preferably 1 nF.

The source impedance in combination with the shunt capacitance of the cable forms a low-pass filter whose −3 dB cut-off we can calculate from:

$$f_{-3\,\text{dB}} = \frac{1}{2\pi CR}$$

However, we would like the high frequency roll off within the audio band to be far less than 3 dB, so we need to know what $f_{-3\,\text{dB}}$ corresponds to a given amount of loss at a given frequency, which we can find from the following formula:

$$f_{-3\,\text{dB}} = \frac{f_{(\text{dB limit})}}{\sqrt{\dfrac{1}{10^{\text{dB}/10}} - 1}}$$

As an example of using this equation, we find that for 0.1 dB roll-off at 20 kHz, we require $f_{-3\,\text{dB}} = 131$ kHz. It should be noted that this formula is only valid for a single HF *CR* or *LR* network. For a single LF *CR* or *LR* network, the formula becomes:

$$f_{-3\,\text{dB}} = f_{(\text{dB limit})} \sqrt{\frac{1}{10^{\text{dB}/10}} - 1}$$

Using this formula, 0.1 dB roll-off at 20 Hz corresponds to $f_{-3\,\text{dB}} = 3$ Hz.

For quick reference, the following table describes the response of 6 dB/octave high-pass and low-pass filters in terms of multiples and sub-multiples of their cut-off frequency:

HPF	*f*/7	*f*/6	*f*/5	*f*/4	*f*/3	*f*/2	*f*	2*f*	3*f*	4*f*	5*f*	6*f*	7*f*
dB	–17.0	–15.7	–14.1	–12.3	–10.0	–7.0	–3.0	–0.97	–0.46	–0.26	–0.17	–0.12	–0.09
LPF	7*f*	6*f*	5*f*	5*f*	3*f*	2*f*	*f*	*f*/2	*f*/3	*f*/4	*f*/5	*f*/6	*f*/7

As an example, if we wanted to investigate a high-pass filter one octave above its cut-off frequency (2*f*), we would see that the response is –0.97 dB.

Once we begin to cascade stages, both HF and LF cut-offs will begin to move towards the mid-band, which is why it was suggested in Chapter 2 that 1 Hz was a good choice for an LF cut-off.

For '*n*' stages, each with identical LF cut-off, the cut-off frequency of each *individual* stage is related to the cut-off of the composite amplifier by:

$$f_{-3\,\text{dB (individual)}} = f_{-3\,\text{dB (composite)}} \cdot \sqrt{2^{1/n} - 1}$$

Applying this formula to a three-stage ($n = 3$) capacitor coupled amplifier, we now find that our 3 Hz cut-off for the entire amplifier requires each stage to have a 1.5 Hz cut-off. The traditional value of 0.1 μF coupling capacitor into 1 MΩ grid leak gives a cut-off of 1.6 Hz.

It is far more difficult to control HF cut-offs than LF cut-offs, so it is unlikely that we will ever have a series of amplifiers each with identical HF cut-offs, but if we did, the following formula could be used:

$$f_{-3\,\text{dB (individual)}} = \frac{f_{-3\,\text{dB (composite)}}}{\sqrt{2^{1/n} - 1}}$$

We can now determine that for 0.1 dB loss at 20 kHz, driving our minimum 170 pF of capacitance, we need an output resistance of 7 kΩ, and preferably a good deal less.

We now need to make some assumptions about power amplifier input resistance and sensitivity.

Traditional power amplifiers had input resistances of 1 MΩ or more. This is a very good resistance, because it allows a low value of coupling capacitor from the pre-amplifier; 47 nF almost meets our 20 Hz 0.1 dB criterion, but 100 nF is better. Sensitivity is ≥1 V, for the better amplifiers.

More sensitive amplifiers should be driven by a passive 'pre-amplifier' consisting simply of switches, volume control and a very short cable, or they should be redesigned to be less sensitive. In the 1960s, *power* amplifiers often had volume controls, selector switches and even tone controls on their front panels. This trend is now being repeated, as the LP, which is the prime reason for a separate pre-amplifier, wanes in popularity.

More modern valve power amplifiers have a sensitivity of ≈500 mV, and an input resistance of 100 kΩ, requiring >470 nF coupling capacitor. This low resistance seems to be because less care is taken over the heater circuitry, which then induces a hum current into the input wiring of the first stage. This current develops a voltage across the grid leak resistor, which results in audible hum if the amplifier is not shunted by the low resistance source of a pre-amplifier. To avoid this, the grid leak is reduced to 100 kΩ, which gives a 20 dB improvement in hum if the input of the amplifier is left open circuit.

Transistor amplifiers also have sensitivity of ≈500 mV, but have an even lower input resistance ≈10 kΩ, which will require >4.7 μF coupling capacitor.

A 4.7 μF 400 V capacitor is not only far more expensive than a 100 nF capacitor, but it is also an inferior component. Rather than trying to find an acceptable 4.7 μF component, we would do far better to attack the problem from the other end, and increase the input resistance of the power amplifier.

This is actually very easy for a valve amplifier. We simply replace the input grid leak resistor with a 1 MΩ resistor, and attend to the heater wiring if this generates a hum problem. Usually this means twisting the heater wiring and pushing it into the corners of the chassis, but we may need to resort to powering the input valve heaters from a stabilized DC supply.

If the input sensitivity is too high, we could fit a potential divider near the input valve; another solution is to reduce the gain of the input stage, and adjust global negative feedback to suit. The second solution is the best, and is easily implemented because the usual reason for excessive gain is that the input valve is an EF86 pentode. If we connect the valve as a triode (using the original anode load resistor) this will reduce the gain of the stage by a factor of 4, and if we keep the feedback factor constant, then the gain of the final amplifier will also be reduced by a factor of 4.

Theoretically, the cathode bias resistor will need to be recalculated, but in practice, the value should not need to be changed greatly. Because of the cathode feedback, the anode resistance of the valve used as a triode will be high, and the output resistance of the stage will be broadly similar to that using a pentode; therefore, compensation components in the anode circuit do not need to be changed. The global feedback network from the output of the amplifier to the input stage will need to be changed to maintain the correct level of feedback with the new gain. This could be done rigorously by calculating the old level of feedback, and recalculating for the new, but a rough method is simply to divide the resistor values by 4 (corresponding to the gain change from pentode to triode). Similarly, any capacitor value would be multiplied by 4 to maintain the correct RC time constant. See Fig. 6.2.

Attempting to raise the input resistance of a transistor amplifier without adding a buffer will probably result in a large DC offset at the output. Realistically, valve pre-amplifiers should not attempt to drive transistor power amplifiers. It can be done, but it requires careful design, and is akin to designing a bicycle that can be raced on ice (this has been done too!). One possible method would be to use a White cathode follower direct coupled to a common cathode triode stage such as an ECC82.

Having prepared the ground, I am now going to state that valve pre-amplifiers should be designed to drive 2 V into 1 MΩ, even if it means modifying the power amplifier to achieve this match.

If we need a sensitivity of ≈250 mV at the input of the line stage, we now know that we need $A_v = 8$; it is a good idea to have 3 dB to 6 dB more than this, to allow for unusually low recording levels, so $A_v = 12$ would be fine.

This stage will be preceded by the volume control, which we will investigate fully later, but for the moment, we can simply state that it will be a 100 kΩ logarithmic potentiometer, whose maximum output resistance will be 25 kΩ.

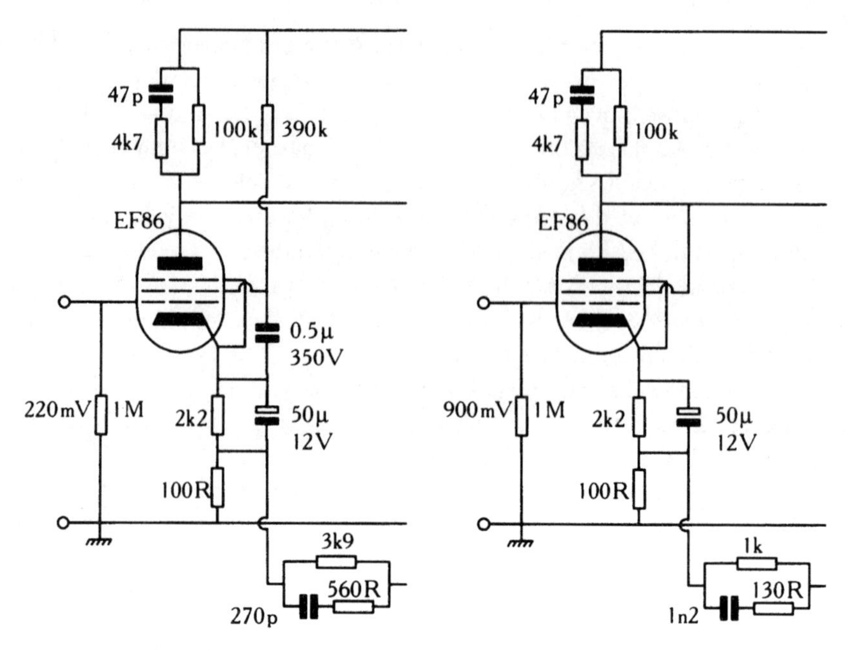

Fig. 6.2 *Modifying a typical pentode input stage for triode operation to reduce noise*

This maximum output resistance may easily be verified by moving the wiper to the *electrical* mid-position of the track. The resistance to each end must be half the total resistance, and assuming zero source resistance, each end is at AC ground. Looking back into the potentiometer, we see the two halves in parallel, and therefore the output resistance is equal to the total resistance of the potentiometer divided by 4. If the wiper is at either end of the track, output resistance will be zero because it is either connected directly to ground, or directly to the (zero resistance) source. Maximum output resistance therefore occurs when the wiper is as far away from *each* end as possible, which is the centre position.

The question of potentiometer output resistance is crucial, because it will form a low-pass filter in conjunction with the input capacitance of the line stage. Using our earlier argument of 0.1 dB HF loss at 20 kHz, we see that the maximum allowable input capacitance of the line stage is ≈50 pF, assuming maximum output resistance of the volume control =25 kΩ.

If the input sensitivity of the stage is ≈170 mV (250 mV −3.5 dB), and we want a signal to noise ratio of ≥100 dB, then the self-generated noise of the stage *referred to the input* will be 170 mV −100 dB = 1.7 μV. The EF86 pentode can manage 2 μV on a good day, so we certainly ought to be able to achieve this using triodes.

Achieving the requirements

We can now pull all our information together to form a full list of requirements for the line stage:

$$\begin{aligned} A_v &= 12 \\ Z_{out} &\ngtr 7\,\text{k}\Omega \\ C_{in} &\ngtr 50\,\text{pF} \\ V_{noise} &\ngtr 1.7\,\mu\text{V} \\ \text{Output coupling capacitor} &= 100\,\text{nF} \end{aligned}$$

A good design is a *simple* design, so we should check to see if the common cathode triode could do the job.

The ECC82, a low μ triode, was investigated. See Fig. 6.3.

$$\begin{aligned} A_v &= 15.5 \\ Z_{out} &= 7.7\,\text{k}\Omega \\ C_{in} &= 30\,\text{pF} \end{aligned}$$

The gain is certainly satisfactory, as is the input capacitance, but the output resistance is a little over the required value. Although the output resistance is higher than we would like, the input capacitance is considerably less than our allowable maximum, and the reduced roll-off at this point will

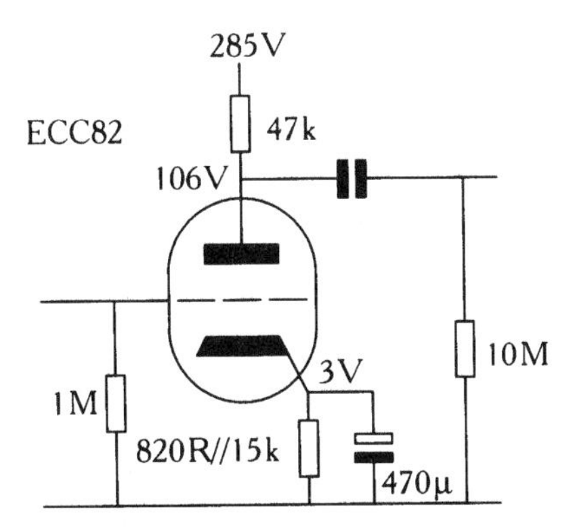

Fig. 6.3 *ECC82 common cathode amplifier as pre-amplifier output stage*

compensate for the slightly increased HF roll-off at the output. Provided that we were very careful about the capacitive loading of the output cable, this would be a satisfactory solution.

Sadly, the ECC82's octal predecessor, the 6SN7, would have an input capacitance of ≈70 pF, because C_{ag} = 3.9 pF and would therefore be unsuitable.

A much better solution would be to incorporate the ECC82 into a μ-follower in order to reduce output resistance. See Fig. 6.4.

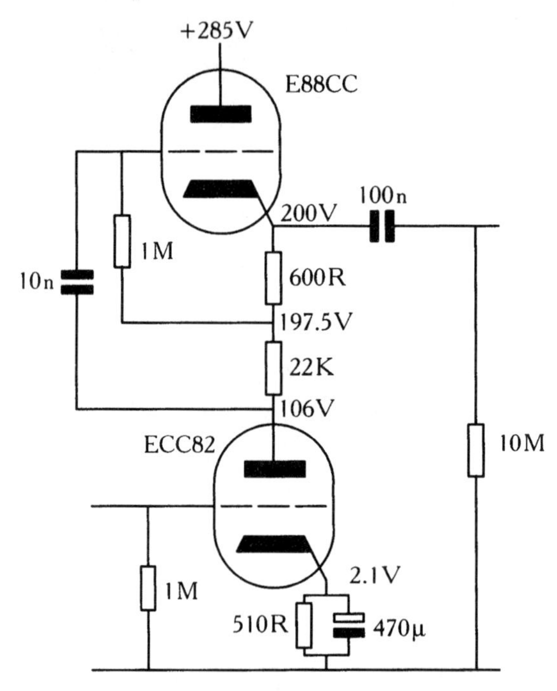

Fig. 6.4 *μ-follower as pre-amplifier output stage*

$$A_v = 19$$
$$Z_{out} \text{ (from upper cathode)} = 1\,k\Omega$$
$$C_{in} = 35\,pF$$

The gain has risen slightly, but noise should still be satisfactory, whilst the output resistance and distortion have been greatly reduced, making this an excellent choice for a line stage.

Unfortunately, this improved performance has been bought at the expense of having a valve with its cathode at ≈170 V, necessitating an extra heater supply to keep V_{hk} within limits. If we can find a good reason for using another μ-follower, or any other stage with an elevated cathode voltage, then this extra heater supply will have been justified.

In these two examples, we have chosen a valve and let it dictate the gain of the stage; in some instances, we may not be prepared to allow that to happen, and we will want to choose the gain.

We could change the gain by adding a potential divider at either the input or the output. Added at the output, it will increase output resistance, which we cannot allow, whereas added at the input, it reduces signal level, but hardly affects noise, thereby degrading the S/N ratio.

Our only remaining option for adjusting gain is negative feedback. We can certainly reduce the gain of a single valve by removing the cathode bypass capacitor, causing series feedback, but this will dramatically raise output resistance, and is therefore not allowed. In order to keep output resistance low, the feedback must be parallel derived, although it may be series or parallel applied. See Fig. 6.5.

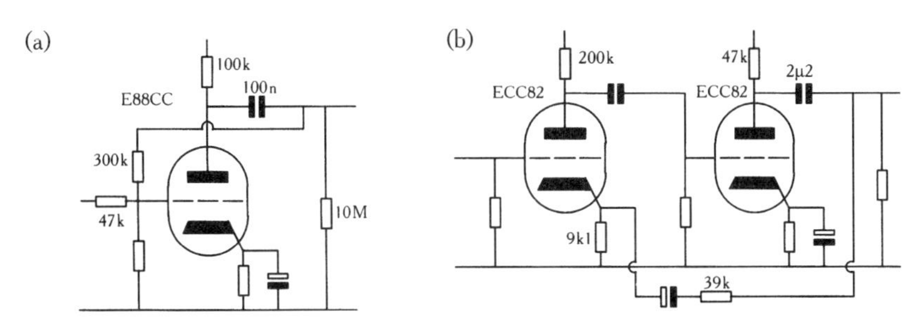

Fig. 6.5 *Shunt feedback versus series feedback*

Both examples use parallel derived feedback to reduce output resistance, but in Fig. 6.5a the feedback is parallel applied, whereas in Fig. 6.5b it is series applied. Both solutions have problems, most of which are linked to the allowable value of the feedback resistor. The feedback resistor is effectively in parallel with the anode load, and is fed by the output coupling capacitor. A low value of feedback resistor will therefore require a larger coupling capacitor and will reduce the (open loop) gain of the stage by reducing the AC anode load resistance. A reasonable first choice for the feedback resistor might be $3 \times R_L$.

For the single valve invertor with $R_L = 100\,\text{k}\Omega$, we might want a final gain of 5, which would require a series input resistor of $47\,\text{k}\Omega$, and a feedback resistor of $300\,\text{k}\Omega$. The input resistance of this stage is $\approx 47\,\text{k}\Omega$, and would heavily load a $100\,\text{k}\Omega$ volume control. We could increase both resistors, making the series input resistor $270\,\text{k}\Omega$, and the feedback resistor $1.6\,\text{M}\Omega$; this reduces the loading on the volume control and further reduces the loading on the output stage. Unfortunately, the noise performance of the stage is now very poor; not only is the input noise of the grid leak resistor no longer being shunted by the source resistance of the volume control, but the series resistor itself will generate noise.

Having dismissed the invertor, we turn to the two valve stage, which has feedback taken to the cathode of the input stage. The grid does not have a series resistor, and the cathode resistance will be quite low, so noise due to the input circuitry will not be compromised, although noise due to R_L will no longer be shunted. (See later in this chapter.)

Having individual input and output valves gives us the opportunity to optimize each stage. V_1 has been biased for low anode current in order to maximize the value of R_k (we will see why this is necessary in a moment), whilst V_2 has been biased for low r_a and good linearity. The capacitor in series with R_{fb} prevents DC from the V_1 cathode from appearing on the output.

Total gain is ≈ 200 before feedback. If, as before, we need a total gain of 5, then we can find β, the feedback fraction, from:

$$\beta = \frac{1}{A} - \frac{1}{A_0}$$

$$= 0.195$$

The following approximation ignores the feedback current in R_k due to V_1, and also ignores the loading of r_k on the feedback network, but it is sufficiently accurate for our present purpose.

$$R_{fb} \approx R_k\left(\frac{1}{\beta} - 1\right)$$

$$\approx 39\,\text{k}\Omega$$

This is a significant loading of the output stage, which will decrease the open loop gain of the stage and increase distortion, *before* the feedback is applied.

The loading of R_{fb} is always a problem in amplifiers of this type, and is reduced by increasing the required gain (reduces β), or by increasing R_k. Alternatively, we can accept the low value of R_{fb}, and arrange to be able to drive it by adding a cathode follower to the output of the amplifier. See Fig. 6.6.

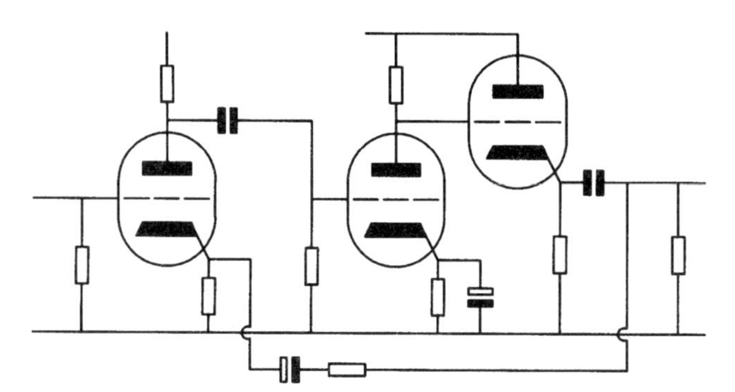

Fig. 6.6 *Adding a cathode follower to allow series feedback*

We have now achieved our objective, but we needed three valves and an extra heater supply for the cathode follower, and we have not even considered the problem of ensuring stability of the completed amplifier.

Unless we specifically need negative feedback to perform frequency correction to the signal, such as tone controls, our best solution is to use the μ-follower.

The volume control

The volume control is not simply a 100 kΩ log pot that was found in the scrap box. It is an essential part of a pre-amplifier and should be treated with the same care as any other part of the design.

The human ear has a logarithmic response to sound pressure level, so if we want a volume control that has a uniform perceived response to adjustment throughout its range, we need to use a logarithmic potentiometer. This is the root cause of all our problems.

It is not a problem to make a linear potentiometer; all we need do is to deposit a strip of carbon of uniform width and thickness onto an insulator, put terminals at each end, and arrange for a contact to scrape its way round. If we don't bother with a casing, it is known as a skeletal type. In an attempt to produce a logarithmic law, the coating thickness is made variable; in deference to audio sensibilities, a pressed metal screening can is fitted, and two potentiometers are ganged together on one shaft onto which we can fit a big, shiny, spun aluminium knob. Making the coating thickness continuously variable would be expensive, so the logarithmic law is approximated by a series of straight lines. See Fig. 6.7.

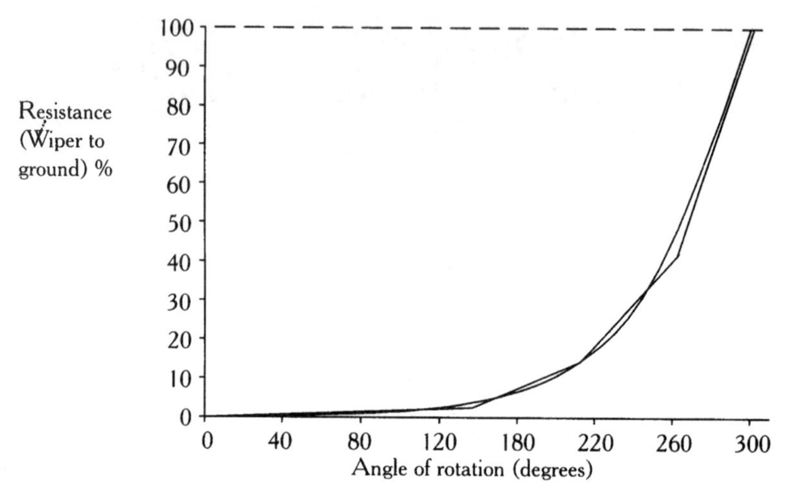

Fig. 6.7 *Approximation of logarithmic law by straight lines*

It is amazing how good a fit to the ideal logarithmic curve can be made using only four different resistance tracks, but it will come as no surprise to learn that this still results in steps in the response as the knob is rotated. We also expect the mechanically linked potentiometers to produce identical levels of attenuation all the way from 0 dB to 60 dB. Some of them are remarkably good, but the carbon track potentiometer's natural habitat is buried in the undergrowth of a television.

One useful fudge is to use a conductive plastic linear potentiometer with a law faking resistor added between wiper and earth. This does not produce a true logarithmic law, but it is far better than a pure linear potentiometer.

The ideal volume control would have identical attenuation (in dB) for a given amount of rotation whether that rotation was at the bottom of the range or the top. The following QBASIC program allows you to investigate the effects of the law faking resistor. Using the program, the ideal value for the law faking resistor seems to be ≈0.83 of the value of the potentiometer, and lower values actually make performance worse! Unfortunately, the stratagem only improves the top part of the range, but if levels can be optimized so that only the top 12 dB of attenuation are commonly used, then it works very well.

```
CLS
P = 1
PRINT "This program calculates the step size resulting from shunting"
PRINT "the output of a linear potentiometer with a law faking resistor. "
PRINT
PRINT "How many steps of resolution do you want to investigate";
INPUT N
DIM LOSSDB(N)
PRINT "What value of potentiometer will you use";
INPUT R
PRINT "What value of law faking resistor will you use";
INPUT L
PRINT
PRINT " LOSS"; TAB(15); " STEP"
DO UNTIL P = N
A = P * R / N
UPPER = R - A
LOWER = A * L / (A + L)
LOSSDB(P) = ((86.8589 * LOG(LOWER / (LOWER + UPPER))) \ 1) / 10
P = P + 1
LOOP
P = 2
DO UNTIL P = N + 1
CLICK = LOSSDB(P) - LOSSDB(P - 1)
PRINT ABS(LOSSDB(P)); "dB"; TAB(15); ((10 * CLICK) \ 1) / 10; "dB"
P = P + 1
LOOP
```

A development of this idea is to provide multiple taps along the linear conductive plastic track as part of the design. *Tapped linear track attenuators* have excellent channel matching and log law; they often have 'Penny & Giles' written on the outside, and *feel* nice when used. It should be noted that these are not potentiometers and that measuring resistance to endstops from the wiper will produce confusing results.

If quality is paramount, and we can accept a control that is not continuously variable, we can use a switched attenuator; this has conventional resistors connected to a switch in order to control volume. Adherence to the log law can now be perfect, as can channel balance. Unfortunately, soldering many resistors onto a switch to make switched attenuators takes considerable time, but commercial attenuators for audio are now commonly available. Switched attenuators are also available with the resistors fabricated directly onto the ceramic substrate of the switch wafer, and their performance is excellent.

The practical disadvantage of the switched attenuator is that we can only have as many different volume levels as switch positions. Although rotary switches are available with 30 positions, as opposed to the more usual 12, this still limits us to 26 or 27 positions by the time that we have fitted the endstop. For a normal volume control, we often want an 'off' position, followed by a –60 dB position, and after that we want uniform steps all the way up to 0 dB. We have already used 6 of our 30 positions, so 60 dB divided by 24 steps = 2.5 dB per step. This is too coarse, and commercial attenuators coarsen the lower levels to allow finer control at the upper levels, but this still only brings the basic step size down to 2 dB.

If we don't mind wiring individual resistors onto switches ourselves, we can do rather better than 2 dB steps, and there are three ways of going about this:

- We design our pre-amplifier so well that all incoming signals arriving at the volume control are at precisely the same level. Additionally, the output of the pre-amplifier is perfectly matched to the power amplifier, and is only able to overload the amplifier on the last few steps of the volume control. This is perfectly feasible, and will allow us to use 1 dB steps. We will usually also need a mute switch in addition to the volume control.

  This type of arrangement is actually surprisingly convenient to use.
- We indulge in mechanical sophistication and digital logic. If we remove the manufacturer's endstop on the rotary switch, we revert to a full 30 positions. We fit a floating traveller to the shaft of the switch which can be picked up, in either direction, by a fixed traveller on the shaft. The floating traveller then engages with an external endstop, and we achieve about 56 'clicks' of the switch from end to end, because the switch now turns almost two revolutions. We now add an optical sensing arrangement that senses the point when the rotary attenuator passes from maximum attenuation to zero attenuation. This simultaneously switches a relay, that adds attenuation equal to the maximum attenuation of the rotary attenuator, *plus*

one step. It is most important that the relay control logic either knows which direction the control is moving, or that it is started up in the correct state, otherwise, instead of achieving a smooth transition, volume will change from –30 dB to 0 dB, or vice versa.

This is not a trivial solution.

- We accept that the control will be coarse, and we add a trimming control that allows the level to be trimmed by ±3 dB in fine steps, perhaps 0.25 dB. This results in two controls for volume, and is not quite as nice to use as the first two solutions, but it does allow fine control *and* adequate range without extensive metalwork. Commercial passive 'pre-amplifiers' of this form are available.

If the volume control is going to be part of a complete pre-amplifier, then it is not unreasonable to suggest that the first method of achieving a switched attenuator be used, since we can always add additional attenuators on individual inputs to achieve correct matching.

Thirty-way stud switches are ideal for making switched attenuators. They are about the same size as a standard wafer switch but the contacts are studs moulded into the plastic of the wafer, and a small wiper travels across these and the centre ring. Bought new, they are expensive, but you may be able to find some in old laboratory equipment at an electronics junk shop. Haggle.

With care, they can be dismantled and their silver contacts cleaned so that they are as good as new, despite being over 30 years old. To do this, carefully remove each wafer from the switch. Scrape the varnish off the two centre screws that hold the travelling wiper mechanism together, and carefully undo them, taking care not to put stress on the wiper itself. Gently prise everything apart, taking extreme care of the plastic barrel that supports the travelling wiper, as it is very fragile. Pour a little silver polish onto clean cartridge (artists' drawing) paper, and rub the wafer, stud side down, on it until it is clean. Do the same for the wiper. Give the whole thing a squirt with electronic cleaning solvent to remove any traces of polish, and reassemble without touching the contacts with your fingers. Sweat contains acid and will tarnish the contacts.

Assuming that we have a pair of multi-way switches available to make a volume control, we need to be able to calculate the values of resistors required. We could do this by hand, but a programmable calculator or computer makes life much easier. There are three basic forms of attenuator that can be used. See Fig. 6.8.

Figure 6.8a is similar to our carbon track attenuator in that it has a ladder of resistors from which we take the appropriate tapping; this is the form used in most commercial switched attenuators.

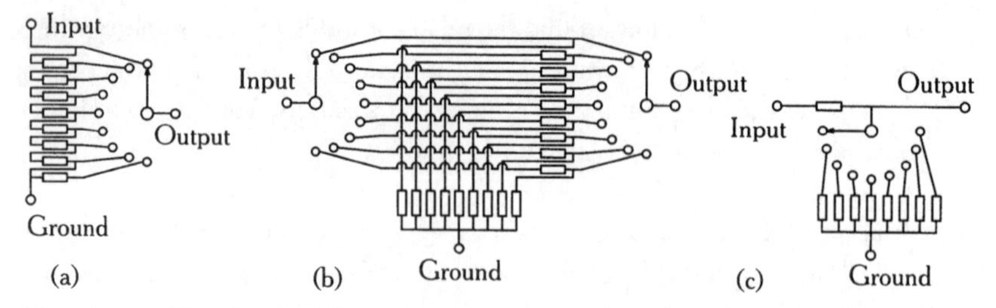

Fig. 6.8 *Basic stepped attenuators*

NB. Just because a volume control has detents, this *does not* guarantee that it is a true switched attenuator; it could be a carbon track potentiometer in masquerade. Real switched attenuators tend to be quite large. A quick test, rather than dismantling it in the shop, which might have you thrown out, is to measure the resistance of the lower arm of each gang at the settings of near maximum attenuation with a digital multimeter. If there is any measurable difference between gangs, it is likely to be a carbon track potentiometer.

Figure 6.8b uses individual potential dividers for each setting of volume which dramatically reduces the number of soldered joints in the signal path at the expense of twice as many switches and resistors. Additionally, the resistors must be close tolerance to maintain a good logarithmic law where each and every step is of equal value.

Figure 6.8c uses one series fixed resistor and a selection of shunt resistors in order to achieve similar performance to Fig. 6.8b at lower cost.

Input resistance is no longer constant, and the series resistor must be equal to the maximum desired output resistance, which we have already established as being 25 kΩ in a valve pre-amplifier. When this attenuator is set to maximum attenuation, the input resistance is equal to the series resistor, and so the input resistance falls to 25 kΩ, whereas the previous attenuators had a constant input resistance of 100 kΩ for a maximum output resistance of 25 kΩ.

The following QBASIC programs will generate the resistor values for the attenuators in Fig. 6.8a, b, c. QBASIC is the version of BASIC that comes with the MSDOS operating system found in many PCs. They are not miracles of programming, but they are quick and easy to use, and can easily be modified for different versions of BASIC, or be run on a programmable calculator.

The programs ask for the load resistance across the wiper, this is the grid leak resistor of the following valve. It is tempting to try to use the potentiometer as the grid leak, but this is poor practice and can cause noise problems, it is also unnecessary since the programs account for its loading in designing the attenuator.

This program finds values for the circuit of Fig. 6.8a. The final value given by this program is connected between the last usable switch contact and ground, it is often convenient to use one of the spare contacts on the switch as a ground terminal.

```
CLS
A = 0
B = 0
N = 0
PRINT "This program calculates individual values of resistors between"
PRINT "taps of the circuit in Fig. 6.8a."
PRINT "How many switch positions can you use";
INPUT S
PRINT "What step size (dB)";
INPUT D
PRINT "What value of load will be across the output of the
potentiometer";
INPUT L
PRINT "What value of potentiometer is required";
INPUT R
DO UNTIL N = S - 1
Y = ((R - L / 10 ^ (-A / 20)) + SQR((L / 10 ^ (-A / 20) - R) ^ 2 + 4 *
R * L)) / 2
C = R - Y - B
PRINT A; "dB "; C; "ohms"
B = B + C
A = A + D
N = N + 1
LOOP
PRINT A; "dB "; R - B; "ohms."
```

The following program is for the circuit of Fig. 6.8b.

```
CLS
A = 0
N = 0
PRINT "This program calculates upper (X) and lower (Y) arms of"
```

```
PRINT "individual potential dividers for the circuit of Fig. 6.8b"
PRINT "How many switch positions can you use";
INPUT S
PRINT "What step size (dB)";
INPUT D
PRINT "What value of load will be across the output of the potentiometer";
INPUT L
PRINT "What value of potentiometer is required";
INPUT R
DO UNTIL N = S
Y = ((R-L / 10^(-A / 20)) + SQR((L / 10^(-A / 20)-R)^2 + 4
*R*L)) / 2
X = R-Y
PRINT A; "dB "; "Y="; Y; "ohms "; "X="; X; "ohms"
A = A+D
N = N+1
LOOP
```

The final program calculates shunt resistors. Note that it *never* achieves zero attenuation, and therefore predicts the minimum unavoidable loss through the volume control and grid leak resistor (basic loss). In effect, this volume control must be considered to be a fixed attenuator plus a variable attenuator.

```
CLS
N = 0
PRINT "This program calculates shunt resistors for the circuit of Fig. 6.8c."
PRINT "How many switch positions can you use";
INPUT S
PRINT "What step size (dB)";
INPUT D
PRINT "What value of load will be across the output of the potentiometer";
INPUT L
PRINT "What value of series resistor is required";
INPUT R
B = ((-100*LOG(L / (R+L))*8.686) \ 1) / 100
PRINT "Basic loss ="; B; "dB, added shunt is infinite"
PRINT "Added attenuation: "
A = B
DO UNTIL N = S-1
```

```
A = A+D
C = (R*10^(-A/20))/(1-10^(-A/20))
Y = 1/(1/C-1/L)
N = N+1
PRINT N*D; "dB, shunt= "; Y; "ohms"
LOOP
```

It may be that you want to build a balanced pre-amplifier. In which case, the volume control should be balanced too. Curiously, some engineers, who should know better, think that this means placing identical controls in *each* path of each channel. This is wrong. The attenuators cannot be perfectly matched, and will therefore unbalance the signal, which degrades the rejection of unwanted noise.

The correct way to construct a balanced volume control using minimum components is to use a configuration based on Fig. 6.8c. If you use the computer program to determine values, remember that the series resistor that the program uses is *twice* the value of the resistor in each leg. This form of attenuator has the disadvantage of a high output resistance when set for a sensible input resistance, and in combination with the input capacitance of the following stage, this will cause HF loss if ignored. Note that tapped track linear attenuators cannot be used as the shunt element. See Fig. 6.9.

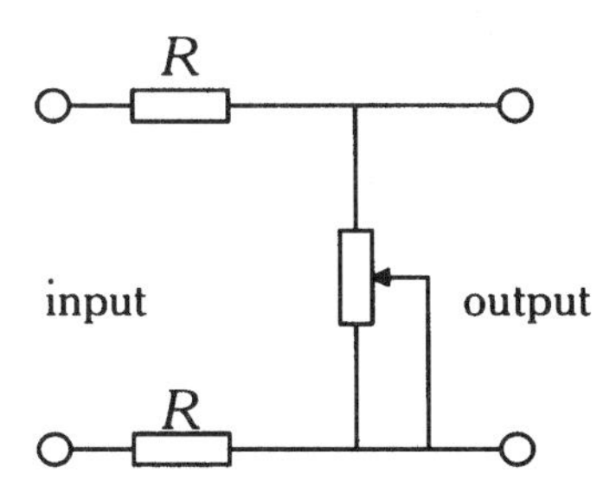

Fig. 6.9 *Balanced volume control*

Input selectors

It is quite likely that we will have a number of alternative sources to the pre-amplifier, such as LP, CD, digital TV or radio, etc. These will need to be selected to the volume control, and we may wish to be able to record them onto tape, so we will need a tape loop. See Fig. 6.10.

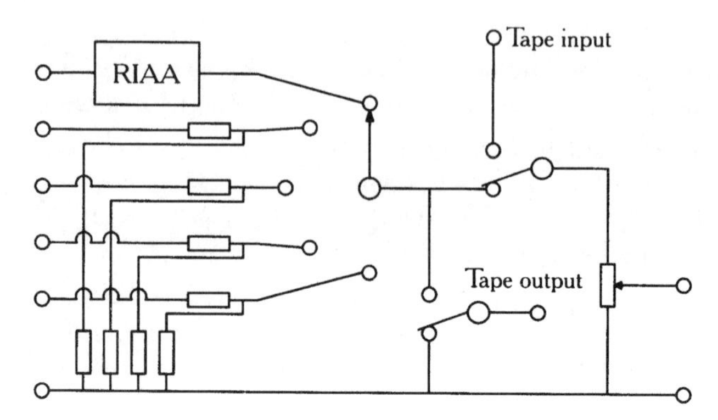

Fig. 6.10 *Input selectors with switchable tape output*

It will be seen that the arrangement is very simple. Incoming signals are routed through the input selector to the volume control via the tape monitor switch to enable off-tape or source monitoring.

An unusual feature is the inclusion of a switch in the *output* to tape. All of the sources to the selector will be low resistance, and perfectly capable of driving the cable capacitance to the (powered) tape machine, and so a tape buffer is not required. However, a switched-off tape machine will present a non-linear load to the source in the form of unbiased transistor junctions, and may cause distortion. Most of the time, we will not be recording the source, so the tape output is switched off.

As the circuit stands, the rotary selector switch could suffer from crosstalk due to capacitance between adjacent contacts. On high quality traditional pre-amplifiers, this problem was solved by having two switches, one selected the source and the other deselected the short circuit to ground on that source. Unfortunately, such wafer switches are no longer available, but a good alternative is to use alternate contacts as inputs on a standard wafer switch, and connect the unused contacts to ground, which then screens the signal contacts. See Fig. 6.11.

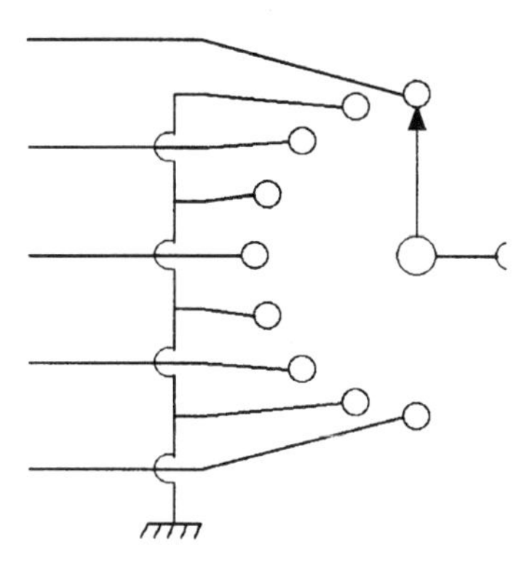

Fig. 6.11 *Reducing capacitive crosstalk on input selector by guarding*

Switch quality

Switches are not created equal, and suffer from a number of known defects. Stray capacitance has already been mentioned, but there are other problems.

Ideally, contact resistance should be zero, but this is never achieved. Contact resistance will rise as a result of: oxidation of the contacts, insufficient contact pressure, and contact wear. In an effort to prevent oxidation, gold plated contacts are sometimes used, despite the fact that the resistance of gold is considerably higher than that of silver or copper. Another method is to accept oxidation and use silver plating, as the silver sulphide that forms is still a conductor. It should be noted that the phenomenon of contact resistance is complex, and that low level signal switching should only be done by switches expressly designed for that purpose.

All switches will have some leakage resistance, which will be specified by the manufacturer, and will be worsened by atmospheric damp.

Push-button mechanically interlocked switches should not be considered. They feel nasty to use, and are unreliable because they are generally of poor quality. Additionally, they are awkward for the amateur to use because cutting the required neat rectangular hole is not easy. By contrast, the rotary wafer switch only needs a drilled or punched hole which is then obscured.

The ideal signal switch is the mercury-wetted relay. This has a droplet of mercury to 'wet' the contacts and ensures minimum, and constant, contact resistance. Because the relay has to be sealed to prevent the (poisonous) mercury from escaping, the contacts do not oxidize; the contacts do not scrape together, so wear is minimized.

Mercury wetted relays generally have a single closing contact, and must be mounted correctly to ensure that the mercury is in the correct position. If we have a series relay on each source, we could precede it with a shunt relay to ground that would normally be closed, and would ensure almost perfect attenuation of unwanted sources. In order to protect the source from the short circuit, a 1 kΩ series resistor is normally fitted before the shunt relay; this also improves attenuation. See Fig. 6.12.

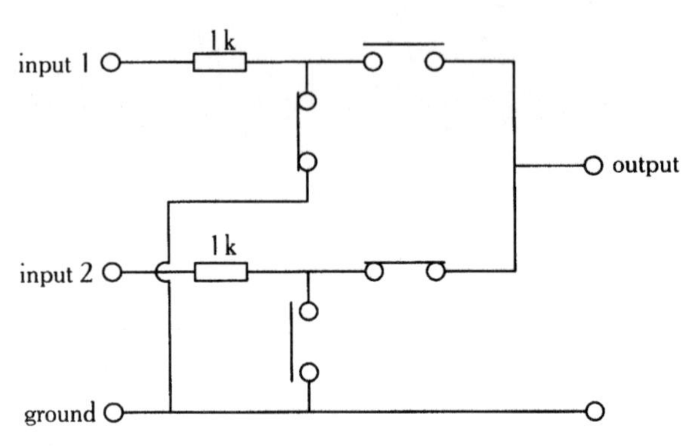

Fig. 6.12 *Shunt and series switching using relays eliminates crosstalk*

These relays would then be mounted as close as possible to the input sockets and could be connected directly to the tape switching relays, which would also be close to their sockets. All of the signal switching is then as direct as possible, and the only signal wires that need come from the rear panel would go to the volume control, as the front panel switch would only carry DC. It makes sense to choose 5 V relays, since these can then be powered by a 5 V regulator fed from the same transformer winding as the heaters. Since we are using 5 V, we could consider using logic gates to drive the relays, and use momentary action push-buttons on the front panel to select routings, or even remote control!

The *only* disadvantage to mercury-wetted relays is their cost.

The RIAA disc stage

RIAA is the abbreviation for 'Recording Industry Association of America', and is the worldwide standard for equalization of 'microgroove' records, as opposed to the numerous standards for 78s. Because the RIAA standard was not invented in Europe, but a worldwide standard *was* needed, the IEC invented an LP equalization standard that was almost identical. The only difference is that the IEC standard recommends bass cut on replay only, with a −3 dB point at 20 Hz (7950 μs) in order to reduce rumble. Most manufacturers of quality pre-amplifiers assume that their products will be complemented by equally good turntables, and that rumble will not be a problem, so they ignore the IEC recommendation. Their equalization is therefore RIAA.

There is considerable pressure to modify RIAA pre-amplifiers to include a low frequency roll-off because:

- Valve power amplifiers are susceptible to output transformer core saturation if high amplitude signals are applied at low frequencies (<50 Hz).
- Bass reflex loudspeakers are easily overloaded at frequencies below their acoustic roll-off because there is negligible damping of cone motion. Bookshelf loudspeakers tend to roll off below 100 Hz, whereas free-standing loudspeakers could improve this to 50 Hz, or less.
- Vinyl records contain low frequency (<20 Hz) noise due to warps and rumble.

It is therefore argued that these problems could be avoided by implementing some form of LF roll-off within the RIAA stage. One possibility is to implement the IEC 7950 μs recommendation, but a more sophisticated approach is to build a pre-amplifier incorporating a properly designed high-pass filter with a final slope of 12 dB/octave or more, set at ≈10 Hz.

The author firmly believes that neither of the preceding electrical approaches is correct, and that RIAA equalization should be reserved solely for correcting the record equalization applied by the manufacturer at the time of cutting. CD players do not add a 10 Hz high-pass filter to solve the problems of poorly designed loudspeakers or questionable output transformers, so why adulterate LP? Warps and rumble are mechanical problems and should therefore have mechanical solutions, not electrical 'fixes'.

The mechanical problem

Fortunately, a 12 dB/octave high-pass mechanical filter is unavoidably formed by the compliance of the cartridge suspension and the effective mass of the arm plus cartridge. The low frequency arm/cartridge resonance may be found using the standard resonance equation:

$$f = \frac{1}{2\pi\sqrt{CM_{total}}}$$

where: C = cartridge vertical compliance
M_{total} = total effective mass

Typical values might be:

Cartridge mass:	5 g
Mounting hardware (screws and nuts):	1.5 g
Arm effective mass:	12 g
Total mass (M_{total}):	18.5 g
Cartridge vertical compliance (C):	15×10^{-6} dyne/cm

The previous figures refer to a unipivot arm designed for a moving-coil cartridge with its outer body removed, and result in a resonant frequency of 10 Hz.

It has been suggested that a higher frequency (12–15 Hz) should be set as this will be more effective in reducing low frequency noise, and this is quite true. However, we live in a practical world and dramatically reducing arm effective mass (to raise the resonant frequency) inevitably produces a flimsy structure only suitable for cartridges that do not transfer much vibration into the arm. Unfortunately, such cartridges are high compliance, and we are back where we started. Additionally, even setting the resonant frequency as low as 10 Hz means that the reproduced response (when RIAA equalized) is likely to be –1 dB at 20 Hz, depending on damping.

Even if the arm/cartridge resonant frequency is correct, the mechanical high-pass filter will only operate correctly if the resonance is correctly damped. The general principle is that the moving pick-up arm is fitted with a paddle which is forced to move through a viscous liquid, thus damping the motion of the arm. Ideally, the damping should be applied at the headshell because this reduces the energy transferred from the cartridge into the arm, and particularly reduces excitation of unavoidable high frequency structural resonances; but damping near the pivot, as required by almost all unipivots, can damp the low frequency resonance equally well.

Mechanical damping has to be set by trial and error, and commonly far too much damping is applied – the fluid is either too viscous, or there is too much of it. One way to set damping is to play a badly warped record with no damping applied and observe cartridge movement as the warps are traversed. If the cartridge appears to bounce relative to the record surface, add a little fluid and try again. Use as little damping as possible, as too much will *increase* low frequency noise and cause tracking problems at higher frequencies as undistorted stylus displacement is wasted tracking warps rather than recorded signals.

Setting the mechanical high-pass filter correctly has two major consequences. First, it means that we no longer need an electrical high-pass filter, but more importantly it means that stylus cartridge vertical deflection is considerably reduced and distortion generated by the cartridge falls.

Designing the RIAA stage

If power amplifiers are let down by their phase splitters, then the Achilles' heel of the pre-amplifier must surely be the RIAA disc stage. The stage has to satisfy so many contradictory requirements simultaneously that its design and execution is fraught with problems.

When we investigated power amplifiers, we looked at some classic designs to see how the goals were achieved. There are *no* classic RIAA stages, they varied from mediocre to plain awful.

This was not always due to incompetence on the part of the designers; they had poor quality components, and could not use semiconductor regulation of power supplies as is habitual today. However, the main factor was that there was no incentive to design superb disc stages, because the signal leaving the turntable was not very good. Good turntables and cartridges were available, but suitable arms and plinths were rare.

Determination of requirements

As before, we need to investigate the requirements of the RIAA stage, which will help define the design.

1 *Low noise and no hum.* We have to admit that valves are not as quiet as the latest generation of low noise IC op-amps, but they can be made to be quiet if we use DC heater supplies. Pentodes are complete non-runners, and we will need to be careful in our use of triodes.
2 *Constant input resistance and capacitance.* This might seem obvious, but a lot of designs have failed to appreciate this requirement. Cartridge manufacturers design for a specific loading of resistance and capacitance which they use to equalize mechanical deficiencies of the generator system.

This is particularly noticeable for moving magnet designs such as the older Shure and Ortofon models.

3 *Accurate RIAA*. It is quite unbelievable how many designs have incorrect RIAA equalization (ancient and modern). This is either down to a failure to use the correct equations, or to appreciate the loading conditions.

4 *Low sensitivity to component variation*. Valves wear out, and as they do so, their anode resistance rises. When a valve is replaced, the new value of C_{ag} may not be the same as the old valve. Neither of these effects should noticeably affect the accuracy of RIAA equalization.

5 *Good overload capability.* But what capability is necessary? Using a Tektronix TDS420 digitizing oscilloscope, the dynamic range of LPs was investigated in conjunction with a high quality record playing system. The TDS420 was first used in 'envelope' mode to find the maximum output of the cartridge, and monitored an entire day of listening to music. The largest musical peaks were found whilst playing a Mobile Fidelity pressing of Beethoven's 9th Symphony. Before equalization these peaks rose to +16 dB above the nominal 5 cm/s level, but clicks due to dust or scratches rose to about twice this level at +22 dB. See Fig. 6.13.

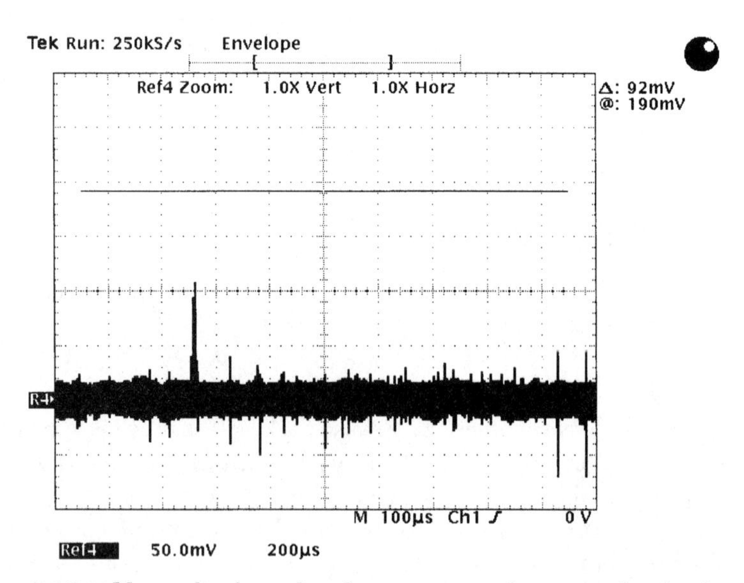

Fig. 6.13 *Unequalized enveloped music output from cartridge (peaks are dust/clicks)*

Individual clicks were then captured, and it was found that the vinyl/tip mass resonance was being excited, and that this produced a damped oscillation at 56 kHz for this particular (moving-coil) cartridge. See Fig. 6.14.

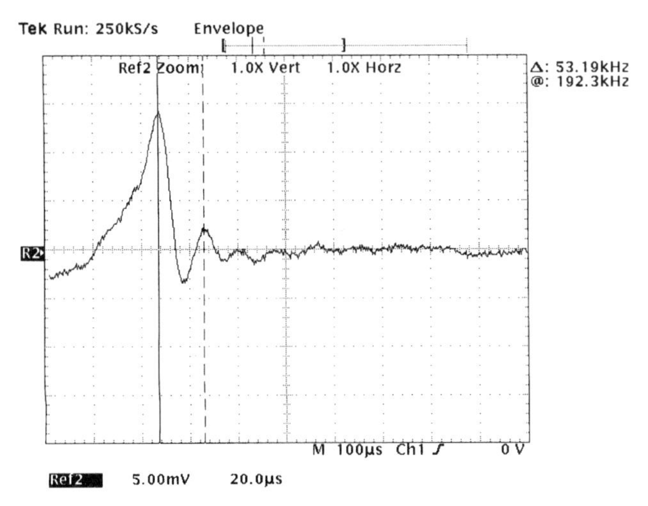

Fig. 6.14 *Unequalized output from cartridge showing excitation of vinyl/tip mass resonance*

If these ultrasonic signals were to overload the pre-amplifier, they would generate intermodulation products that would come back down into the audio band and make the clicks much more noticeable.

We should now allow for variable cartridge sensitivity of about 6 dB; if we need more than this, we should reconfigure the disc stage.

A good design should not operate permanently at its limits, so a further 6 dB margin is desirable, to give a total of 28 dB in the audio band, *rising* to 34 dB or more at ultrasonic frequencies. Very few pre-amplifiers of any age achieve this requirement and low noise simultaneously.

Worn/old discs will generate more ultrasonic energy than a new disc. This may be due to dirt ground into their groove, or because they were played by a cartridge that mistracked causing wall damage as the stylus flailed helplessly from side to side of the groove.

Inadequate ultrasonic overload margin is the reason why some pre-amplifiers will make worn records sound unplayable, but a good pre-amplifier is able to extract the best from the disc.

6 *Low distortion.* This is an obvious requirement, and is linked to 5.
7 *Low output resistance.* Ideally, the stage should be able to drive cable capacitance, so that it can drive a tape machine, or be sited within the plinth of the turntable.
8 *Low microphony.* Valves are always microphonic; but it is possible to make them worse. A low value of anode load combined with high anode current reduces noise, but increases the *power* gain of the stage, which causes microphony to rise. Ultimately, we have to mechanically isolate the valves. This is less of a problem than it seems, since most of the structural resonances of the electrodes are above 1 kHz, and a mechanical filter to deal with this is easily made.

Now that we know our requirements, we can consider what topology would suit.

The constant input resistance and low noise requirements eliminate shunt feedback. Low noise also rules out the pentode. We are therefore left with a combination of triode stages having active equalization determined by series feedback, or with passive equalization. Each of these contenders may be further broken down into performing the equalization 'all in one go', or splitting it over a number of stages.

To see how we need to tackle the problem of RIAA equalization, we need to define RIAA equalization. The equalization is specified in terms of time constants: 75 μs, 318 μs and 3180 μs. The equation that generates the gain required by the RIAA replay equalization is given below:

$$G_s = \frac{(1 + 318 \times 10^{-6} \times s)}{(1 + 3.18 \times 10^{-3} \times s)(1 + 75 \times 10^{-6} \times s)}$$

where $s = j\omega$, and $\omega = 2\pi f$.

This is not a friendly equation, and a computer program is the easiest way of subduing it. Running the following computer program yields the data shown in the following table, and calculates the required response of the equalizer *only*, a perfectly pre-equalized signal passed through a perfect equalizer would yield an amplitude response of 0 dB, and a phase response of 0° for all frequencies.

```
CLS
OPTION BASE 1
DATA
0,10,20,50.05,70,100,200,500.5,700,1000,2000,2122,5000,7000,10000,
20000,50000,70000,100000,200000
FOR R = 1 TO 20
```

```
READ F
W = 2 * 3.1415927# * F
A = .000318
B = .00318
C = .000075
REALU = 1 - W ^ 2 * B * C + W ^ 2 * A * (B + C)
IMAGU = W * (A - B - C - W ^ 2 * A * B * C)
LOWER = (1 - W ^ 2 * B * C) ^ 2 + W ^ 2 * (B + C) ^ 2
MAG = SQR(REALU ^ 2 + IMAGU ^ 2) / LOWER
GAIN = (((19.91102 + 8.68589 * LOG(MAG)) * 1000) \ 1) / 1000
PHASE = ((572.96 * ATN(IMAGU / REALU)) \ 1) / 10
PRINT F, GAIN, PHASE
NEXT R
```

Note that correct syntax is vital, and the numbers following the DATA statement must all be typed in on one line even though they cannot be fitted onto a single screen (or page) width.

Frequency	Gain (dB ref. 1 kHz)	Phase (degrees°)
0	19.911	0
10	19.743	–10.4
20	19.274	–20
50.05	16.941	–40.6
70	15.283	–48.4
100	13.088	–54.8
200	8.219	–59.6
500.5	2.643	–52.6
700	1.234	–49.7
1000	0	–49
2000	–2.589	–55.9
2122	–2.866	–56.9
5000	–8.210	–72.1
7000	–10.816	–76.8
10 000	–13.734	–80.6
20 000	–19.620	–85.2
50 000	–27.541	–88.1
70 000	–30.460	–88.6
100 000	–33.556	–89
200 000	–39.575	–89.5

Note that although the amplitude response has been normalized to a 1 kHz reference, the phase response has been left untreated.

From the table we can see that considerable gain is needed at low frequencies, whilst HF attenuation is specified to continue indefinitely, which excludes the series feedback 'all in one go' topology, because its gain cannot fall below unity. Although this failing can be exactly compensated after the feedback amplifier, it does mean that the response before compensation is rising, which means that ultrasonic overload capability within the amplifier is being compromised. Later in this chapter, the requirement for RIAA gain to fall indefinitely will be questioned, but the compromised ultrasonic headroom is still a disadvantage.

Because the 1 kHz level is ≈ 20 dB below the maximum level at LF, any 'all in one go' passive network must have a *minimum* of 20 dB of loss, and probably more, since the network will have the grid leak resistor of the following valve in parallel with it, which will cause additional attenuation. It will be found that it is extremely difficult to design a pre-amplifier of acceptable noise and overload capability using this network, so this topology can also be excluded.

If we should decide to use either of the two previous topologies, the relevant formulae are given in the definitive paper by Lipshitz.[1]

Of the four possible networks that Lipshitz gives, these reduce to two for passive equalization. Of these two, only one has a capacitor in parallel with the lower arm of the network. This feature is important because it allows us to account for stray and Miller capacitance, and it is therefore the only feasible network in a valve pre-amplifier. See Fig. 6.15.

The relevant equations for this *passive* network are:

$$R_1C_1 = 2187\,\mu\text{s}$$

$$R_1C_2 = 750\,\mu\text{s}$$

$$R_2C_1 = 318\,\mu\text{s}$$

$$C_1/C_2 = 2.916$$

These numbers are exact and have not been rounded.

Remember that any grid leak resistor in parallel with the lower arm of the network, or non-zero output resistance of the driving stage, will change the effective value of R_1 as seen by the network. Therefore, the values for the network must be calculated using the Thévenin resistance seen by that network.

Likewise, any stray, or Miller, capacitance will need to be subtracted from the calculated value of C_2.

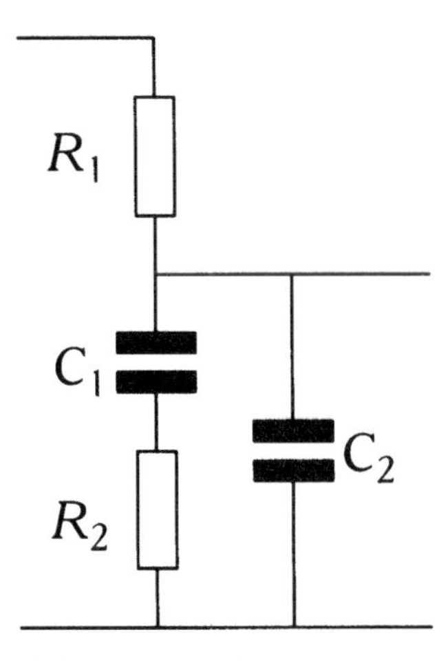

Fig. 6.15 *Passive RIAA de-emphasis network*

For any 'all in one go' topology other than the above network, it is essential to refer to the Lipshitz paper, and read it thoroughly before embarking on design.

We are now left with only two possibilities for equalization: split active and split passive, and we must define how we would split the equalization. Fortunately, there is only one rational way to split the equalization, and that is to pair the 3180 μs with the 318 μs, but to perform the 75 μs separately.

The 75 μs time constant defines a low-pass filter whose –3 dB point is at ≈2122 Hz and rolls off at 6 dB/octave thereafter. This would be an ideal filter to use early in the pre-amplifier since it will allow HF overload capability after that stage to rise at 6 dB/octave above cut-off, which is *exactly* what we need.

It is usual to perform the 75 μs time constant passively following the input stage, which has the advantage of ensuring that the resistance seen by the cartridge is constant with frequency, other than input capacitance. See Fig. 6.16.

The reason for the choice of a passive network is that a series feedback amplifier cannot make $A_v < 1$, and a shunt feedback amplifier would have noise problems. Additionally, although it was not noted earlier, a feedback

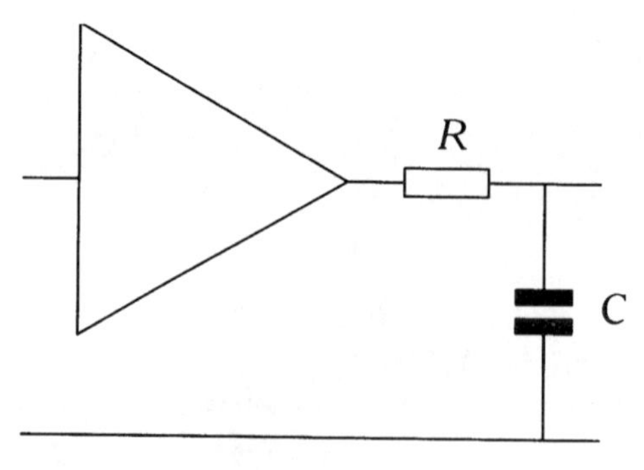

Fig. 6.16 *Split RIAA de-emphasis*

amplifier attempting this response would find its output stage faced with a heavy capacitive load. This capacitive load would demand a large current at HF, and would be equivalent to changing the AC loadline to a far lower value of load resistance, which would result in additional distortion *before* the feedback loop is closed.

It should be noted that *all* of the observations so far are equally relevant to discrete semiconductor or IC-based pre-amplifiers.

The 3180 μs, 318 μs pairing defines a shelf response with a level variation of exactly 20 dB. Using IC op-amps it is equally convenient to perform this actively or passively, but with valves it is more convenient to use passive equalization.

We have now defined the optimum way of achieving RIAA equalization in a valve pre-amplifier.

We shall use passive 75 μs, followed by passive paired 3180 μs, 318 μs over several stages of triodes. All we need now do is to define the topology and operating conditions of each stage, and calculate component values. We are now able to draw a block diagram of the pre-amplifier. See Fig. 6.17.

Drawing the block diagram was useful, because it has further defined the design by requiring a minimum of three stages. Note that the block diagram has completely ignored practicalities such as coupling, or decoupling, capacitors and grid leak resistors. Nevertheless, it represents a simplicity of design to which we should aspire, i.e. DC coupling throughout. This ideal is achievable, but it is not ideal for the novice constructor; we will be a little more cautious in our first design.

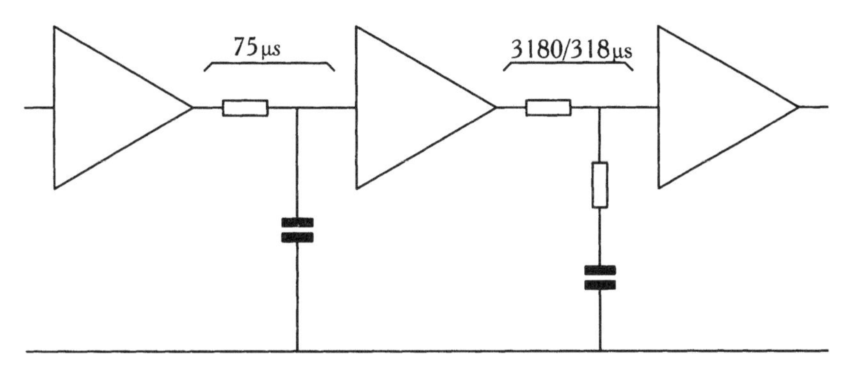

Fig. 6.17 *Block diagram of RIAA pre-amplifier*

Noise and input capacitance of the input stage

Bearing in mind our previous requirements for low noise, the first stage is the *crucial* stage, and must have low noise above almost every other requirement. This is reasonable, because even +34 dB ref. 5 mV is only 700 mV_{pk-pk}, so linearity ought not to be a problem.

Designing for low noise usually means wringing the utmost gain out of the first stage such that noise considerations in succeeding stages are irrelevant. This would imply using a high μ triode such as the ECC83, but with a typical gain $A_v = 70$, this would result in an input capacitance of ≈120 pF including strays, but load capacitance >100 pF tends to cause a peak in the response of a moving-coil step-up transformer.

Many moving magnet cartridges are designed to be loaded by a specific capacitance, older Shures and Ortofons needed 400–500 pF, but more modern cartridges tend to need 250 pF. Once we include pick-up arm wiring capacitance and connecting cable capacitance to the 120 pF contributed by the ECC83, the loading capacitance seen by the cartridge could rise to 300 pF. If necessary, we can easily add a capacitor across the input socket so that the cartridge sees the correct total capacitance, but it is much harder to remove unwanted capacitance.

The ECC83 is probably now out of the running, unless we are prepared to rewire the arm (which might not be such a bad idea), and we are back to the E88CC with a lower gain and lower shunt capacitance; even the ECC82 is usable provided that we take a little care over the noise the next stage.

The earlier generation of octal-based triodes are almost certainly forbidden because of their excessive C_{ag}. The 6SL7GT, which with $\mu = 70$, is the predecessor of the ECC83, and has C_{ag} = 3.4 pF. With a typical gain of 50, this would result in an input capacitance, including strays, of 180 pF. This is now extremely close to our 250 pF limit, and could only be achieved by mounting the entire RIAA pre-amplifier directly below the pick-up arm mounting so that the internal wires of the arm connected directly to the grid.

Mounting the RIAA pre-amplifier onto the plinth, directly below the arm mounting, has enormous advantages in terms of input capacitance, rejection of induced noise, and microphony. It also makes the turntable completely non-standard, and may not even be physically possible, due to limited space, or limited weight carrying ability; a suspended sub-chassis turntable will not take kindly to a pound or two (500–1000 g) of pre-amplifier being bolted to the arm mounting. Conversely, a turntable that is directly mounted on a very heavy plinth, such as a Garrard 301, will hardly notice the extra mass, and would be ideal.

Both the E88CC and ECC82 have an additional advantage in that r_a is low, and, as we will soon see, this helps noise performance. Additionally, a low r_a will be a small proportion of the total resistance that defines the 75 μs roll-off, which then satisfies our earlier requirement of reduced sensitivity to component ageing and changes.

Noise in the input stage is not only determined by the valve, but also by the associated resistors, of which R_L is by far the most important. See Fig. 6.18a.

In order to be able to calculate the noise performance of the stage, we need to redraw the circuit as a simple equivalent circuit, which will make analysis easier. See Fig. 6.18b.

We have replaced the output of the valve with a perfect Thévenin voltage source, and r_a has been included. A moving magnet cartridge can be represented as a resistor in series with an inductor, and since a Thévenin source has zero resistance, we could replace it with a short circuit, and redraw the circuit yet again. See Fig. 6.18c.

We are now in a position to add some noise sources to our equivalent circuit. See Fig. 6.18d.

The derivation of this final equivalent circuit was taken in many steps because the final circuit bears very little resemblance to the original circuit. Before embarking on complex calculations, we can make some important, and useful, observations.

All of the noise sources (with their associated resistances) are in parallel, so a source of zero resistance will short circuit *any* other source, provided that there is no additional series resistance. Typically, we aim for $R_g \approx 100\, r_a$ and

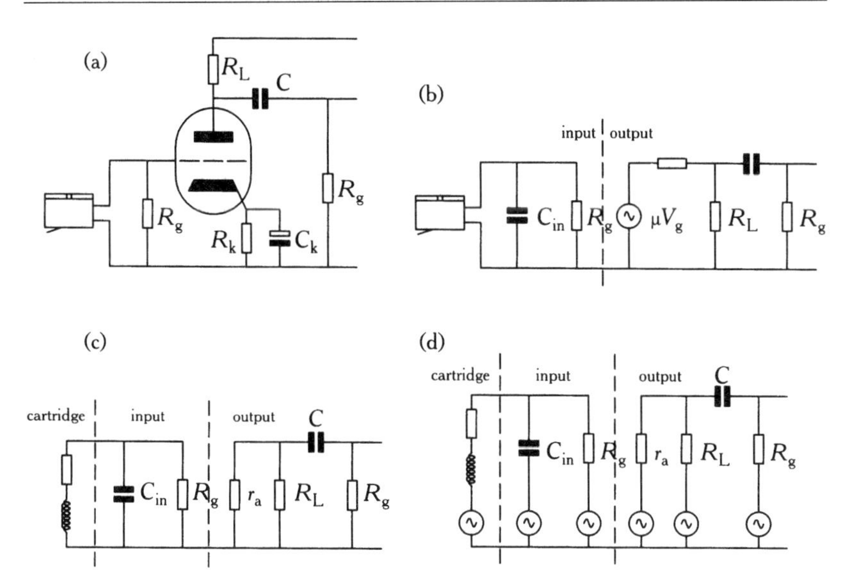

Fig. 6.18 *Noise in the input stage*

$R_L \approx 10r_a$, so r_a will tend to shunt these other sources. This should make the contribution of R_g insignificant, so that any convenient value of R_g could be used, but the series coupling capacitor will reduce the shunting effect of r_a. The reactance of this capacitor is:

$$X_c = \frac{1}{2\pi fC}$$

For a typical grid leak of 1 MΩ, we might use a coupling capacitor of 100 nF to give a –3 dB frequency of 1.6 Hz. If we assume that the lowest noise frequency of interest is 20 Hz (and this is debatable), then we find that at 20 Hz, X_c = 80 kΩ. This is such a high value that it will nullify any possible shunting effect by r_a, until X_c falls to a value lower than r_a.

The result of this is that the usual choice of coupling capacitor does *not* allow r_a to shunt the noise from the grid leak resistor at frequencies below 1 kHz. The resistor therefore produces noise whose amplitude is inversely proportional to frequency ($1/f$ noise), but that rises to the maximum theoretical thermal noise for that value of resistor ($v_n = \sqrt{4kTB}$).

To prevent this excess noise, we may therefore decide to use a value of coupling capacitor sufficiently large that r_a is able to shunt R_g at *all* frequencies, which would require a value ≈10 μF. This would also reduce the noise generated by the capacitor itself, which we had not previously considered. This is a large capacitor, and DC coupling is preferable if possible, but the technique has been used in a number of commercial pre-amplifiers.

Assuming that we have dealt with the grid leak resistor and the coupling capacitor, we are left with the anode load resistor R_L, and the valve itself, which leaves us with a simple equivalent circuit. See Fig. 6.19.

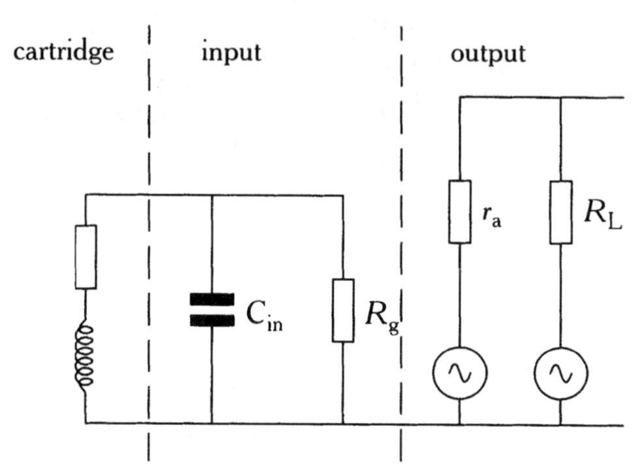

Fig. 6.19 *Final equivalent circuit for noise sources in the input stage*

R_L will generate thermal noise, and if it is a film resistor, it will also generate *excess* noise. Excess noise is generally specified by manufacturers in terms of μV/V of applied DC. We will therefore investigate a typical stage. See Fig. 6.20.

The DC voltage across $R_L \approx 200$ V. A typical 100 kΩ 2 W metal film resistor generates 0.1 μV/V of excess noise, so 20 μV would be generated in this circuit. The thermal noise of a resistor is given by:

$$v_n = \sqrt{4kTBR}$$

where: k = Boltzmann's constant = 1.3805×10^{-23} JK^{-1}
T = absolute temperature in K ≈ °C + 273
B = bandwidth in Hz
R = resistance in Ω

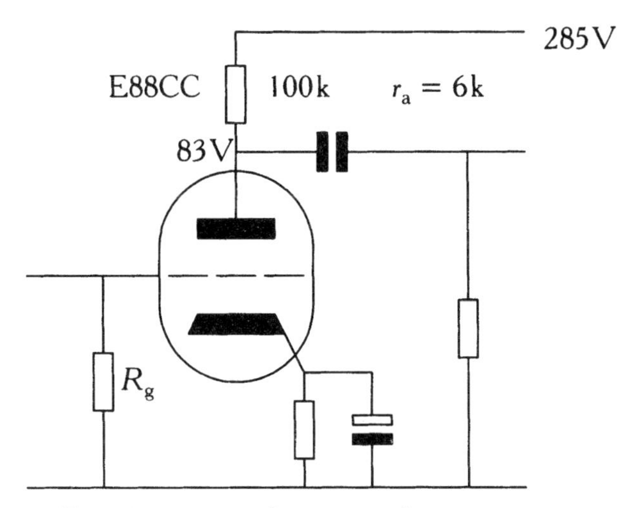

Fig. 6.20 *Typical input stage for noise analysis*

For an ambient temperature of 40°C (313K), with a bandwidth of 20 kHz, this is more conveniently expressed as:

$$v_n = 1.86 \times 10^{-8} \sqrt{R}$$

Using this equation, we find that a perfect 100 kΩ resistor would generate 5.9 μV of thermal noise. In this instance, the resistor's thermal noise has been greatly exceeded by its excess noise. To find the total noise of the resistor, we must add the individual noise *powers*, which, if we remember that $P = v^2/R$, means that:

$$v_{\text{noise (total)}} = \sqrt{v_1^2 + v_2^2 + v_n^2 \ldots}$$

This gives a total noise for the resistor of 21 μV, and was rather tedious, but it demonstrates two points:

- For wirewound resistors we need only calculate the thermal noise (no excess noise).
- For metal film resistors we need only calculate the excess noise. (This simplification works because in practical circuits, as the voltage across the resistor falls, so does its value.)

Now that we have simplified the noise sources in the resistor, we can see how they will be shunted by the r_a of the valve, and we redraw the circuit. See Fig. 6.21.

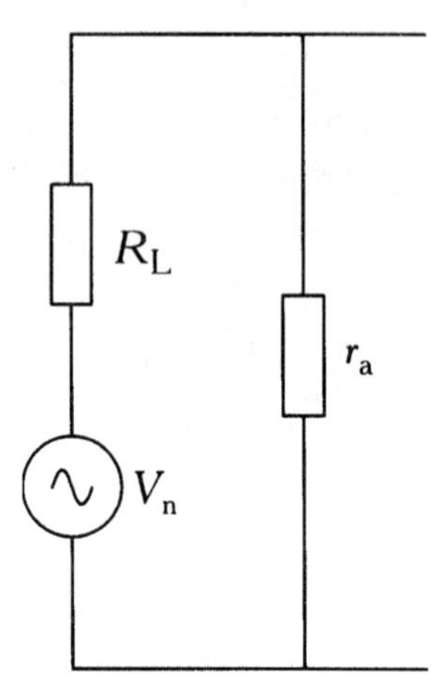

Fig. 6.21 *Effect of r_a on noise produced by R_L*

It is now easy to see that the circuit is a potential divider, and that the actual contribution of resistor noise to the circuit is equal to the open circuit resistor noise multiplied by the attenuation of the potential divider. In our example, this reduces the noise of the resistor from 21 μV to 1.26 μV. It should be noted that if R_k is left unbypassed, r_a will rise dramatically and will no longer be able shunt resistor noise.

If we divide the noise voltage by the gain of the stage $A_v = 29$, we can find the *input referred noise*, which is 43 nV.

Triodes and noise

We should now consider the noise generated by the valve itself. For triodes, there is a handy little approximation:

$$r_{(\text{equivalent noise})} = \frac{2.5}{gm}$$

This says that the *white* noise generated by the valve is equivalent to the thermal (white) noise generated by a perfect resistor r_{eq} at the *input* of the valve. For our example, $gm \approx 5.3$ mA/V, so the equivalent noise resistance would be 470 Ω.

The input voltage noise produced by the valve is therefore ≈400 nV and swamps the 43 nV noise produced by the anode load resistor (as it should in a good design), and we need not sum the noise powers of the valve and the resistor.

We could calculate the thermal noise generated by R_g and the cartridge, but this would be insignificant compared to the valve noise. In any event, we do not have a choice about any of these values since they are set by the cartridge.

Alternative input valves

Although we can calculate the theoretical noise performance of a triode, it does not relate particularly well to practice, and so the exercise might seem futile. However, if we already had a pre-amplifier whose noise performance was deemed acceptable, we could calculate its noise performance and compare it with the predicted noise performance of a new design. This would be extremely useful if the existing cartridge had a worn stylus and a new cartridge of differing sensitivity was being contemplated. Since noise is generally more significant for moving-coil cartridges, the possibilities for the new cartridge are:

Output voltage	**Considerations**
Significantly reduced (–6 dB)	Transformer essential: noise questionable
Identical (0 dB)	Choice: Retain transformer and improve noise, or trade noise against transformer problems
Significantly increased (+6 dB)	Transformer questionable: A quiet valve could make it unnecessary

The assumption is made that the dominant noise source is the input valve, and the *gm* at its operating point is compared with proposed alternatives. Relative noise is found using:

$$N_{(\text{dB})\text{relative}} = 10 \log \frac{gm_{\text{current valve}}}{gm_{\text{proposed valve}}}$$

Example

The pre-amplifier was initially designed to be used with a moving-coil cartridge plus 1:10 step-up transformer to raise the input to the pre-amplifier to 2 mV$_{RMS}$ @ 5 cm/s. A possible replacement cartridge is claimed to produce >500 μV at the same recorded velocity, so it might be possible to dispense with the input transformer. The virtually unobtainable Loctal-based STC 3A/167M and Magnoval-based WE 437A single triodes were designed for telephone repeater amplifiers, and operated with 40 mA of anode current have a *gm* of 47 mA/V. In the initial pre-amplifier, the input valve had a *gm* of 5.3 mA/V:

$$N_{(\text{dB})\text{relative}} = 10 \log \frac{5.3}{47} = -9.5\,\text{dB}$$

The proposed input valve is 9.5 dB quieter, but the proposed cartridge is 12 dB lower output, so the signal to noise ratio will be degraded by 2.5 dB. In practice, the step-up transformer will introduce some loss, 1 dB is typical, so the true deterioration in signal to noise ratio is likely to be around 1.5 dB – hardly noticeable. Although the previous valves are almost extinct, a triode connected E810F can easily achieve a *gm* of 50 mA/V, and rather more stocks exist, so this might be a practical possibility.

Alternatively, we can increase *gm* by connecting a number of devices in parallel, since the noise falls by a factor of $\sqrt{n}$. The LM394 super-match transistor is an extreme example of this technique, as it contains a pair of composite transistors each made of 100 individual devices to give a 20 dB improvement. Paralleling 100 E88CC is somewhat impractical, but a worthwhile if somewhat modest improvement of 4.5 dB can be gained using three devices in parallel. Note that the input capacitance has trebled, and this loading will certainly cause HF peaking if a moving-coil input transformer is used.

Unfortunately, the previous examples all demonstrate an important point. Whilst we may improve noise by a better choice of input valve, or valves, we pay dearly for quite small improvements, since obtaining a high *gm* is expensive and invariably current hungry. It is always better to present the input stage with a healthy signal, rather than hope to cleanly amplify a weak one...

Triodes versus pentodes

If we were to calculate the signal to noise ratio for our example stage using E88CC, for a typical input signal of 2.5 mV, we obtain an answer of 76 dB. RIAA equalization will improve this figure to 98 dB, which is a superb figure. It is also incorrect. The problem is that we have only accounted for white noise, and have not accounted for $1/f$, or flicker, noise. Unfortunately, there is no way of predicting flicker noise for a valve as it is highly sample dependent, although its level is directly related to the level of thermal noise for a given valve.

The EF86 low noise pentode, operating under very similar conditions, but with a noise bandwidth from 25 Hz to 10 kHz, produces an input noise of 2 μV (Mullard specification). Adjusting this figure to a 20 kHz bandwidth raises the figure to 2.8 μV, which means that the EF86 is 17 dB noisier than the theoretical thermal noise of a triode. In practice, a triode will also generate $1/f$ noise, which will slightly erode its superiority over the pentode.

Valve noise summary

Despite all the previous caveats, we can make some useful generalizations that will save unnecessary calculation when designing for low noise:

- Pentodes are noisier than triodes.
- Sample variation of both types is large.
- To render the noise of R_L insignificant, there must be no feedback at the cathode, since this reduces the shunting effect of r_a. This is also true for a μ-follower, even though omitting C_k has no discernible effect on gain. The cascode has $r_a \approx \infty$, and so the noise from R_L must always be considered.
- Maximize *gm* for low noise, either with a single excellent valve, or with a number of lesser valves in parallel. (This maxim holds true whatever the device.)
- Maximized *gm* invariably raises the input capacitance of the input stage and precludes moving-coil step-up transformers.
- Excess noise generally dominates in film resistors passing DC. Wirewound and bulk foil resistors do not have excess noise.
- A very large (typically 100 times normal) coupling capacitor will allow r_a to shunt the noise generated by the grid leak resistor of the following stage, and will itself produce low noise. Even better, we should DC couple if possible.

Together, these noise and input capacitance considerations all but eliminate the ECC83 and other high μ valves from the input stage of an RIAA disc pre-amplifier.

Strays, and implementing RIAA

From our previous arguments we need three individual stages to make up the RIAA complete stage, and we now know that our input valve is likely to be an E88CC, or better. A cascode or a μ-follower are both possibilities, but we shall initially use a common cathode triode stage for simplicity. The second stage can be the same, but the third will need to be a cathode follower for reasons that will become apparent later. We can now draw a circuit diagram for the complete RIAA stage. See Fig. 6.22.

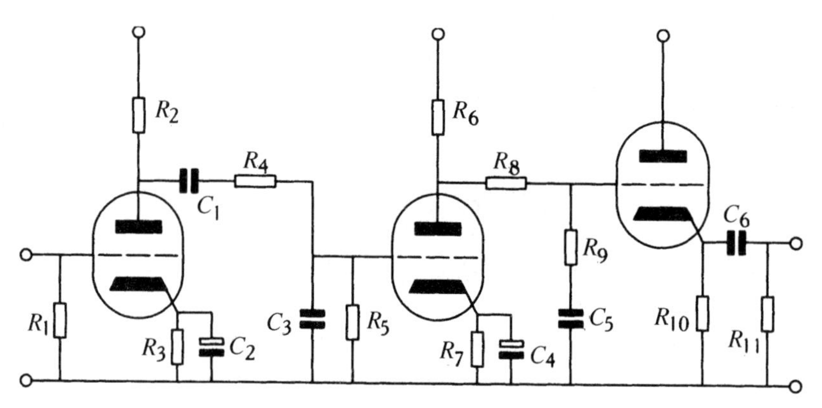

Fig. 6.22 *Basic RIAA pre-amplifier*

The 75 μs HF loss is formed by the combination of R_4, R_5 and C_3, whereas the 3180 μs, 318 μs pairing is formed by R_8, R_9 and C_5. The calculation of these components is simple, but we must remember to account for hidden components. The hidden components are the output resistance of the valve and the Miller input capacitance in parallel with strays.

Calculation of component values for 75 μs

For the DC conditions chosen for our common cathode triode input stage, $r_a = 6\,\text{k}\Omega$, this is in parallel with the $100\,\text{k}\Omega$ anode load resistor, so $r_{out} = 5.66\,\text{k}\Omega$.

The gain of the stage is 29, and C_{ag} = 1.4 pF, so the Miller capacitance will be 30 × 1.4 pF = 42 pF. In addition to this, the cathode, the heaters and the screen are at earth potential, and will be in parallel with this capacitance. $C_{g-k+h+s}$ = 3.3 pF, and we ought to allow a few pF for external strays. A total input capacitance of 50 pF would be about right.

To calculate the capacitor needed for the 75 μs time constant, we need to find the total Thévenin resistance that the capacitor sees in parallel. See Fig. 6.23.

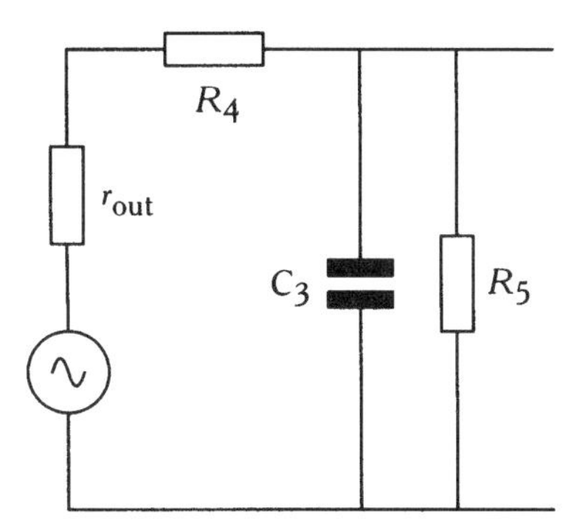

Fig. 6.23 *Determining 75 μs RIAA values*

For the moment, we will ignore C_1, but this will be accounted for later. C_3 sees the grid leak resistor R_5 in parallel with the series combination of the output resistance of the preceding valve and R_4. As is usual, we will make the grid leak as large as is allowed, so R_5 = 1 MΩ.

We are now free to choose the value of R_4. We need r_{out} to be a small proportion of R_4, otherwise variations in r_a will upset the accuracy of the equalization, but too large a value of R_4 will form an unnecessarily lossy potential divider in combination with R_5. At HF, the capacitor C_3 is a short circuit, and so the additional AC load on the input valve will be R_4. 200 kΩ is a good value for R_4, and it has the bonus of being available both in 0.1% E96 series and 1% E24 series (very few E24 values are common to the E96 series). In combination with R_5, this will give an acceptable loss of 1.6 dB, whilst not being an unduly onerous load for the input stage.

The capacitor now sees 200 kΩ + 5.66 kΩ in parallel with 1 MΩ, which gives a total resistance of 170.58 kΩ. Dividing this value into 75 μs gives the total value of capacitance required, =440 pF. But this network is loaded by the second stage which already has 50 pF of input capacitance from grid to ground, so the *actual* capacitance that we need is 440 pF – 50 pF = 390 pF; a 390 pF 1% capacitor would be fine. (Cynics will think, 'How convenient that this came out as a standard value.' Considerable effort was expended in nudging values so that this desirable state of affairs occurred!)

We ignored the effect of the coupling capacitor C_1, but this must have some effect on the Thévenin resistance seen by the capacitor. We could use such a large value that its reactance was negligible compared to the 200 kΩ series resistor, but a more elegant method is to move its position slightly. See Fig. 6.24.

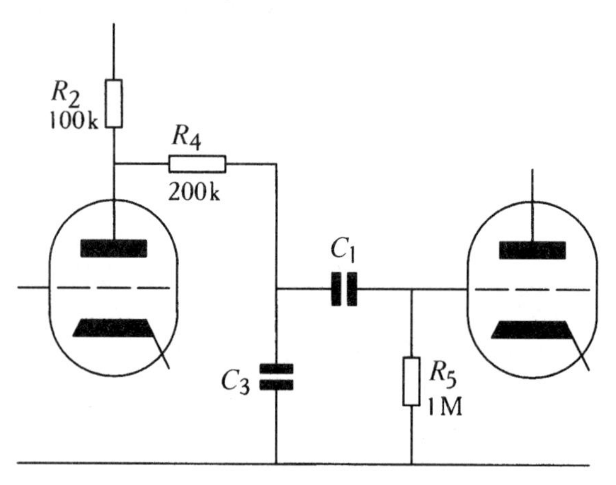

Fig. 6.24 *Moving the coupling capacitor to reduce interaction*

The capacitor now only has to be negligible compared to 1 MΩ. 75 μs corresponds to a –3 dB point of ≈2 kHz, so it is at this frequency that the values of other components are critical. At 2 kHz, a 100 nF capacitor has a reactance of ≈800 Ω, which is less than 0.1% of 1 MΩ. If we had not moved the capacitor, we would have needed a value of 470 nF simply to avoid compromising RIAA accuracy.

There is little point in using a very large coupling capacitor in an effort to reduce noise at low frequencies, since the 200 kΩ series resistance of R_4 will swamp the output resistance of the input valve and nullify its shunting effect on the grid leak of the second valve.

The following program allows you to experiment with resistor values to achieve a convenient capacitor value and also gives the loss at 1 kHz.

```
CLS
X = 1
PRINT "This program finds the value of capacitor required for 75 us CR in the
PRINT "circuit of Fig. 6.24 and calculates the loss at 1kHz. "
PRINT
DO WHILE X > 0
PRINT "What value is the series resistor PLUS the output resistance, in Ohms, "
PRINT "input 0 to stop";
INPUT X
IF X = 0 THEN END
PRINT "What is the value of the grid-leak, in Ohms";
INPUT Y
R = (X * Y) / (X + Y)
C =. 000075 / R
PRINT
D = (((C * 10 ^ 12) * 10) \ 1) / 10
PRINT "C ="; D; "pF"
A = 1 / (2 * 3.14159 * 1000 * C)
L = ((8.68589 * LOG((Y / (X + Y)) * (A / SQR(R ^ 2 + A ^ 2))) * 10) \ 1) / 10
PRINT "Loss ="; -L; " dB "
PRINT
LOOP
```

3180 μs, 318 μs equalization and the problem of interaction

The second stage is direct coupled to the cathode follower, and so we do not need to worry about interaction between a coupling capacitor and the 3180 μs, 318 μs pairing. This is just as well, since 3180 μs corresponds to 50 Hz, which is close to our 1 Hz cut-off; these time constants are sufficiently close that they would interact significantly.

The other reason for using a cathode follower is its low input capacitance. Any stray capacitance across the 3180 μs, 318 μs pairing will cause an additional high frequency roll-off. In the 75 μs network, we were able to incorporate the value of stray capacitance into our calculations, but in this instance this is not possible, and it is therefore essential that stray capacitance is so small that it can be ignored. The full equation for the input capacitance of a cathode follower is:

$$C_{input} = C_{ag} + (1 - A).C_{gk}$$

To a good approximation, the gain = $\mu / (\mu + 1)$, for an E88CC, $\mu \approx 32$, so the gain $A_v = 0.97$. $C_{ag} = 1.4$ pF and $C_{gk} = 3.3$ pF. The C_{gk} term is entirely negligible at 0.1 pF, and so the input capacitance is virtually independent of gain at 8 pF, including an allowance for strays.

The equations that govern the 3180 μs, 318 μs pairing are delightfully simple, $CR = 318 \times 10^{-6}$, and the upper resistor = $9R$, whilst the loss at 1 kHz for this network is 19.05 dB. See Fig. 6.25.

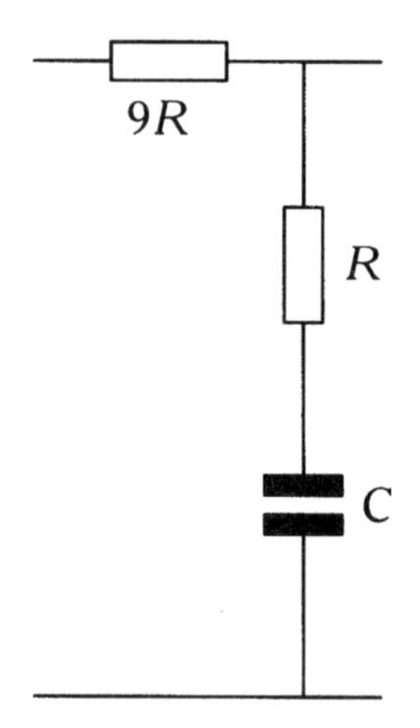

Fig. 6.25 *3180, 318 μs RIAA values*

We should now check whether 8 pF shunt capacitance is sufficiently small not to cause a problem. To do this, we need to employ a slightly circular argument.

We first say that it will *not* cause any interaction. If this is true, then the frequency at which the cut-off occurs will be so high that C in the network is a short circuit. If it is a short circuit, we can replace it with a short circuit and calculate the new Thévenin output resistance of the network. Since the ratio of the resistors is 9:1, the potential divider must have a loss of 10:1, and the output resistance is therefore one-tenth of the upper resistor. If we assume that our upper resistor will again be 200 kΩ (neglecting r_{out} of the previous stage), the Thévenin resistance that the stray capacitance sees *at HF* is 20 kΩ; combined with 8 pF, this will give an HF cut-off of 1 MHz.

As a rough rule of thumb, once the ratio of two interactive time constants is ≥100:1, the response error caused by interaction is inversely proportional to that ratio, and a ratio of 100:1 causes an error of ≈0.1 dB.

In our example, the ratio of 1 MHz to the nearest time constant of 318 μs (500.5 Hz) is 2000:1, so we can safely ignore interaction and go on to accurately calculate the values for the 3180 μs, 318 μs pairing.

If we were driving the network from a source of zero resistance, ideal values for the resistors would be 180 kΩ and 20 kΩ, since these are both members of the E24 series, and the capacitor could then be 16 nF with only 0.6% error. Unfortunately, our source has appreciable output resistance, so we will choose 200 kΩ as the upper resistor and must accept whatever values this generates for the lower two components.

Since we have used an identical stage to the input stage, output resistance is 5.66 kΩ, making a total upper resistance of 205.66 kΩ. The lower resistor will therefore be 22.85 kΩ, and the capacitor 13.92 nF. 22.85 kΩ can be made out of a 23.2 kΩ 0.1% resistor in parallel with a 1.5 MΩ 1%. 13.92 nF can inconveniently be made out of a pair of 6.8 nF in parallel with a 330 pF. We can now draw a full diagram of the pre-amplifier stage with component values. See Fig. 6.26.

Awkward values and tolerances

It will be found that equalization networks and filters *invariably* generate awkward component values and that much manoeuvring is required to nudge them onto the E24 series. This is usually wasted effort, since, although 0.1% resistors are readily available, capacitors are only readily available in 1%. Therefore, for best accuracy, we measure the value of the largest capacitors on a precision component bridge (or perhaps an *accurate* digital multimeter), and add an additional capacitor to achieve the required value.

For the 13.92 nF capacitor needed earlier, we might measure the 6.8 nF capacitors, and find that they were actually 6.74 nF, so we would actually need a 430 pF, rather than 330 pF. This is not a problem, but suppose we had manoeuvred the values so that exactly 10 nF was required, but when the

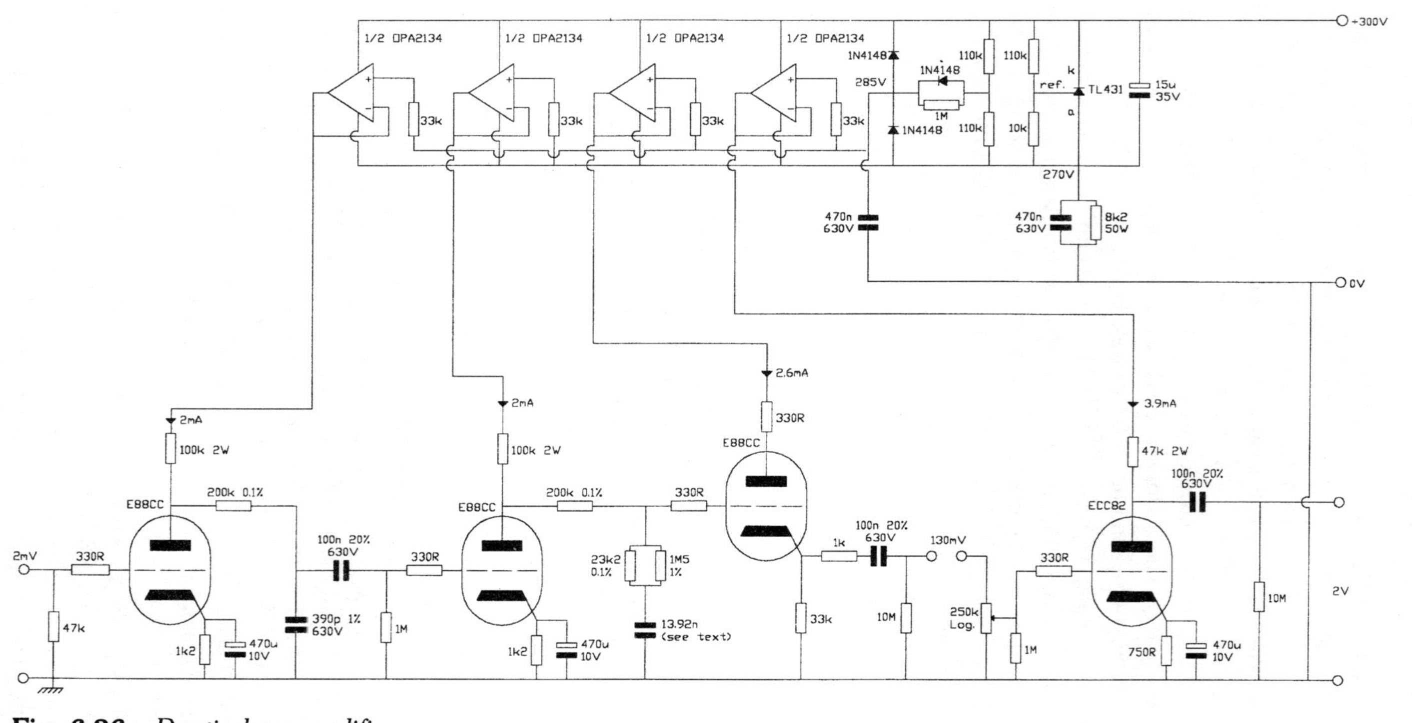

Fig. 6.26 *Practical pre-amplifier*

capacitor was measured, it was found to be 10.1 nF? We can hardly file a bit off the end!

If extreme accuracy is needed, and a component bridge is available, it is better to design for values to be the nearest available value *plus* its tolerance, and then add capacitance to make up the final value. This also neatly fits our 100:1 rule for making perfect capacitors by bypassing with a smaller value.

It is worth considering component tolerance at this point. As a rule of thumb, if we combine a close tolerance component with a loose tolerance component, then the resulting component will *still* be close tolerance, *provided* that the ratio of the values is greater than the ratio of the tolerances. Clearly, the close tolerance component should be the larger component, whilst the smaller component can be looser tolerance.

As an example, we need a 22.85 kΩ resistor to a close tolerance. We choose 23.2 kΩ 0.1%, and parallel it with 1.5 MΩ 1%. 1.5 MΩ/ 23.2 kΩ = 65:1; this is greater than the 10:1 ratio of the tolerances, so this combination will be fine.

Conversely, for the 13.92 nF capacitor needed earlier, the ratio of the primary component to the trimming component is 16:1, so even a 430 pF 10% would be fine. We could probably only buy a 1% component, so there is no need to measure it.

Just because we have adjusted component values on test to meet our exact required value *does not* mean that we now have zero tolerance components. Real components drift with time and temperature, so our values will change. What we have done is to remove calculation error, and real value equals calculated value, which places us in a better starting position for overall tolerance due to drift.

The author's prototype

The prototype of the previous pre-amplifier design was built in a case bought from an electronics junk shop; it cost £3, and originally contained a modem. A new front panel was fitted to mask the holes in the original panel, and the case was sprayed with car paint. See Fig. 6.27.

Originally, the author had intended to fit the RIAA board to the turntable, but when a second Garrard 301 was acquired, this scheme was rejected, and so there are two separate signal boards within the pre-amplifier. See Fig. 6.28.

Fig. 6.27 *The author's prototype*

Fig. 6.28 *The author's prototype (internal view)*

The volume control was a commercially made switched attenuator (found unused in a junk shop for £1!), and the input selector switch mounted near the back panel is a stud switch with alternate contacts earthed to reduce crosstalk. Phono sockets were provided in addition to a 5-pin DIN for the moving-coil disc input, which was wired to be balanced and floating (see later).

The dual colour LED in the middle of the front panel has its green LED lit by the permanently applied heater supply, and the red LED is in series with the lower leg of the HT sink resistor for the op-amps. Switching the pre-amplifier on therefore results in an orange glow similar to the colour of a valve heater, but a pure red glow would indicate LT failure.

Although the PCBs are ground planed (see Chapter 7), the pre-amplifier was initially wired with 30 strand PVC insulated extra flexible wire, but this was later replaced with PTFE sleeved 0.7 mm solid silver. The change gave the best £30 improvement that the author has heard for some time.

Balanced working and pick-up arm wiring

Balanced working is a technique commonly used in broadcast and recording studios to protect audio signals from external electromagnetic interference, and is particularly important for low level signals such as microphones.

A balanced source is simply one where each terminal of the source has balanced impedances to ground. Frequently, the only path to earth from the terminals is via stray capacitances, and the source is then *floating*. Connecting cables for balanced systems therefore have *two* identical signal wires, or *legs*, to maintain this balance, plus an overall screen. The input stage of the following amplifier must have its stray impedances carefully balanced to ground, and is either based on a differential pair (cheap) or a carefully designed transformer (best).

When we immerse the connecting cable in an electromagnetic field, an identical noise current is induced into both wires. The series resistance of the cable is the same on each leg, and the shunt capacitances and resistances to ground are also equal, so the noise current develops a voltage of identical amplitude and phase on both legs at the amplifier input. The common mode signal is then rejected by the amplifier, whereas the wanted audio signal is differential mode, and is amplified.

Typically, a moving-coil cartridge produces $\approx 200\ \mu V$ at 1 kHz 5 cm/s, but before RIAA equalization, the level at 50 Hz is ≈ 17 dB lower at $28\ \mu V$. Achieving our goal of inaudible hum on a signal at this level is not trivial, and we need all the help we can get. The cartridge is inherently a balanced device, so why unbalance it?

We should immediately rewire the output cable of our pick-up arm to maintain this balance by throwing away any coaxial cable. The connecting cable should be replaced by a twisted pair, with overall screen for each channel.

A construction used by the author has a twisted pair of 0.7 mm solid core silver (PTFE sheath), covered with a braid electrostatic screen. Both cables are then threaded down one overall braid screen which also serves to hold the cables together. The braid should not have voids, and most aerial cable is therefore unsuitable. Broadcast quality video cable, or multicore umbilical cable, are both ideal sources of non-voided braid. Once the plastic outer sheath has been removed, the braid will easily concertina off the inner conductors. Finally, the cable should be sleeved with nylon braid to prevent the noise that results from the cable touching a metal part.

Phono plugs should *not* be used for connecting this cable to the pre-amplifier, as they are unbalanced connectors, and a professional quality metal bodied 5-pin DIN plug is ideal, although the cable entry will need to be enlarged. Ideally, the screen should be connected to mains earth at the pick-up arm end, but this is not quite so critical in a balanced system.

Within the arm tube, most pick-up arms twist all four (thin, non- screened) wires from the cartridge together, because this makes the harness far easier to handle. Crosstalk between channels, and hum rejection, can be improved by twisting channels individually as they pass down the arm tube, but reverting to the four wire twist (required for low friction) as the wires pass through the bearings to the output cable. Because this modification primarily affects longitudinal currents, it tends to be of more value to pre-amplifiers with balanced inputs, but is still worthwhile on unbalanced ones. Martin Bastin (of Garrard modification fame) reports that he has been using this method for years.

Balanced wiring is particularly beneficial for moving-coil cartridges, and will help hum rejection even if the pre-amplifier is unbalanced.

Basic pre-amplifier compromises

If we really want to achieve a significant improvement on the basic pre-amplifier, we must look closely at the fundamental design, and reconsider some of the compromises that were initially made.

- Rejection of unwanted signals was not maximized.
- Individual anode currents were set quite low in order to minimize total current consumption, so that the pre-amplifier could be powered from an associated power amplifier. This meant that *gm* for each stage was low, and noise was not minimized.

- Metal film resistors were used in the anode load, resulting in excess noise (although most of this was shunted by r_a).
- The individual stages were kept simple, but linearity was therefore not optimum.

A balanced pre-amplifier

The author feels that the following balanced pre-amplifier has finally reached an evolutionary dead end. Although the ideal would be to keep the signal balanced all the way from balanced cartridge or DAC to the push–pull output stage, this precludes active crossovers (active crossovers are quite complex enough without making them balanced). Nevertheless, a balanced pre-amplfier makes a great deal of sense if passive loudspeakers or electrostatic headphones with a dedicated amplifier are to be used, so the design has been updated. See Fig. 6.29.

The input stage

In order to reap the full benefits of balanced working, a moving-coil step-up transformer for 3 Ω cartridges was especially designed for this pre-amplifier by Sowter Transformers of Ipswich. Correctly terminated, the first batch of type 8055× had a frequency response that was flat ±0.1 dB from 12 Hz to 100 kHz, whilst the HF phase response was pure delay ±1 ° to 50 kHz.

The 8055x transformer also has an electrostatic screen between primary and secondary, and its stray capacitances to ground have been balanced, resulting in excellent rejection of common-mode noise on the connecting wires from cartridge to pre-amplifier.

The first stage has an improved transistor cascode constant current sink to enhance common-mode rejection. Although a 'ring of two' circuit could have been used as a sink for the first stage, each transistor would operate at a very low voltage, which not only makes the circuit more susceptible to RF overload, but also narrows the depletion region within the transistor, causing increased output capacitance. These two factors demanded the use of a subsidiary negative supply, and once this had been accepted, a superior constant current sink using RF transistors could be used, thus making a virtue out of a necessity.

The input transformer is used as the grid leak resistance for the 5842 differential pair. The load seen by the transformer is determined purely by the loading resistor across the secondary windings, and for best sound quality this

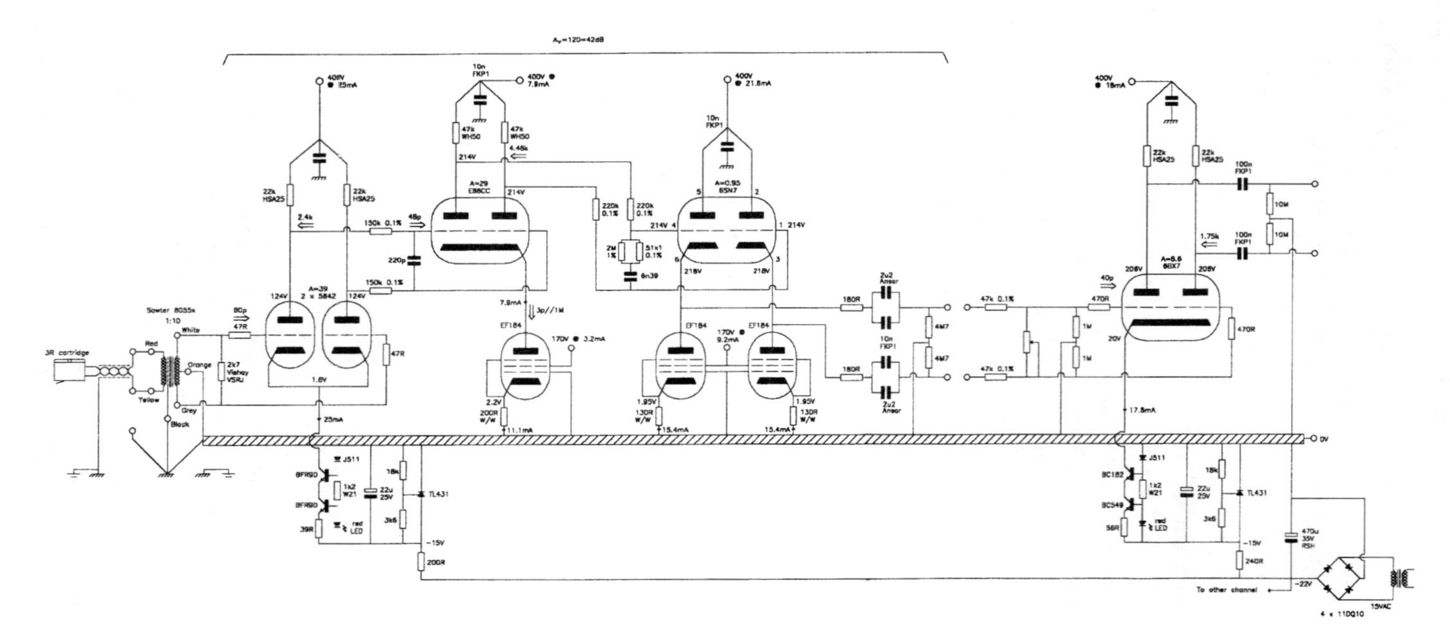

Fig. 6.29 *Fully balanced pre-amplifier*

should be a bulk foil resistor. Because the differential pair is made of valves in individual envelopes, pairs can easily be found that give matched anode voltages, and a DC balance control is superfluous.

If a moving magnet cartridge were to be used, we would need two equal value grid leaks, each of half the value required as a load by the cartridge. See Fig. 6.30.

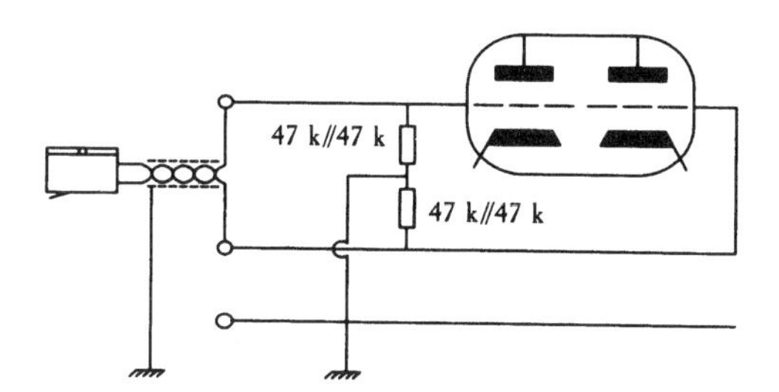

Fig. 6.30 *Arrangement of grid-leaks in balanced input*

One point that may not be immediately apparent is that by connecting the cartridge in balanced fashion to both inputs of the stage, the cartridge or transformer sees the input capacitance of the valves in series, so the capacitance seen is *half* the input capacitance of one valve. See Fig. 6.31.

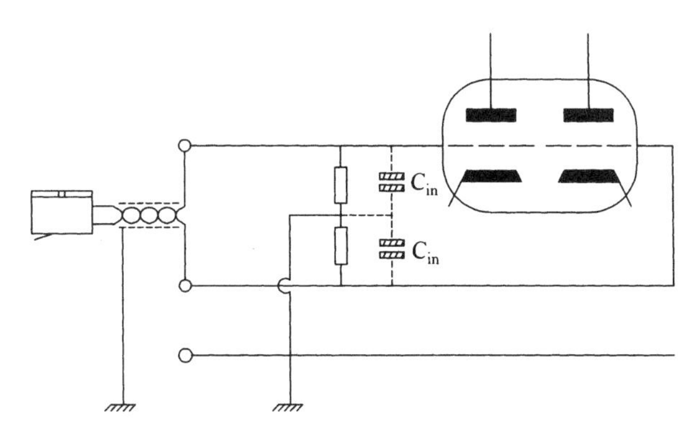

Fig. 6.31 *Reduction of Miller capacitance as seen by cartridge due to balanced connection*

The second stage and 75 μs time constant

In order to direct couple the first stage to the second, the cathode of the second stage must be at an elevated voltage, and a constant current sink seems inevitable. The excellent performance (≈1 MΩ||3 pF), simplicity, and cost of an EF184 constant current sink was extremely attractive, and the noise penalty insignificant. At 20 kHz, 3 pF has a reactance of ≈ 2.7 MΩ, so common-mode rejection within the audio band will largely be determined by the matching of the two halves of the E88CC and stray capacitances.

Because the second stage valve is direct coupled to the first, the second stage does not need grid leak resistors, and we therefore avoid the 1.6 dB *excess* loss that we suffered in the basic pre-amplifier's 75 μs network. Every little helps.

The 75 μs time constant is achieved in a balanced fashion, and the shunt capacitor should be mounted directly onto the valve base with short leads in order to minimize stray capacitance. Similarly, the bodies of the series resistors must be as close as possible to the valve pins, which means that they also perform the function of grid stopper resistors. The best way to understand the equalization is to redraw the circuit as two unbalanced networks. See Fig. 6.32.

The values for R and C are calculated exactly as before, but we observe that we could break the centre tap of our added capacitors away from ground, which leaves two capacitors in series. These can be replaced with a single capacitor of half the value, and a noisy ground is now less able to inject noise into the audio signal. An additional advantage is that there is now no DC across the capacitor, so a lower voltage rating may be used if necessary.

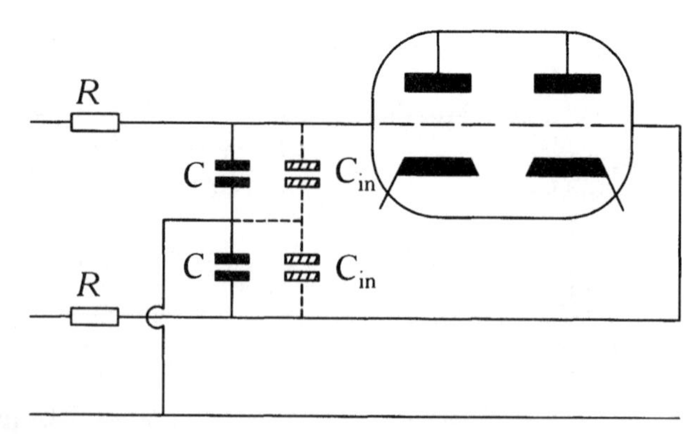

Fig. 6.32 *Implementing 75 μs in balanced mode*

3180 μs, 318 μs pairing and cathode follower

Since this pairing is achieved in a balanced fashion, the value of the capacitor is halved, and it has virtually no DC across it, which makes it much easier to find close tolerance components. Because of the balanced 3180 μs, 318 μs pairing, twin cathode followers are necessary, resulting in a balanced output from the disc stage. The cathode followers have a rather lower V_a than is ideal for linearity, so constant current loads were substituted for resistive loads to remedy this, since an EF184s in its socket, plus a current programming resistor, is cheap. 2.2 μF output coupling capacitors and high I_a in the cathode followers enable the RIAA stage to drive semiconductor equipment gracefully.

The line stage and volume control

A balanced switched attenuator precedes the 6BX7 (see earlier in this chapter for design details). The low μ of the 6BX7 minimizes Miller capacitance and r_{out}, making it ideal as a line stage. The 6BX7 needs a good constant current sink because its low μ would otherwise render CMRR quite poor.

The 5842 RIAA pre-amplifier

Background

Shortly after publication of the first edition of this book, the author's transmission line loudspeakers finally acquired their long awaited active crossovers tailored to individual drive units, enabling a beautifully flat (±2 dB) on-axis frequency response from ≈100 Hz to 20 kHz. (The lower limit of reliable measurement was limited by the separation of the bass driver from the port, and the quasi-anechoic FFT measurement system used.) Unfortunately, even the author baulked at the idea of a balanced active crossover, so the utopia of keeping signals balanced from LP cartridge and (potentially differential output) DAC was reluctantly abandoned. Consequently, the question arose, 'How could the basic pre-amplifier be significantly improved so as not to embarrass the (rather splendid) active loudspeakers?'

The perfect pre-amplifier

The best way of improving any pre-amplifier is to remove it.

It was previously argued that the ideal sensitivity for a power amplifier is $2\,V_{RMS}$, and since this is the maximum possible output of a standard CD player, the two are perfectly matched. All pre-amplifiers degrade sound quality, so why not dispense with an expensive impediment, and connect signal sources directly to the power amplifier? We need only add a volume control and input selector switch to the power amplifier to have the same facilities as before, but with much higher quality.

Modern CDs are carefully mastered so that peaks reach within 1 dB of the $2\,V_{RMS}$ maximum, but older CDs might leave 3 to 4 dB of headroom. Therefore, it might be worthwhile to increase the sensitivity of the power amplifier by 3 dB from $2\,V_{RMS}$ to $1.4\,V_{RMS}$ in order to be sure that older CDs are able to take advantage of the full power available.

The power amplifier is now perfectly matched to CD, but what about other sources?

- *DAB, DAT, DVD, MD, etc.:* All these digital sources tend to produce $2\,V_{RMS}$, or you play them through an outboard D/A which produces $2\,V_{RMS}$, so there isn't a problem.
- *¼" analogue tape:* If you can afford the tape costs, then you can afford a professional machine. +8 dB ref. 0 dBu ≈ $2\,V_{RMS}$, so this isn't a problem. (Incidentally, the industry standard Studer A80 has a superb transport, but less than superb audio electronics, and they are now becoming available at very reasonable prices, making them ideal candidates for valve conversion.)
- *Analogue cassette:* Despite the (discontinued) efforts of Nakamichi, cassette is not a hi-fi medium, and you wouldn't want to achieve full volume for fear of hearing all that tape noise.
- *'Hi-Fi' VCR:* No, it isn't. The 'hi-fi' audio is recorded on an FM carrier using the rotary video heads. Unfortunately, head switching causes a momentary phase change of the FM carrier on replay, which is demodulated as a noise spike. Powerful noise reduction is employed to reduce these spikes, which occur at 50 Hz, but it cannot remove them. Record a classical concert, replay it at a realistic level, and prove this to yourself. Once again, you won't want to achieve full volume with this source.
- *NICAM TV sound:* Although limited to a dynamic range of ≈81 dB, this is potentially quite a good source. Unfortunately, most TV tuners and

VCRs have indifferent D/A convertors, resulting in additional noise, so you probably won't want to achieve full volume with this source. (Some of the older tuners, such as the excellent Arcam D150, used oversampling filters that encoded the data into an SPDIF data stream, which can then be fed to an outboard D/A – see the Appendix for details of the required equalization.)

- *FM radio:* Potentially quite a nice source if the broadcasters could be persuaded not to cane the Optimod (most can't). Many tuners produce quite a low audio level, and stereo Leak Troughlines are the worst offenders. Solution: modify the tuner.
- *LP:* Perhaps you don't have one of these, in which case, there is no problem. Most RIAA stages do *not* produce $2\,V_{RMS}$ at maximum level, and they ought.

Comparison of LP cartridge and CD signal levels

Although CD is specified as providing a maximum possible output of $2\,V_{RMS}$ for an undistorted sine wave, LP cartridges usually have their output voltage specified referred to a recorded (sinusoidal) velocity of $5\,cm\,s^{-1}$ at a frequency of 1 kHz. How can we reconcile these entirely different methods?

The LP is not nearly as tightly specified as CD, and maximum recorded level is totally dependent on the skill of the cutting engineer. Recorded level (and quality) may be reduced by as much as 6 dB in order to allow 40 minutes rather than 20 to be recorded on one side. RIAA equalization further complicates matters, but when equalized, LPs typically have peaks (measured with a Peak Programme Meter) reaching 12 dB above the nominal $5\,cm\,s^{-1}$ line-up level.

The significance of this is that we could now consider a cartridge specified to produce a nominal $2\,mV_{RMS}$ at $5\,cm\,s^{-1}$ as capable of producing musical peaks of $8\,mV_{RMS}$, which when multiplied by the 1 kHz gain of a suitable RIAA pre-amplifier, could give a signal level directly comparable with that of CD.

Pre-amplifier RIAA errors

Errors due to Miller capacitance

We previously argued that the only logical third stage for a valve RIAA pre-amplifier was a cathode follower. This was because it allowed DC coupling and had a low input capacitance, which eliminated interaction and errors to the 3180 μs, 318 μs pairing. If we can tolerate interaction, and have a means of predicting and solving the problem, then this might allow a little more freedom of design choice.

If we want to achieve levels from LP comparable to those from CD, we *must* increase the gain of the RIAA stage. Higher μ valves cause Miller capacitance problems, so the only practical way of increasing gain (without increasing the number of valves) is to substitute a common cathode amplifier for the final cathode follower, which immediately introduces two new problems:

- The stage must have its input AC coupled, so there is interaction between the new LF roll-off introduced by the coupling capacitor and the 3180 μs time constant, causing LF response errors.
- Because the new stage has a gain >1, Miller capacitance becomes significant, and the equalization network will be loaded by a far larger stray capacitance than before, causing HF response errors.

The 75 μs problem

Extended foil polystyrene capacitors were chosen for the equalization networks, since this form of construction lowers self-inductance and ESR. Unfortunately, the commercially available types have a voltage rating of only 63 V_{DC}, so the interstage coupling capacitor between the first and second stage has been forced to revert to its more traditional position, ensuring interaction with the 75 μs equalization.

The Computer Aided Design (CAD) solution

The previous problems can only be sensibly investigated with the assistance of iterative CAD AC analysis. We start by calculating values in the normal way (assuming no interaction), then use CAD to predict the effects of interaction on frequency response using a sweep between 2 Hz and 200 kHz. Assuming that a problem is revealed, we adjust individual component values to seek improvement. Although this sounds laborious, it can actually be quite quick, provided that we think about how and where we make our adjustments.

The easiest way to analyse the circuit is to feed it with a record equalized RIAA signal so that correct replay equalization produces a flat frequency response. Depending on the package you use, it may be possible to directly insert the record RIAA equation (simply the replay equation inverted), or a circuit using perfect components can be simulated to produce the same result.

We have five variables that must be juggled to produce the correct result, so some simplification is needed. We can start by analysing a design that does not have interaction, and then gently modify it, introducing more interactions until we reach our final design. The 3180 μs, 318 μs pairing is most affected by interaction, so we should change these components first.

3180 μs, 318 μs pairing manipulation

- The shelving loss at very low frequencies, <20 Hz, caused by adding the interstage coupling capacitor can be cured by reducing the upper resistor value in the potential divider.
- A shelved response (where the levels of frequencies above 1 kHz are at a constant level different from the levels of frequencies below 250 Hz) can be cured by changing the lower resistor value in the potential divider. If the higher frequencies are at too high a level, then the divider is not attenuating sufficiently, so the lower resistor must be reduced in value, and vice versa.
- A peak in the response centred near 500 Hz can be cured by increasing the capacitor value, whereas a dip can be cured by reducing capacitor value. This result is not quite so easily deduced, but a larger capacitor would increase the time constant, lowering the frequency at which the potential divider takes effect, so that attenuation begins earlier than it should, resulting in a dip in the final response.

The final two adjustments are highly interactive, and an increase in one will always require a proportionate decrease in the other to maintain the correct time constant. It is usually easiest to optimise the resistor first. The model should be tested down to 2 Hz, and the LF roll-off adjusted to emulate a simple 6 dB/octave filter, then optimized for minimum deviation between 20 Hz and 20 kHz.

75 μs/3.18 μs manipulation

Allen Wright[2] points out that at the time of cutting, RIAA pre-emphasis cannot continue indefinitely and a final time constant of ≈3.18 μs is commonly added to prevent excessive amplitude at supersonic frequencies from damaging the (probably Neuman) cutting head. Unfortunately, the value of this time constant varies between cutting head manufacturers, and the less common Ortofon heads use a time constant nearer to 3.5 μs. Nevertheless, it seems reasonable to accept that an electrical 3.18 μs time constant has been deliberately added at the cutting stage in addition to the inevitable mechanical losses within the cutting heads themselves.

The justification for adding a 3.18 μs time constant to the replay equalization has nothing to do with amplitude response, but a great deal to do with group delay and transient response. Uncorrected, the 3.18 μs time constant slightly changes the phase of frequencies above 5 kHz, so that they no longer arrive at the same time as lower frequencies (unequal group delay), and this distorts the transient response. We cannot compensate for the HF cutter resonance, and we probably don't have the data to compensate for the cartridge response, but we can certainly compensate for the hidden 3.18 μs time constant.

An ideal RIAA pre-amplifier should therefore include a final time constant of 3.18 μs, which is physically easily included by adding a resistor in series with the capacitor producing the 75 μs time constant. See Fig. 6.33.

Setting the value of the resistor is considerably more difficult because there are so many other HF roll-offs within the amplifier, most notably the input capacitance of the 12B4-A loading the 3180 μs, 318 μs pairing. Mainly, the additional resistor needs adjustment, but very minor adjustments of the 75 μs capacitor are likely. The model should be tested up to at least 300 kHz, and adjusted to emulate a simple 6 dB/octave smooth roll-off, then further optimized for minimum deviation between 20 Hz and 20 kHz. It may be even necessary to make minor changes to the 3180 μs, 318 μs pairing ...

Production tolerances and component selection

Once the circuit model had been derived and optimized, it was possible to investigate the effects of component tolerance errors. There is little point in specifying close tolerance components, if others with looser tolerances are able to upset performance.

The computer determined the 20 Hz–20 kHz frequency response 10 000 times, each time with random changes in all component values within their manufacturer's tolerance. This technique is known as Monte Carlo analysis, and provided sufficient runs are used, it produces a likely worst case spread of frequency response, in this case ±0.3 dB, using standard component values, without deliberate preselection to obtain optimum values.

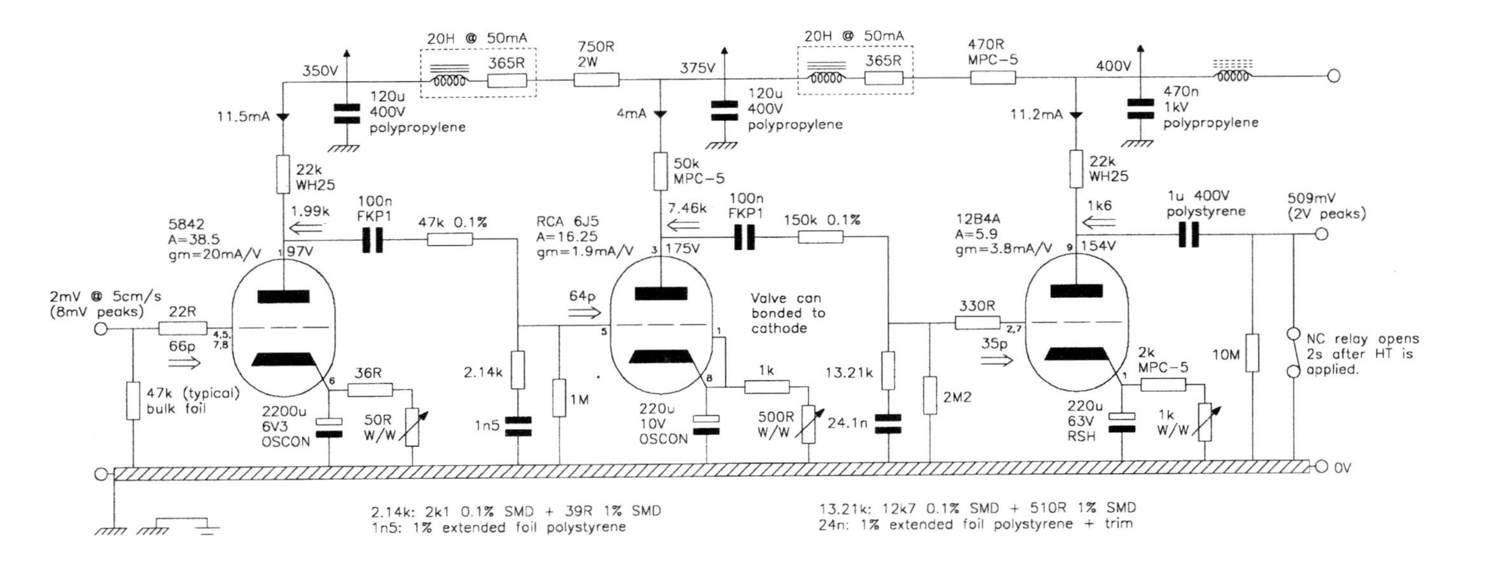

All resistors 1% SMD unless stated.
RIAA includes 3.18us.
2Hz to 280kHz, −3dB.

Errors (20Hz − 20kHz).
Design error: <0.02dB
Monte Carlo spread: +/−0.3dB
All valves halved gm: +0.26dB, −0.19dB

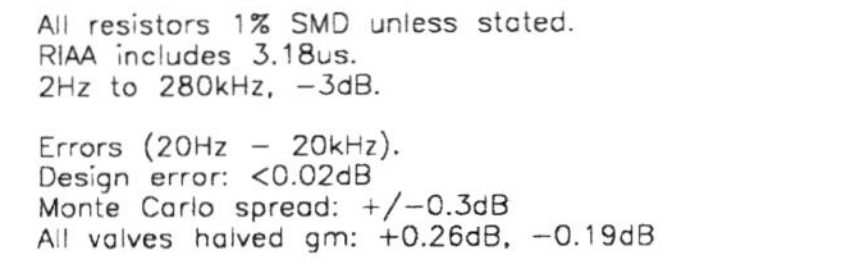

Fig. 6.33 *5842 RIAA pre-amplifier*

Measuring real world RIAA deviation is not very practical – the author's MJS401D can give reliable results to 0.05 dB, but perhaps you don't have a laboratory? The problem can be circumvented by preselecting capacitors using a (much cheaper) component bridge, whilst a 4½ digit DVM allows selection of 0.1% resistors. Even without component preselection, the error with new valves is likely to be well within ±0.3 dB, and preselection can reduce errors to below 0.1 dB.

Errors due to the valves

Finally, we should assume that all of the valves have reached the end of their life, and that their *gm* has halved, doubling r_a (assuming that μ stays constant). Doubling r_a increases output resistances and reduces Miller capacitances due to reduced stage gain. The overall effect is to produce a slightly bumpy tilted response across the entire audio band with 0.45 dB loss at 20 kHz referred to 20 Hz. A less extreme reduction in *gm* to ⅔ of nominal value reduces the error to 0.25 dB.

Despite being designed to minimize the effects of valve tolerances, it is mainly the valves that determine RIAA errors; however, there is even a way around this. Intriguingly, samples of a given type of valve tend to have the same *gm*, *provided* that their anode current is forced to be the same. Thus one valve might need V_{gk} = –2 V to achieve 12 mA, and another might need V_{gk} –3.6 V, but both then have similar *gm*. Therefore, this RIAA stage has variable cathode resistors to adjust bias and set anode voltages to the design values, thus ensuring correct *gm* and r_a, and minimizing RIAA errors.

Some pre-amplifiers using high μ valves, such as the ECC83, have been found to sound audibly different with different makes of valve, giving rise to the belief that a Siemens ECC83 is better (or worse) than a Mullard, when it was actually differing r_a and C_{gk} causing clear errors in RIAA equalization.

Valve choice

Valve choice by *gm* and μ

Despite the design effort required, the 5842 RIAA pre-amplifier is conceptually very similar to the RIAA stage in the basic pre-amplifier.

When the first edition of this book was written, the future availability of valves was in doubt, so 'safe' design choices were made to ensure that a valve could be replaced in the future. There is now sufficient demand for valves that there is an ever increasing number of valve vendors, and we can be rather more adventurous in our valve choice.

A good design (whether pre-amplifier, or power amplifier) should have a graduated transition of μ from high μ in the input stage to low μ in the output stage. This is because high μ in the input stage ensures that noise in succeeding stages is irrelevant, and low μ in the output stage maximizes voltage swing and minimizes r_a, thus minimizing r_{out}. In addition, input stages designed primarily for low noise must maximize *gm* at almost any cost.

As mentioned previously, *gm* can be increased by connecting devices in parallel, but this is not ideal because the valves are unlikely to be matched, so we need a single valve with high *gm*.

Type	**Achievable *gm* (mA/V)**
E810F (triode connected)	≈50
3A/167M, 437A	≈42
5842, 417A	≈20
EC86, PC86, EC88, PC88	≈11
ECC88/6DJ8, E88CC/6922	≈8

The values given in the table are somewhat lower than manufacturer's values because they reflect usable values that can be achieved in a real design. Although the very high *gm* valves are attractive, they are expensive to buy and use (because of the high I_a required), so the plentiful 5842 (μ = 40) was chosen for the input stage.

The dual 6SN7 and single 6J5 (μ = 20) have an excellent reputation, and are ideally suited for the second stage. The third stage valve was chosen equally easily, simply because there are so very few low μ small triodes available – the 6BX7 (μ = 8) and 12B4-A (μ = 6) were obvious choices.

Refining valve choice by heaters

The 6BX7 is current hungry (I_h = 1.5 A), and the 6SN7 is not much better (I_h = 0.6A). Together with the 5842 (I_h = 0.3A), a total of 2.7A would be required from a 6.3 V regulator. This is achievable, but awkward, and a series 300 mA heater chain would be far easier. Fortunately, the 6J5 (I_h = 0.3 A) is suitable, and the 12B4-A can be used as a 12 V 300 mA heater, so the final line-up was 5842, 6J5, 12B4-A.

Apart from the relaxed requirements of the heater regulator, a series heater chain has many other advantages, which are detailed in Chapter 4, not least of which is reduced sensitivity to RF noise.

Summary

The 5842 RIAA stage has been designed to surpass all of the author's previous designs, and is possibly the last RIAA stage that the author will build, since it is hard to see where a substantial improvement can be made. Having optimized the circuit, it is essential to optimize construction, and the prototype is sandwiched between two large (200 × 100 × 40) heatsinks to allow cooling of anode loads and of the 5842 input valves. Initially, a PTFE PCB was considered, but many components were not ideal for PCB mounting, so hard-wiring was finally chosen, allowing easy modification at a later date.

References

1. Lipshitz, Stanley P. (1979) On RIAA Equalization Networks. *Journal of the Audio Engineering Society*, June, Vol. 27, No. 6, pp. 458–481.
2. Wright, Allen (1994) *The Tube Pre-amp Cookbook*.

Further reading

Dove, Steve (Sept. 1980 to Feb. 1982) Designing a Professional Mixing Console. *Studio Sound*.

Fletcher, Ted (Dec. 1981) Balanced or Unbalanced? *Studio Sound*.

Fletcher, Ted and Dove, Steve (Oct. 1983) Development of a Digitally-controlled Console. *Studio Sound*.

Gayford, M. (1994) *Microphone Engineering Handbook*. Focal.

Morrison, J.C. (1993) Siren Song: A Phono Pre-amplifier for Hedonists. *Sound Practices*.

Talbot-Smith, Michael (ed.) (1994) *Audio Engineer's Reference Book*. Focal.

7

Construction and safety

We have looked at how to design circuits using valves, but successful design does not finish with a circuit diagram and a component list.

Once the design leaves the realms of theory, and becomes real components that cost money, practical considerations must be accounted for if the circuit is to work as designed. Whilst good construction will not save a poor design, poor construction can certainly ruin a good one.

In this chapter, we will look at construction techniques, but it should be emphasized that the methods given are not inflexible and are simply a guide; if you can think of a better method for dealing with a problem, then do so.

Comments on electrical safety appear frequently throughout this chapter, and are based on good commercial practice. Since you will not be building, or modifying, equipment in order to sell it, you may feel that you do not need to make so much fuss about safety. Nothing could be further from the truth. Most domestic fires are caused by faulty electrical appliances, and having seen many people's homes, the author is not surprised.

If you build something to a standard that you know to be lower than a reasonable commercial standard, do not be surprised when you find that your house insurance is invalid. It's your home, and your life.

Mechanical layout

Valve amplifiers invariably use a number of large components that need to be positioned relative to one another such that the connecting wires between each component are as short as possible, but that the components do not interfere with each other.

The large components are generally: mains transformers, output transformers, power supply chokes, power supply capacitors, and valves. One way of deciding how to position them is to cut out pieces of paper of the same size and shuffle them around a piece of graph paper.

Even better, the components can be shuffled around the chassis using an engineering drawing package on a computer, with the bonus that a template of the layout can be printed with all fixing holes precisely positioned, saving errors in marking out.

It is vital to make the chassis large enough!

Achieving neat construction on a cramped chassis requires a great deal more skill and patience than on a spacious chassis. There are many considerations that must be taken into account, and it is vital that this stage is not rushed.

Heat

Component position and modes of cooling

Heat is the enemy of electronics. At best, it shortens component life and causes components to drift in value. At worst, it causes fires. And we intend to use valves, which are deliberately heated.

Some components, such as output valves, will be hot, and must be allowed to cool properly. Power valves should be separated from one another by a spacing of 1½ envelope widths or more, and placing output valves in the middle of the chassis is not likely to be a good idea. Output transformers and chokes are usually quite cool, so they can move towards the centre of the chassis. Mains transformers are generally warm, and it is usually best to mount them at one end of the chassis. See Fig. 7.1.

Some small components such as power resistors and regulator ICs unavoidably generate significant heat. Resistors are commonly mounted on stand-offs to allow an unimpeded air flow, and regulators are often fitted with a small finned aluminium heatsink. Both of these strategies are flawed because they attempt to lose heat by convection to the air enclosed by the chassis.

Convection cooling only works if there is a free flow of cooler air past the hot component. Once the cooling flow stops, the hot component is surrounded by still air, which is a good insulator, and its temperature quickly rises. Eventually, the still air begins to lose its heat by conduction to the

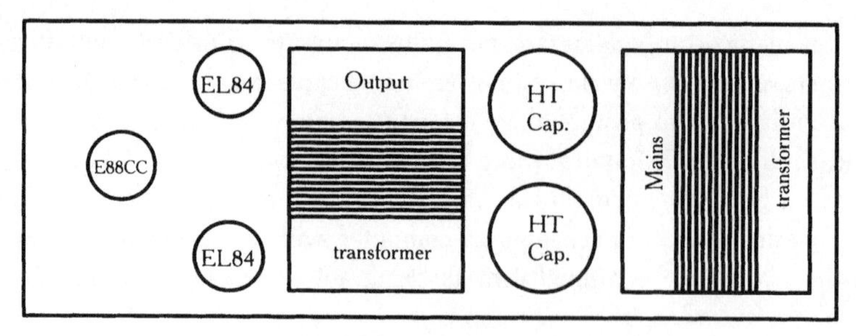

Fig. 7.1 *Layout of power amplifier chassis*

surrounding chassis, and an equilibrium results with a high internal air temperature and a hot component.

A high air temperature within the chassis is undesirable because:

- The components causing the high air temperature are unnecesarily hot, and even though they were designed to withstand heat, their working life is unnecessarily reduced.
- Electrolytic capacitors are extremely sensitive to heat, and their working life is halved for each 10°C rise in temperature.
- Components having a critical value, such as in equalization or biasing networks, will drift way from their optimum value as a consequence of heating from the air.

The best way to cool components is to thermally bond them directly to the chassis, thus ceramic cased resistors should be eschewed in favour of aluminium-clad or TO-220 types which then provide the bonus of convenient tags for other components. With a little ingenuity, even components designed for free air cooling, such as small diodes and valves, can be successfully thermally bonded to the chassis.

A snug fitting black anodized aluminium P clip around a valve envelope will absorb the radiant heat from the anode, which can then be efficiently conducted to the chassis or heatsink, thus cooling the valve. The clip is formed by rolling a strip around a cylinder of somewhat smaller diameter than the valve and opening it out until it grips the valve envelope snugly.

The chassis has a large area and can usually lose heat to the surrounding air quite efficiently. It may seem unnerving to touch a chassis with hotspots due to local heatsinking, but this technique minimizes the internal air temperature, and thus minimizes the heating of sensitive components.

The chassis cools by convection, and this can be assisted by improving the air flow past it. Although most designers recognize the importance of providing adequate ventilation by providing holes in the top of the chassis, the underneath is sometimes neglected. To provide adequate convection, air must be free to enter the chassis as well as to exit; thus, the ideal solution is to make the *entire* underside of the chassis from perforated steel or aluminium, and support it on feet ≥20 mm high.

If necessary, individual components may be cooled even more efficiently by bonding them to a finned heatsink bolted to the outside of the chassis, as is common on transistor amplifiers. Although this technique most obviously springs to mind when considering large power amplifiers or power supplies, precision pre-amplifiers should have their internal temperature rise minimized in order to prevent equalization networks drifting in value.

Wire ratings

Wires have resistance, and as current flows through them, they will self-heat. Excessive current is therefore a fire hazard, and it is important to ensure that the wiring is rated appropriately for the current to be passed. In general:

Conductor diameter (mm)	**Maximum current (A)**
0.6	1.5
1	3
1.7	4.5
2	6

It should be realized that these figures are not absolute, since they depend on the surrounding temperature and ventilation. Component catalogues are good sources of information on the suitability of a particular wire.

Arcing and insulation breakdown should not be a problem at the voltages found within most valve amplifiers, but it is still advisable to maintain 2–3 mm separation between conductors with a high voltage between them, unless each conductor is insulated. As an example, hardwiring with bare wire touching the surface of a capacitor with wound polypropylene tape insulation is not advised, but sleeved wire touching the same capacitor would be unlikely to constitute a safety hazard.

Induced noise

Electromagnetic induction

Almost all of the larger components either radiate a magnetic field or are sensitive to fields. Not all of the primary flux reaches the secondary in a transformer. The leakage flux will induce currents into wires such as valve grids, whether this is significant will depend on the signal level and source impedance at the grid; output valves will be less of a problem than the input stage. Transformers can couple into one another, and hum can be produced by the mains transformer inducing current directly into the output transformer.

Fortunately, the cure for this is reasonable simple. Electromagnetic fields decay with the square of distance, so as the distance between offending items is increased, the interference falls away rapidly. However, simply increasing the physical gap between two adjacent chokes from ¼" to 1" does not materially reduce the interference, because the chokes are typically 3" cubes, and the spacing that applies for the inverse square law is the distance between choke centres, which has only changed from 3¼" to 4", resulting in only 3.6 dB of theoretical improvement.

Rotating transformer cores by 90°, so that the coil of one transformer or choke is not aligned with the other, is far more effective, and typically results in an immediate 25 dB of practical improvement. Even better, if one coil is driven from an oscillator whilst the interference developed in the other is monitored, careful adjustment of relative angles can often gain a further 25 dB.

Pre-amplifiers using choke interstage smoothing should be tested for optimal alignment (because HT chokes are particularly leaky) and should ideally use oil-filled chokes, not because the oil confers any advantage, but because the metal case needed to contain the oil provides electrostatic screening, thus reducing capacitive coupling at HF. Similarly, output transformers should always have their coils at 90° to mains transformers, and if possible be adjusted for optimum alignment.

Although smoothing chokes are gapped, and therefore particularly leaky, they do not generally have an appreciable AC voltage across them, so their AC leakage is low, and they can often be used to screen output transformers from the mains transformer. The exception to this rule is the choke input power supply, which has a substantial AC voltage across it, and will induce currents into surrounding circuitry.

Input valves are very sensitive to hum fields, and should be placed at the opposite end of the chassis to the mains transformer. Induction from push–pull output transformers is less of a problem, because output transformers are operated at a lower flux density, more care is taken over their design, and leakage flux is minimized. However, output transformers for single ended amplifiers are unavoidably gapped, and therefore leaky, so must be kept away from input valves.

Electrostatic induction

Electrostatic coupling is capacitive coupling. If we minimize the capacitance between two circuits, we will minimize the interference. Remembering the equation for the parallel plate capacitor:

$$C = \frac{A \cdot \varepsilon_0 \cdot \varepsilon_r}{d}$$

We should aim to attack all parts of this equation to minimize capacitance. This means keeping wires short and away from other wires. Short wires are only possible if the layout is good.

One approach that is not immediately obvious is that sleeved wires cause higher stray capacitance than self-supporting bare wires because $\varepsilon_r \geq 1$.

The input valve is sensitive, but we can screen this from electrostatic induction by covering it with the traditional earthed metal screening can. Unfortunately, this restricts ventilation, and the valve becomes significantly hotter, dramatically reducing its life. But the previously mentioned P-clip can be made sufficiently wide that it screens the valve quite effectively whilst simultaneously cooling and increasing valve life.

Safety

Although we will look at safety in detail later, one point that should be made now is that electrolytic smoothing capacitors should not be exposed. The voltage on the can of the capacitor is indeterminate, but is generally near to the potential on the negative terminal. Although the can could be bonded to chassis if the negative terminal is at 0 V, this invariably causes a hum loop, so the can is usually insulated from the chassis. The capacitor therefore has only a single layer of insulation between the HT supply and the outside world, but safety calls for either a double layer of insulation, or one layer plus an earthed metal shroud. (See later for explanation of Class I and II equipment.)

Practicalities

The desired layout has to be achievable, and this will inevitably mean compromises. Not only must it be possible to construct the project in the first place, but it should be made so that it can be maintained or modified once completed. Wiring and components should therefore be accessible, although some wiring, such as heater wiring, is most unlikely to be changed, and can usually be obscured with impunity.

Aesthetics

Designing and building equipment from first principles takes time and effort, and you will want to be proud of the results, so try to put some thought into the design. The finished result does *not* need to look like a collision between a rat's nest and a supermarket trolley. See Fig. 7.2.

Fig. 7.2 *Early stage of construction of 10 W amplifier (note layout)*

The chassis

The purpose of the chassis is to mechanically support the components and to *enclose all the dangerous voltages*, thus eliminating the risk of electric shock. The safest form of chassis for the home constructor is the totally enclosed earthed metal chassis.

A traditional method of construction was the folded aluminium chassis, and classic designs included beautiful engineering drawings complete with exact dimensions, folding lines, and all holes marked and dimensioned. The author can only assume that there were many more folding machines available in the early 1960s, and that all constructors had access to a full mechanical workshop. Interestingly, none of these drawings made provision for a cover at the bottom of the chassis, so live parts could be contacted if the chassis was lifted whilst power was applied.

Steel is not suitable for the chassis of valve amplifiers. Steel is magnetic, and will allow leakage flux from transformers to flow through the chassis and induce currents into the pins of the valves. If a steel chassis is unavoidable, induction into the chassis can be greatly reduced by fitting a non-ferrous gasket between transformers and the chassis; 1.5 mm Paxolin is ideal.

Although it is sometimes possible to buy an undrilled folded aluminium chassis, it needs to be 1.6 mm (16 swg) thick in order to be able to support the weight of the transformers. Although a pre-folded chassis might seem convenient to use, it is always awkward to drill holes in the sides of the chassis, because it is difficult to support the metal whilst it is being drilled.

A far better alternative is to make the chassis out of separate pieces. See Fig. 7.3.

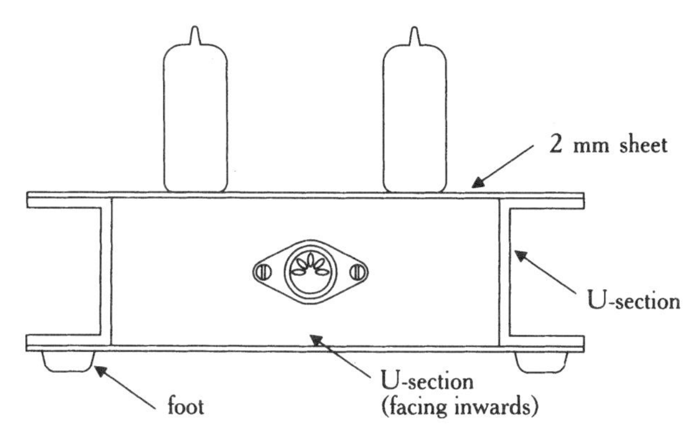

Fig. 7.3 *Making a chassis from U-section extrusion and sheet*

The top plate, to which most of the components will be fitted, is made of 2 mm aluminium, which is readily available either as off-cuts from an aluminium stockholder or from one of the electronic factors. Whilst it is tempting to use even thicker metal, many of the holes will be cut using chassis punches that can be damaged by thicker metal. Additionally, most valve holders were designed for 1.6 mm chassis, and whilst they can tolerate 2 mm, clearances become limited if the plate is thicker.

The front, back and sides are made from U-section aluminium extrusion cut to length. The sides have the U-section facing out, thus providing convenient handles with which to lift the chassis, and the front and back fit into the remaining space between the sides. The whole construction is then fastened with engineering screws and nuts. This form of construction has many advantages over the folded chassis:

- The chassis can easily be made to any convenient size using hand tools. It need not even be rectangular!
- Cutting holes in the chassis is now easy, because each surface can be properly supported whilst it is being worked upon.
- If modifications are required later (not uncommon), individual parts can be replaced if necessary.
- Aluminium U-section tends to be quite thick, making it a good heatsink, and threaded holes may be tapped into it, which is often convenient.
- If access is needed at one edge, that piece of U-section can be temporarily removed.
- Looking at the bottom of the chassis, the U-sections are rigid load bearing members to which feet and the safety cover plate can be easily fixed (which should ideally be perforated, to allow a cooling air flow). Fitting a cover plate to the bottom of a folded chassis is usually rather more difficult.

There are only two minor disadvantages. First, the total top area is a little larger than the folded aluminium chassis, because some space is wasted at the sides by the outward facing U-section. Second, for safety, each separate piece of aluminium should be reliably earth bonded to the top plate with star washers at one of the fixing points.

Metalwork for poets

Tools

Many electronics enthusiasts hate metalwork. If they were to take a longer and more thoughtful look at their pet hate, they would realize that what they actually hate is attempting to do metalwork with *poor tools*. An experienced craftsman can produce good work even with poor tools, but would rather use the best. Beginners do not have this level of expertise, and need all the help that they can get, so they *need* good tools. Buy the best that you can't quite afford; good tools will last a lifetime, and will not only be cheaper in the long run, but they will be a pleasure to use.

Marking out

This is where the mistakes are made, so don't rush; ten minutes saved here could cost hours later.

Metal is traditionally marked out using a *sharp* engineer's scriber, in conjunction with a 150 mm engineer's try square, and a clean 300 mm grey steel rule. See Fig. 7.4.

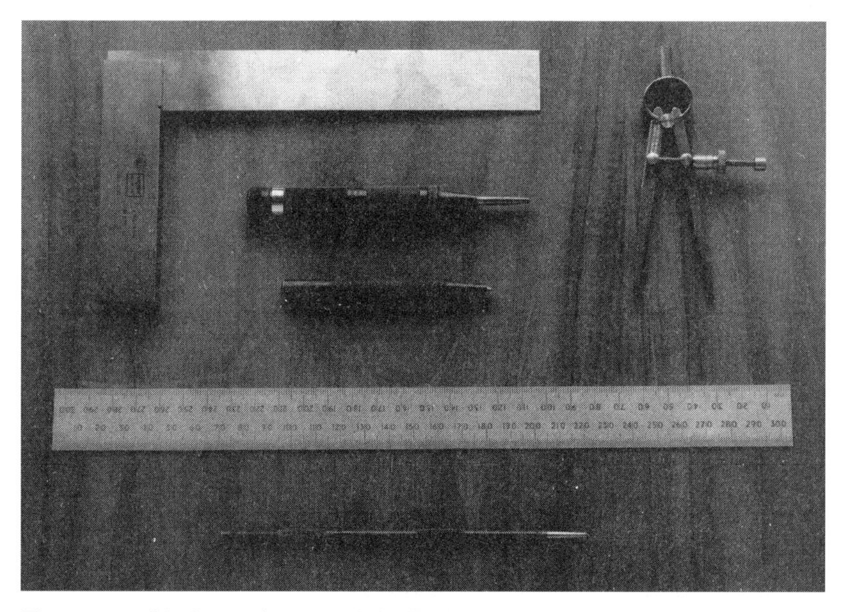

Fig. 7.4 *Clockwise from top left: Engineer's 150 mm try square, dividers, scriber, 300 mm steel rule, manual centre punch, autopunch*

With the best will in the world, your marking out will never be perfect, so choose a reference edge from which to measure, and only use the try square on this edge to minimize errors.

The centrepoint of drilled holes is marked by the intersection of two lines. With all the construction lines that you will need, there will be a lot of these intersections, so when you are marking the position of a hole, use a scriber to lightly draw a circle around the intersection, of roughly the same diameter as the hole. This will prevent you from drilling holes in the wrong place, and may stop you drilling the hole oversize. If you have a pair of dividers, use these to accurately mark the size of larger holes – any mistakes in marking out will instantly become apparent.

You will often need to cut irregularly shaped holes for transformer connections. Cross-hatch the metal to be removed with a marker pen to avoid confusing construction lines with cutting lines.

Alternatively, you can rely on the precision of a printer to do your marking out for you. Most modern printers print with a resolution of at least 300 dots per inch (dpi), and often far more, and are thus much better than you at positioning lines, so an engineering drawing package can easily produce a precise template that just needs to be punched through to ensure perfect marking out.

At this point it is still possible to correct mistakes, so offer up the various parts to be fitted, and check that the marking out looks sensible.

If it still looks good, use a centre punch to lightly indent the centres of all the drilled holes. The modern punch is the automatic spring loaded punch, whereas the older hand punch has to be lightly struck with a hammer. Although the autopunch is superficially attractive, it is actually less accurate than the hand punch.

Whichever punch you use, it must be *sharp*, and should be ground to an included angle of 90°C.

Centre punching sheet metal is noisy. Noise can be minimized, and accuracy increased, by supporting the area to be punched directly above a leg of the bench or table. This work is being done for the pleasure of your ears, so wear earmuffs.

Drilling round holes

Drilling triangular holes in sheet metal is remarkably easy; drilling round holes in the correct position takes a little more care.

There are four basic requirements for drilling round holes in the correct position:

- Do not attempt to drill to the final size using only one drill. Use a *pilot* drill (≤2 mm) to drill the first hole, and then enlarge it with larger drills. This ensures that the larger drills follow the correct course. If the pilot should drift off course, use a round needle file to enlarge the hole until it is in the correct position, and then continue drilling.

- Use sharp, correctly ground drills – *never* attempt to re-sharpen drills (if you genuinely know how to do this properly, you shouldn't be reading this chapter). The best way to buy drills is to look through a model engineering magazine for the mail order suppliers – most of them will send you a free catalogue. All of them stock sets of drills by quality manufacturers, but individual drills are expensive, so buy a set of 1 mm to 6 mm in 0.1 mm steps – another from 1 mm to 12 mm in 0.5 mm steps is useful. These are virtually all the drills you will ever need, and they will come in a protective steel box with each size marked. See Fig. 7.5.

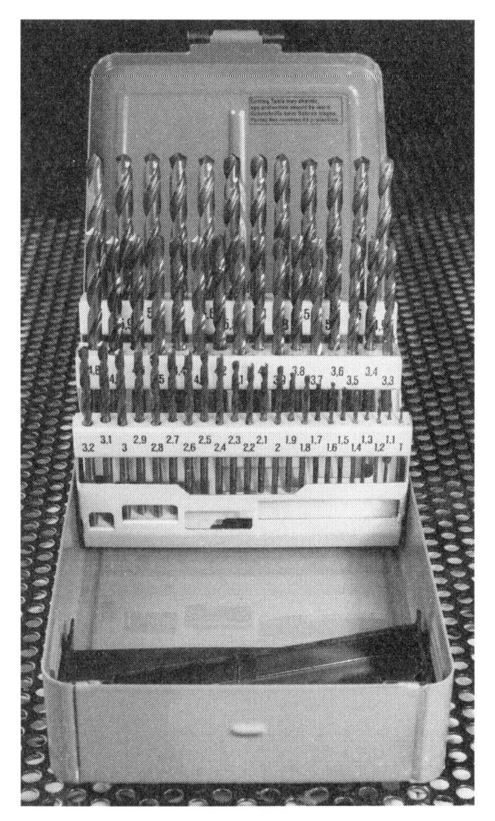

Fig. 7.5 *Set of engineers twist drills as supplied in steel case*

- Use the correct drill speed. A 1 mm drill needs to run fast to clear the swarf or it will break, so it should run at 2500 rpm, whereas a 12 mm drill should be run at the slowest possible speed, 200 rpm or lower. This is not nearly as much of a problem as it used to be for the amateur, because the better quality power drills have a two-speed mechanical gearbox and electronic speed control.

- Use a drill stand bolted to your bench. Again, this is much easier than it used to be, because drill manufacturers have standardized on a 43 mm collar to grip the drill in the stand, so a choice of stands is available. Alternatively, small drill presses are now available at remarkably low prices, and although they will not withstand comparison with a genuine workshop drill press, they are quite good enough for amateur work. If you have never used a drill in a stand before, you will not believe how much easier it makes your work!

Deburring

When you drill a hole in metal, there will always be a small burr on the upper surface, and a larger burr on the lower surface, which will need to be removed. All metalworkers have their preferences, but a rose countersink in a handle, or a specialized deburring tool, are the author's favourites. See Fig. 7.6.

Fig. 7.6 *Top: Rose deburring tool. Bottom: 'Wiggly' deburring tool*

At a pinch, you can use a large drill without a handle to deburr holes, but you will find that the flutes of the drill will tend to cut the surface of your skin as you grip it.

Drill safety

Work tidily. A tidy workshop will inspire careful work but an untidy workshop is dangerous.

A drill mounted in a stand, bolted to a bench, invariably means that the operator bends, or sits down to carry out the work. Drilling produces swarf which flies out from the drill, this swarf may be hot (although if it is hot whilst you are drilling aluminium, then something is wrong). You are now at eye level to the swarf, so wear safety glasses. Your ears are near your eyes, and drills are noisy, so also wear ear muffs.

When you drill sheet metal, the work will try to vibrate, and if it is allowed to do so, the drill will snatch at the work, and you will suddenly find the work spinning on the end of the drill. This is most alarming, and can be very dangerous.

There are a number of precautions for preventing drill 'snatch':

- Clamp the work. If the work is firmly clamped near the drilled hole, then the drill cannot cause it to spin and rip your fingers open. Unfortunately, this may not necessarily prevent the work being partly pulled up the shaft of the drill and being deformed.
- Use the correct drill speed. Too slow is far better than too fast.
- *Always* support the work on a piece of wood. An off-cut of chipboard, or similar, is ideal because it is *flat* and supports the work evenly. Always position the drill above a fresh area of board so that the work is supported directly below the drill. In time, your drilling block will become a Swiss cheese, but by then you will have some more off-cuts from your new loudspeakers.
- Make sure that the work is firmly in contact with the drilling block by deburring the underside of each hole immediately after it is drilled.
- If you must hold the work by hand, press down firmly, and avoid having your fingers near a part of the work that would cut you if the drill snatches. If the worst comes to the worst, do not try to fight the drill; it is much stronger than you are, let go, and switch the drill off. Ideally, you should have a foot operated stop switch, but few amateur workshops can afford this level of sophistication.
- A little lubrication will work wonders. The traditional *cutting fluid* was a 50/50 mix of lubricating oil and water with a healthy dash of washing-up liquid. This aids cooling and cutting, but it is extremely messy, and rusts tools. Aerosol lubricants are not quite as good, but far less of a problem. Only a little is required; too much will spray yourself and the surroundings.

Sheet metal punches

You should not attempt to drill holes larger than 9 mm in sheet metal, it is simply asking for trouble.

The solution for round holes is to use a sheet metal punch. Although these are not often available from high street shops, electronics factors and engineering suppliers stock them. See Fig. 7.7.

Fig. 7.7 *Selection of chassis round punches*

The punches are of a two-part construction that are drawn together by an Allen bolt. Provided that they are kept well greased, they cut a beautifully neat hole, and last for years. Standard sizes are:

- $\frac{3}{8}$": Imperial potentiometers and rotary switches, grommets for small cables.
- $\frac{1}{2}$": Large toggle switches, 32 A loudspeaker terminals, larger grommets.
- 16 mm: DIN sockets.
- $\frac{3}{4}$": Most B9A valve sockets (modern ceramic sockets may be larger), some cable clamps.
- $1\frac{1}{8}$": Phenolic International Octal sockets (again, check the size of ceramic types).
- 35 mm: Large capacitors.

Larger punches need quite a large hole for the bolt, so there is no reason why you should not use the ⅜" punch to cut the bolt hole for the 35 mm punch.

If, as was suggested earlier, you have used a pair of dividers to draw the exact position of the finished circle, you can now align the punch before tightening up on the Allen bolt and cutting the hole.

Note that although chassis punches produce very little burr, they do slightly deform the surface from which the cut began. It is therefore usual to punch from the inside of the chassis to the face side to avoid this deformation being visible. Additionally, the pressure of the supporting face on the work can mark decorative surfaces (such as brushed anodized aluminium), but this can be avoided by placing a thin cardboard washer between the punch supporting face and the decorative surface.

Making small holes in fragile thin sheet

We often need to mount parts separated by a very thin insulating spacer, and although the material can be cut neatly to size with a scalpel or sharp scissors, making clean small holes for screws to pass through is a problem. This quandary can often be solved by a paper punch. Mark the position of the hole, take the bottom off the punch, tip out the chads, and use the punch upside down. Gripped gently, the cutter will hold the material in place, and the marking out can be clearly seen through the exit hole of the punch. The material can be moved precisely into position, whereupon a beautifully clean hole in the correct position may be punched.

Sawing, and irregular shaped holes

It might be thought that to cut a piece of metal, it is only necessary to take a few wild swings at the work with a hacksaw whilst the room rings to the screech and shudder of the saw.

This is an excellent way to ruin a perfectly good hacksaw blade, deafen yourself, and produce work of an appallingly low standard. Before using a hacksaw, check:

- Is the blade inserted the correct way round? (It should cut on the forward stroke.) See Fig. 7.8.
- Does it have a complete set of teeth? If any are missing, discard the blade, blades are cheap – your time is not.

Fig. 7.8 *Close-up of hacksaw blade (note direction of teeth)*

- Does it have the right number of teeth per inch (TPI)? A saw should have three teeth in contact with the work at all times.
- Is the blade properly tensioned? (The wingnut should be as tight as possible.)

Cutting sheet metal is a problem because a hacksaw blade is insufficiently fine to cut at right angles to the work, so the only way to cut metal sheet is to cut at an extreme angle, and if this means crouching on the floor whilst the work is held vertically in a vice, so be it. A better method is to clamp the work horizontally to the bench, and use a *panel* saw, which looks like a wood saw, but takes a hacksaw blade. See Fig. 7.9.

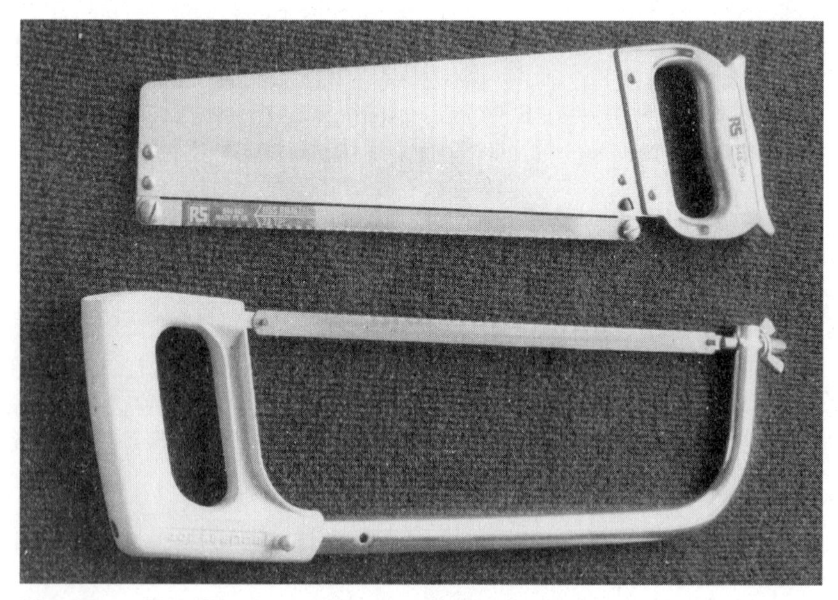

Fig. 7.9 *Top: Panel saw. Bottom: Hacksaw. (Both take identical size hacksaw blades.)*

A drop of lubricant whilst sawing will do wonders for your cutting efficiency.

The best way to cut irregular holes is by hand, using a tension file in a standard hacksaw frame. See Fig. 7.10. The process starts by drilling a hole in the material that is to be removed. The file cuts on the forward stroke, so run the file gently through your fingers to determine the cutting direction. See Fig. 7.11. Now fit the file to the handle end of the frame, but pass the file blade through the hole, before fitting it to the other end of the frame and tensioning it.

With care, the hole can be cut so accurately that very little remedial filing will be needed. Sometimes the frame of the hacksaw will be unable to reach the proposed hole from any direction . . .

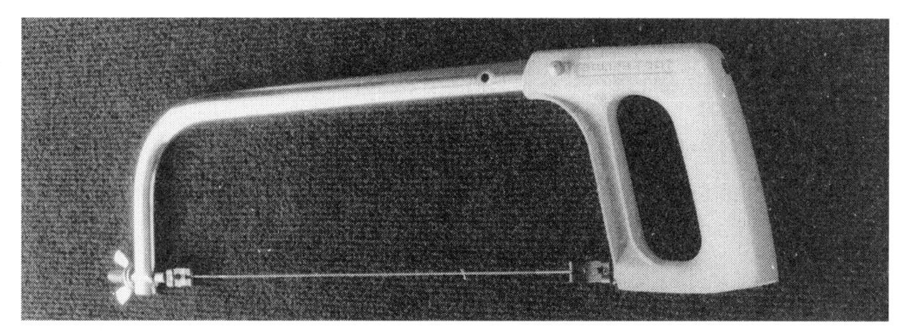

Fig. 7.10 *Tension file in hacksaw frame*

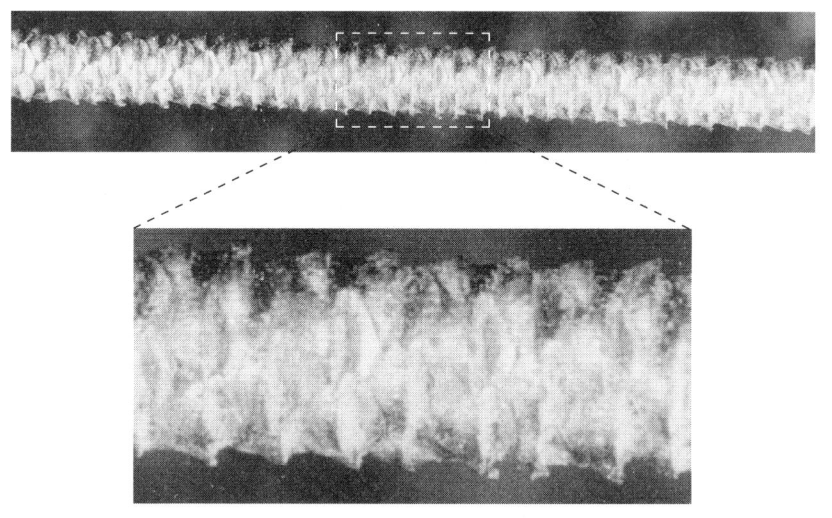

Fig. 7.11 *Close-up of teeth of tension file (note direction of teeth)*

The alternative is to use an electric jigsaw at low speed with a *metal* cutting blade with the finest possible teeth. This is not nearly so easy to control, and is potentially dangerous. It is also extremely noisy, and earmuffs are essential. The drumming of the saw will leave marks on the work unless a piece of thin cardboard is carefully fitted to cover the sole of the saw.

As before, a hole is drilled in the material to be removed and the saw blade passed through. Ensure that the teeth of the blade are *not* touching the edge of the hole, and whilst pressing down firmly, start the saw. When you reach the end of a cut, back the saw off a little before switching off, or the teeth will snatch and the saw will try to jump up from the work.

Finishing

Aluminium can be spray painted, but paint does not stick very well to aluminium, and subsequently chips off. Buying aerosol cans of car primer and top coat is quite expensive, and the fumes are most unhealthy. Nevertheless, this is one way of finishing the chassis.

A much better method, but one that requires rather more planning, is to have the chassis anodized by a professional anodizer. Note that only aluminium can be handed to an anodizer, no foreign substances whatsoever are allowed. Surprisingly, this is actually quite cheap, because the pieces that you will hand over will be very small compared to the main batch that is being anodized. It may mean that you need to wait until a batch of your chosen colour is to be anodized, but the finished result will be far superior, provided that you have prepared the work properly.

Both painting and anodizing show up every imperfection of the underlying surface, so the surface cannot be too well prepared. A 'brushed' finish can be obtained using reducing grades of silicon carbide (often known as 'wet and dry') rubbed along one direction only. If the final stage is lubricated with soap and water, a very smooth finish can be obtained. Alternatively, soap-filled wire wool scouring pads soaked in hot water can be very effective.

It is far better to begin with a good surface, so most stockholders keep aluminium that has one face protected with a plastic film. Keep the film on for as long as possible, and do not allow objects to be dropped on the sheet; aluminium is soft and easily dented. When using the scriber, keep your construction lines to a minimum, and score lightly with a *sharp* scriber; the marks that a blunt scriber makes are much more difficult to remove.

Whether you paint or anodize your chassis, make sure that you really have drilled *all* the holes you need. Drilling holes afterwards is invariably messy, and will easily spoil your finish.

Soldering

Soldering irons

Obviously you need a soldering iron, but what is the difference between them, and why are some so cheap?

The job of the iron is to heat the parts to be soldered to a temperature such that as the solder is applied, it melts quickly and flows to form a perfect joint. Almost anything will do this, but the component may not work afterwards. Two thermal properties characterize the iron: thermal mass and temperature. Thermal mass is simply the mass of the hot part of the iron, and the larger this is, the more difficult it is for the proposed joint to cool it.

A cheap iron determines its temperature by only generating sufficient heat to match its losses to the environment, and to keep the tip hot enough to melt solder. As soon as it is touched to the joint, it begins to cool. If the work is not to cool the iron down so much that it is unable to melt solder, then the iron must contact the joint at a rather higher temperature. A higher thermal mass helps here, but makes the tip clumsy to use. The upshot of all this is that it usually runs *too* hot and burns the flux in the solder, and may well damage the components. It will almost certainly cause tracks on printed circuit boards to lift if used for desoldering.

A better iron is thermostatically temperature controlled, and has an oversized element (typically 50 W as opposed to 12–25 W). The iron is at the correct temperature all the time, and if the joint begins to cool the tip, the thermostat trips and the oversized element has no trouble in restoring the correct temperature. Additionally, temperature controlled irons are generally low voltage (usually 24 V), which makes them longer lasting (the wire of the element is thicker, and less liable to break). Because the tip doesn't overheat, it too lasts longer and is easier to keep clean.

All soldering irons suffer from leakage current between the heating element and the tip. The electrical insulation between the element and the tip must be thin in order to transfer heat quickly and efficiently, but because it is thin and hot, this insulator is not perfect. (Insulators become more leaky as temperature rises.) This leakage current is determined by the electrical resistance of the insulation and the voltage applied to the element, so low voltage irons are far lower leakage than mains irons. This point is significant because semiconductors, and particularly CMOS digital ICs, can be damaged by the leakage current from a mains iron.

Because low voltage irons need a mains transformer and are more complex, they are invariably more expensive than the cheap mains irons. However, they will pay for themselves in time, because elements and bits last far longer, and they are less likely to damage a printed circuit board or an IC.

Remember, it is the *board* that is expensive, not the components. Individual components can always be replaced, but if the board is wrecked, then everything has to be replaced.

Busbars should ideally have a large cross-sectional area to reduce electrical resistance, but this inevitably makes them difficult to solder because they conduct heat away so efficiently that it takes considerable time for the iron to raise a part of the bar to soldering temperature, by which time the nearby polystyrene capacitor or IC has already been destroyed. The solution is to use a larger iron which can heat the bar faster – the author uses a 200 W temperature controlled mains iron.

Tips

The part of the iron that contacts the work is the tip. Old fashioned irons had solid copper tips whose working surface would gradually be dissolved by the solder to become concave, and would then need to be filed flat. Filing the tip was also the accepted way of cleaning these irons.

Modern irons use iron coated tips to protect the copper, and should *never* be filed. The normal method of keeping the tip clean is to wipe it on a moistened sponge (specially made for the purpose by most soldering iron manufacturers). It is most important to keep the sponge moist, and most engineers keep an old washing-up liquid bottle of water near their iron for this purpose. If the tip becomes sufficiently dirty that a quick wipe on the moistened sponge will not clean it, then a wipe on one of the proprietary tip cleaners should do the job. If that fails, then careful scraping with a knife or wire wool will cure the problem. Do not be tempted to use silicon carbide or glasspaper; the heat will melt the glue and make the tip even dirtier.

Tips come in many different shapes, sizes and temperatures. The preferred tip is conical, with an oblique cut across the end to produce an elliptical soldering surface. These tips are usually specified by the width across the minor axis of the ellipse, and a good general purpose width is 2.4 mm. A wider tip will allow you to get more heat into the work, and is better for heavier jobs, but is more clumsy, whereas a fine 1.2 mm tip is excellent for pick-up arm wires, but is unable to heat larger jobs. Ideally, you should have a range of tips, and change them depending on the application. See Fig. 7.12.

Irons that use a magnetic thermostat, such as the industry standard 'Magnastat' iron made by Weller, have tips that are available in different

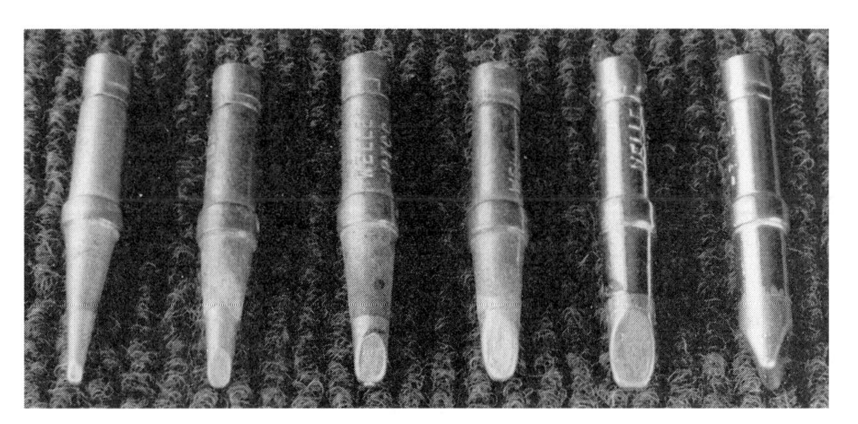

Fig. 7.12 *Selection of soldering iron tips*

temperatures. For most work, a No. 6 (315°C) tip is ideal, but when soldering inside old amplifiers, a No. 7 (370°C) is better at burning the dust away to melt the old solder.

Some irons have their temperature electronically controlled, and can be adjusted during use, but this is a luxury for amateur work, and probably not worth the extra expense.

Solder

The most common solder for electronic use is 60/40 self-fluxing solder. The 60/40 refers to the ratio of tin to lead, and the flux is a chemical, which when heated by the iron, will clean the surfaces to be soldered, and allow a good joint. There are various other solders available, some of which have sufficiently powerful fluxes to enable soldering to aluminium. *Never* attempt to use aluminium solder for electronic work, and once a soldering iron tip has been contaminated by aluminium solder, it should not be used for electronic work.

It is most important to keep your solder *clean*; keep it in an airtight box when you're not using it. Drawing it through a clean cloth before use, to remove surface contamination, will significantly improve the quality of your soldered joints.

In small quantities, solder is expensive, so buy a 500 g reel of solder. In this quantity, there are various different sorts of solder in various thicknesses. An excellent general purpose solder is 0.7 mm low melting point solder with 2% silver content. For surface mount components you *must* use silver loaded solder; otherwise the silver in the plating of the components leaches out and they won't solder. Silver loaded solder will allow you to make far better soldered joints than ordinary 60/40, and you will be seduced by its superior properties.

Soldering

Assuming that you have an iron at the correct temperature, with a correctly sized clean tip, and some clean solder of the appropriate type, how do you ensure that you make a perfect soldered joint?

The surfaces to be soldered must be clean, and the soldering iron tip must be clean and free of dross. Soldering works by the solder combining intimately with the surface metal of the components, and dirt hinders this process. The solder should be applied to the point of contact between the work and the iron such that it melts and flows immediately; the tip of the iron should then be wiped clean on the moistened sponge. See Fig. 7.13.

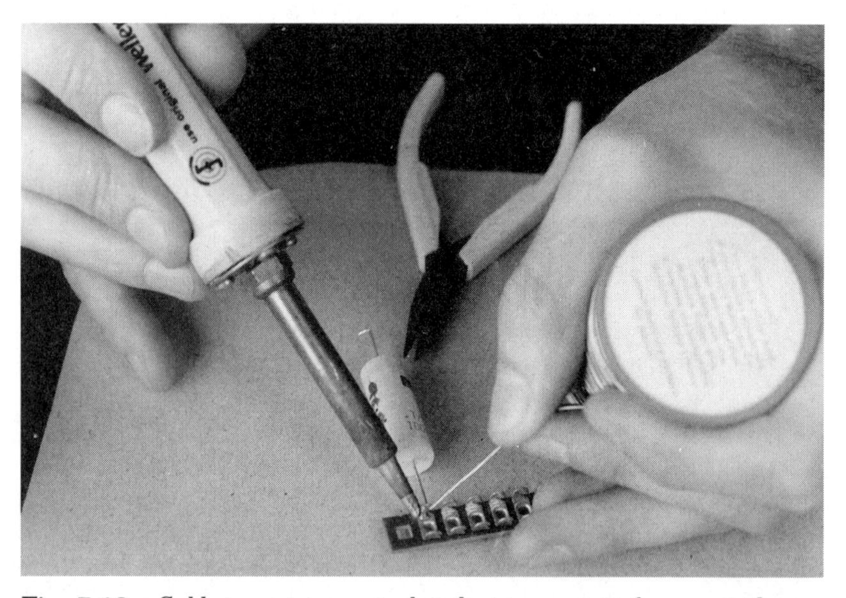

Fig. 7.13 *Soldering a joint; note that the iron contacts from one side, and the solder from the other*

Clean surfaces will *wet* perfectly, and surface tension will cause the solder to flow instantly across the work to form a perfect joint. Dirt will cause the solder to form globules on the surface that do not wet the joint, and defective joints are therefore known as *dry* joints. There are many variations between these extremes, but it can generally be said that good joints are made quickly, whereas dry joints are more likely to occur if the iron is in contact with the joint for more than a second or two.

The best joints are *mechanical* joints. If the parts to be soldered are already unable to move relative to one another, then they will not move as the solder solidifies, and a perfect, shiny, joint should result. If there is any movement whilst the solder is cooling, a dull, dry, joint will result.

The best joint is the *first* joint. Any subsequent resoldering of a joint will degrade the joint because it allows further oxidization of the heated materials. If you are forced to resolder an old joint, remove the old solder and replace with new. The fresh flux will ensure clean surfaces, and the fresh solder will not be contaminated with dross.

Desoldering

Sometimes you will need to desolder a joint. If the joint is mechanical, you will need to cut the wires away, otherwise the prolonged heat whilst bending wires will damage the wire or component. The remaining joint has fragments of wire in solder and must be removed.

The solder can be removed by one of two methods.

Desolder *wick* uses surface tension to wick the solder into copper braid which is then discarded when contaminated by solder. Solder wick must be kept clean if it is to work. This method causes the least damage to the work, but is wasteful and expensive.

The other method is to draw the solder off with a vacuum. In industry, vacuum desoldering stations are common, but they are extremely expensive to buy, and must be maintained and used carefully. The far cheaper alternative is the handheld, spring loaded, solder sucker. See Fig. 7.14.

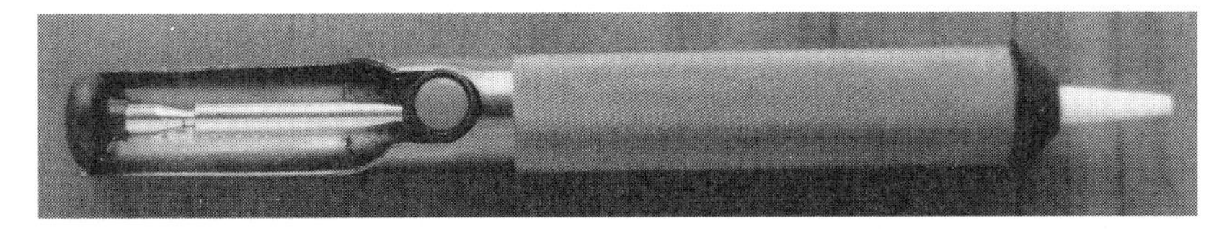

Fig. 7.14 *Manual solder sucker*

This device looks like an oversized pen, and has a PTFE tip that is placed directly in contact with the molten solder, whereupon the trigger is depressed and the sucker sucks up the solder – hopefully. Even handheld solder suckers need to be looked after if they are to work. The inside of the sucker needs to be cleaned periodically, or it will jam, and the PTFE tip eventually becomes damaged and must be replaced, because it is no longer able to seal against the solder to draw it into the sucker.

The main problem with solder suckers is the recoil. The sucker works by accelerating a plunger within the sucker away from the work. The recoil from this drives the PTFE tip into the work with sufficient force to break ceramic stand-offs, or to kick tracks off old printed circuit boards. For this reason, solder suckers should be used very carefully, and you may wish to revert to braid on delicate work.

Once the solder has been removed from a joint, the remaining fragments of wire can be easily peeled away with fine nose pliers. When working on a printed circuit board, care should be taken not to apply any force to the track, or it will lift.

Hand tools

In addition to a soldering iron, solder, and some means of desoldering, you will need hand tools to dress leads and fit components. It is easy to be seduced by all the wonderful pictures of tools in a catalogue, but you will find that for day-to-day use, you only need a very few tools, provided that they are of excellent quality. Good hand tools cost more, but they will last far longer, and will be cheaper in the long run.

Cutters

You need two sizes, one for cutting cable from wires to heavy mains cable, and one for cutting component leads precisely.

RS stock a superb pair of cable cutters (Code 547–470) which are a delight to use. Curiously, vets also sell them (at twice the price) for cutting dog's toe-nails. If you have these, and use them for most work, then the only other cutters you need are Lindström 'Supreme' semi-flush micro cutters. The Lindström cutters seem to last about ten years before becoming too blunt to use, but cutters with tungsten carbide cutting edges are now available, which should last even longer.

Pliers

Again Lindström 'Supreme', short jaw (21 mm). These are for dressing component leads, not for removing your exhaust! A pair of larger pliers is handy too, but you probably already have a pair for dealing with your car or bike.

Wire strippers

There are many different sorts of wire strippers, and personal preference is important. The chosen wire stripper should be able to strip cleanly without nicking, or cutting strands of the wire beneath the insulation. The author strips wire using a scalpel or traditional wire strippers. See Fig. 7.15.

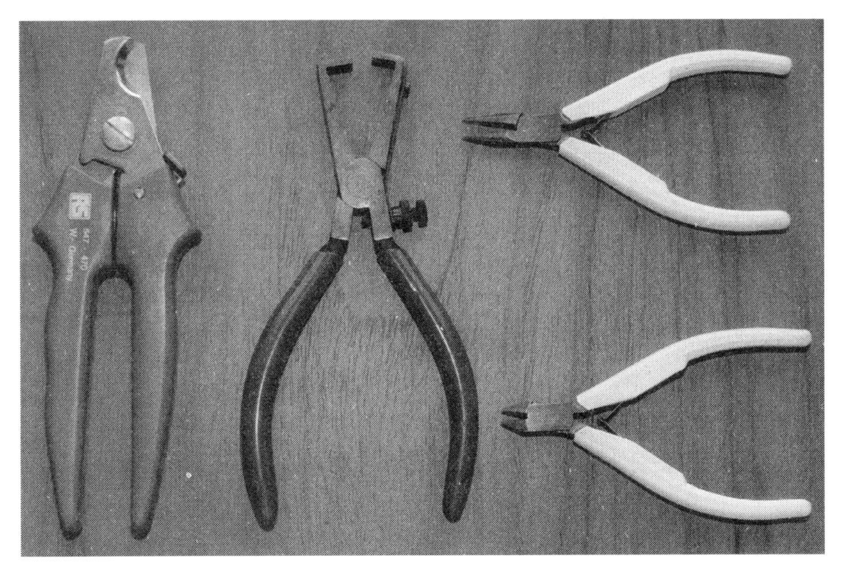

Fig. 7.15 *Clockwise from top right: Micro pliers, micro cutters, wire strippers, cable cutters*

Flat-bladed screwdrivers

You really can't have too many good accurately ground screwdrivers. Stubby screwdrivers are not recommended because they cause inaccurate screwing, and damage screwheads. An extra-long screwdriver with a 5 mm flat blade is extremely useful. You also need a very small screwdriver; it usually has a yellow handle and is about 60 mm long in total, with a blade width of about 1.5 mm.

Supadriv screwdrivers

These are reverse compatible with Pozidriv, but used with Supadriv screws, are far superior; if possible, choose the extra-long version as these are useful for computer monitors. Sizes 0, 1, 2 are useful, 3 and 4 are a luxury. European Pozidriv and Supadriv screws are identified by the additional radial lines between the splines of the screw slots.

Phillips screwdrivers

Oriental electronics is often held together with Phillips screws; sizes 0 and 1 seem to be the most common sizes, but Pozidriv screws (unfortunately without identifying lines) are also used.

Allen (Hex) Keys

Buy a good quality set of long arm Allen keys in Imperial *and* Metric sizes as it is essential to have the correct size. Be careful when using long arm Allen keys; whilst they are extremely useful for chassis punches, they are easily capable of splitting screwheads made of inferior metal. Avoid ball nose keys, although they allow easy access, they are not nearly as robust.

Spanners

Nuts are undone with spanners or nutrunners, *not* with pliers! A set of open-ended BA (British Association) is needed for traditional British valve amplifiers, whilst modern equipment uses metric fasteners. Electronic equipment uses very small nuts, and if the spanner is inaccurately ground, it will slip and chew the nut, so it is essential to use top quality spanners. See Fig. 7.16.

A nutrunner for potentiometer nuts is very useful, and will prevent gouging of front panels. See Fig. 7.17.

A Bahco 6" adjustable spanner will do nicely for everything else if used with care, and is sufficiently good to be used for careful roadside repairs of a Ducati.

Scalpel

Scalpels are extremely useful, and the small handle version is best suited for electronics, with either No. 10 or No. 10A blades. You will find that the

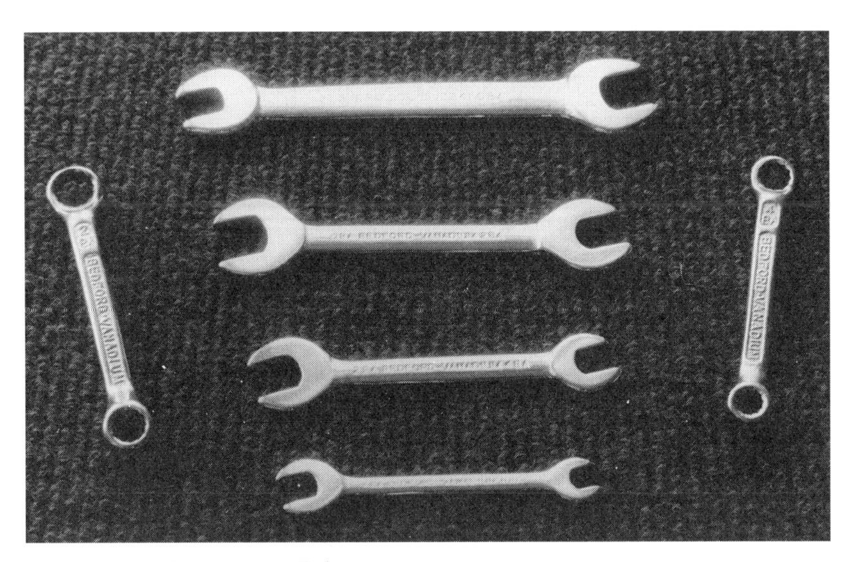

Fig. 7.16 *Selection of BA spanners*

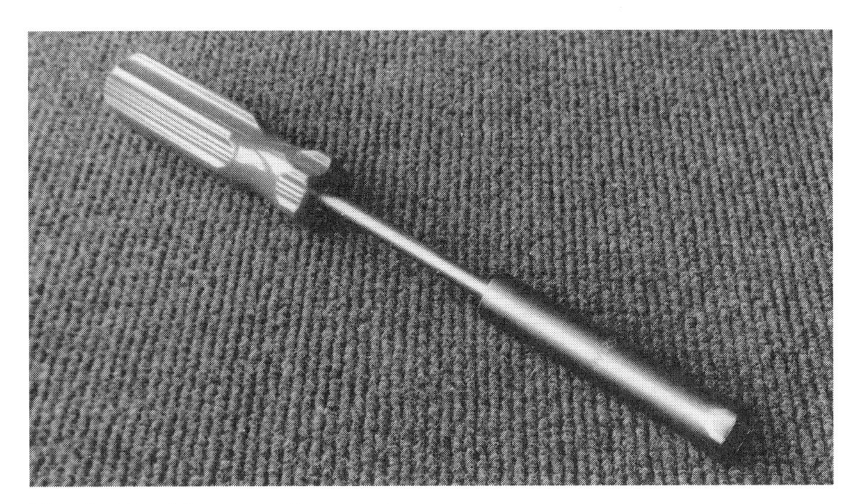

Fig. 7.17 *Nutrunner for potentiometer nuts*

blades lose their edge very quickly, so buy blades in bulk, and be prepared to fit a fresh blade the moment you notice a lack of keenness. Be careful with scalpels; they were designed for cutting flesh. A guard made from layers of heatshrink sleeving can be made to cover the blade when not in use. See Fig. 7.18.

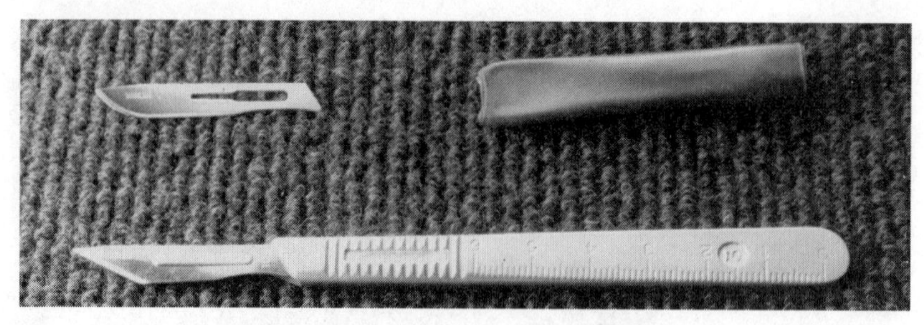

Fig. 7.18 *Scalpel with No. 10A blade and guard*

Heatshrink gun

Electronic tools catalogues stock all sorts of expensive hot air heatshrink guns, but a hot air paintstripper works just as well, and is far cheaper.

Marker Pens

It may sound obvious, but if you label things *before* you take them apart, life becomes so much simpler.

Toolbox

All of these hand tools are precision tools, and they should not be thrown in a toolbox together with old spark plugs and oil filters. Keep them in a clean partitioned box of their own.

AC power wiring and hum

All AC power wiring generates an external field that can induce audible hum into signal wiring. Heater wiring is the obvious problem, because it will unavoidably be close to sensitive signal wiring, but AC mains and the high voltage AC to rectifiers can also cause problems.

The electromagnetic field

The electromagnetic field is due to the *current* flowing in the power wires, which will induce currents in any nearby signal wiring.

Heater wiring is usually taken from a winding on the mains transformer to the nearest valve, and looped through, from one valve to the next, until each valve has heater power. The input valve is the most sensitive stage, so this should be the last in the heater chain, so that the wiring leading to this valve carries the least current.

To minimize the external electromagnetic field, the heater wire should be tightly twisted. This means that although any given twist induces a current of one polarity, the twists either side of it induce opposite polarity, and so the field tends to cancel. This twist should be maintained as close up to the pins of the valve as possible, and when one phase of the heater wire has to cross the valve base and return, as might be the case when looping wiring past an ECC81/2/3 to the next valve, the wire across the base should also be twisted. Admittedly, the return current is less than the outgoing current, but some cancellation is better than none.

Heater wiring leading to valves using B9A sockets such as EL84 etc. is best twisted from 0.6 mm (conductor diameter) insulated solid core wire, which is rated at 1.5 A. When wiring to octal valves, more current may be required, and a thicker wire can be used, which could not have been connected to the pins of a B9A socket. When wiring to valves other than rectifiers, it is useful to use a different colour for each phase, and the author has traditionally used black and blue. When wiring to a push–pull output stage, if the same colour goes to the same pin on each valve, then the hum induced within each valve will be the same phase, and will be cancelled in the output transformer. (This argument assumes that both valves were made by the same manufacturer to the same pattern.)

Valve rectifiers such as the GZ34 will not only have a dedicated 5 V heater supply, but they will also have the incoming high voltage AC, which should be distinguished. The author uses a red twisted pair for the HT, and a blue twisted pair for the heater.

Twisting solid core wire is easy. Cut 4 m, or more, lengths of wire to be twisted, and clamp one end of each wire in a vice. Gently tension each of the two wires, and grip their other ends in the chuck of a power drill. Hold the wires reasonably taut by pulling on the power drill, and switch on. After about 10 seconds, the wire will begin to draw you quickly towards the vice. The wire should now have about 10 twists per inch. Switch off, and *whilst maintaining tension* by holding the wire with your fingers, undo the chuck. The wire will now try to untwist, and if allowed to do so suddenly, will tie itself in knots. Gently release the tension in the wire, and release from the vice. You now have perfectly twisted wire.

Although perfectly satisfactory with commercial sleeved copper wire, the previous method should not be used with silver wire posted down PTFE sleeving as it will tend to break. Twist silver wire by hand.

Electromagnetic fields decay with the square of distance, so heater wiring runs should be as far away as possible from signal circuitry, and only come up to the valve at the last possible moment.

The valve sockets should be oriented so that the pins that receive the heater wiring are as close to the chassis wall as possible, and heater wire should *never* loop round a valve (except for rectifier valves, where hum is not an issue).

The electrostatic field

The electrostatic field is due to the *voltage* on the wiring.

Heater wiring should be pushed firmly into the corners of the chassis, since this tends to null some of the electrostatic field. Heater wiring should not run exposed from one valve to the next. This makes heater wiring awkward, and is why commercial amplifiers sometimes skimp on the quality of their heater wiring, since good heater wiring takes considerable time to put into place.

AC heater wiring should be connected to the transformer in a balanced fashion. Unfortunately, heater wiring *must* have a DC path to HT 0 V in order to define the heater to cathode voltage, and this can be achieved in various ways. See Fig. 7.19.

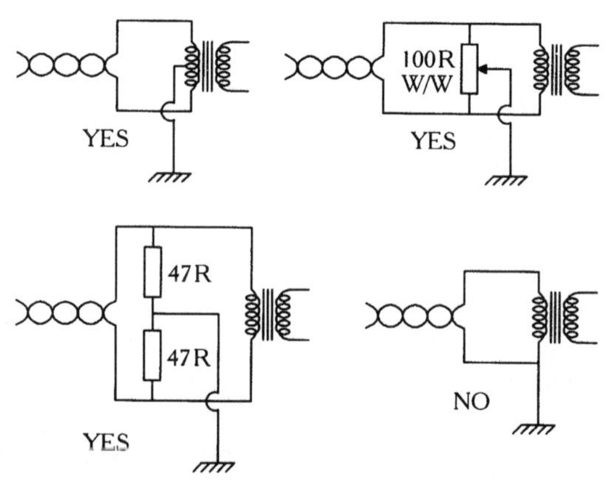

Fig. 7.19 *Grounding heater wiring*

The worst way to define the DC path is to simply connect one side of the transformer winding to 0 V. This ensures that one phase of the wire induces no hum, whilst the other phase induces maximum hum.

The ideal way of defining the DC path is to use a transformer with a centre tap on the heater winding, but if this is not available, fixed or variable resistors can be used to derive a midpoint. Accurately matched resistors used to be rare, so a variable resistor known as a *humdinger* control used to be fitted, and adjusted for minimum hum. Once an LT midpoint has been derived, and connected to HT 0 V, each wire has equal voltage (but opposite phase) hum, and the electrostatic fields tend to cancel.

The previous cancellation cannot be perfect, and for ultimate reduction of heater induced hum, we should screen the heater wiring with braid, and/or use DC heater supplies. Even when using DC heater supplies, it is worth treating the heater wiring as if it were carrying AC, as this will ensure that the finished project has *no* heater induced hum. The output of a DC heater supply will not have a centre tap, so the LT 0 V is often connected directly to the HT 0 V, but a better method is to derive a centre tap using a pair of matched fixed resistors. See Fig. 7.20.

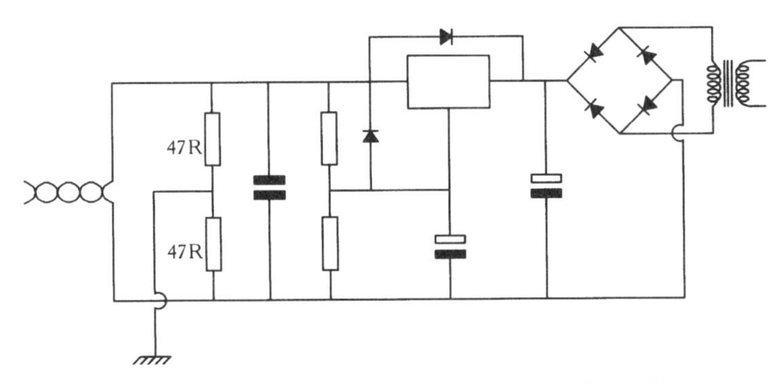

Fig. 7.20 *Balancing regulator noise using a potential divider*

Heater wiring is the first piece of wiring to go into a project; thereafter, it is obscured by signal wiring. Once all the other wiring is in place, it is impossible to replace the heater wiring, so it must be installed correctly. See Fig. 7.21.

Fig. 7.21 *Heater wiring pushed into the corner of the chassis*

Mains wiring

Mains wiring should be as short and direct as possible because the wire is often quite thick, and cannot be twisted well. Mains wiring unavoidably generates considerable interference fields.

Modern semiconductor equipment *sleeves* all exposed mains wiring with rubber or PVC sleeving such that it is moderately safe to rummage inside a piece of powered equipment. Valve amplifiers operate on such high voltages that it is *never* safe to rummage in powered equipment, and even unpowered equipment should be approached with caution. Safety is therefore not greatly improved by sleeving mains wiring, but it is still good practice to sleeve mains wiring with heatshrink sleeving or purpose made rubber boots to fit over IEC sockets or fuseholders.

The *only* approved 3-pin domestic mains connector is the IEC plug/socket, which is familiar to many as the connector at the base of electric kettles. These are available with integral fuseholders which allow much safer construction, and are recommended.

The mains connectors used on classic valve amplifiers are, without exception, outrageously dangerous by modern standards.

Mains switching

To switch a piece of equipment off, all we need do is to break the circuit from the source of power. A mains switch could therefore equally well be inserted in the live or neutral wire, and still perform the job, and this is known as single pole switching. However, a switch in the neutral leaves all internal mains wiring live, and constitutes a shock hazard within the equipment. Single pole switching should therefore *always* switch the live circuit to minimize shock hazard.

Double pole switching switches both the live and the neutral, and ensures safety even if the live and neutral wires are reversed. CD players often use non-polarized 'shaver' plugs on their rear panels, which allow live and neutral to be reversed, so they commonly use double pole mains switching.

Where there is *no possibility* of live/neutral reversal, single pole switching is safer, and more reliable, because failure of the switch ensures a break in the live connection to circuitry. A double pole mains switch has twice as many contacts to fail, and if the neutral fails, the equipment could appear to be safe, even though the mains wiring is connected to live mains and still constitutes a shock hazard.

Fuses

A fuse is a piece of fine wire with resistance connected in series with the circuit to be protected. If excessive current passes through the wire, it heats in accordance with I^2R, and heats sufficiently that it melts, or ruptures. A fuse is a single pole switch and should therefore be connected in the live wire, *before* any other circuitry, such as a mains switch.

It will be appreciated that the fuse wire must lose heat to its surroundings, so a mild overload will allow much of the heat that should have melted the fuse wire to escape, whereas a short duration gross overload will not be able to lose so much heat, allowing the fuse to rupture quickly. As an example, a 13 A fuse to BS1362 (as fitted to a UK 13 A domestic plug), requires 100 A if it is to rupture in 0.4 s, whereas it requires 10 s at 50 A.

To calculate a mains fuse rating, the total power consumption of each individual load on the mains transformer should be summed to find the total load taken from the mains. The current drawn from the mains may now be found, and the fuse rating should be the next rating above this. This calculation contains many sweeping assumptions and approximations, but fuses are not accurate either, so the method will be found to be satisfactory.

Some equipment, particularly toroidal mains transformers, draw large inrush currents, but their working current is much lower. To cope with these requirements, anti-surge or timed fuses are available, which will withstand short overloads. These fuses usually have their values preceded by a 'T' to indicate 'timed'.

Protecting each output of a multiple winding mains transformer is difficult for the following reasons:

- Fuses are very rarely fitted to HT supplies because they offer only very limited protection to the output valves. In a Class A amplifier, the output valves are usually run at their maximum anode rating, and a doubling of anode current will cause damage. However, a fuse may not blow with an overload as small as this, so little protection is offered. Fuses can be fitted to Class AB amplifiers, and are advisable for OTL designs, but their non-constant resistance can cause distortion.
- Fuses are never fitted to heater supplies because heater circuitry is normally so simple as to not warrant a fuse. Failure of heater supplies can cause damage elsewhere in a DC coupled amplifier, as valves switch off and anode voltages rise to full HT volts.
- Grid bias supplies to output valves should never be fused because failure of this supply will destroy the output valves.

Most valve amplifiers do not, therefore, have any fuses other than a fuse on the primary of the mains transformer.

Glass bodied fuses should *never* be used to protect high voltage circuits such as AC mains. A short circuit causes the fuse to rupture instantly and vaporize, thus depositing a conductive metal film on the inside of the glass which continues to pass current and heats the glass. If glass is heated sufficiently, it becomes a conductor in its own right, and so the fuse has failed to protect the circuit. Fuses suitable for high voltage use have ceramic bodies filled with sand to prevent the creation of a continuous conductive film.

Mains and high voltage fuseholders accessible from the outside of the chassis should be of a construction that requires a tool to release the fuse.

Class I and Class II equipment

Class II appliances have all hazardous voltages (>50 V) *double insulated* from contact with the operator, and use a 2-core mains lead. Double insulated means two insulating barriers, one of which may be air, each independently capable of withstanding the shrouded voltages and protecting the operator. It is possible to make a Class II appliance that has exposed metal, but rigorous testing is required to ensure that the appliance meets the full technical standard. A symbol consisting of two concentric squares signifies double insulation.

Class I appliances require only one layer of insulation from hazardous voltages, but this layer must be totally shrouded by a conductive layer *bonded* to mains earth via a low resistance path (see later). It is far easier to make equipment that conforms to Class I than Class II, so amateur equipment should *always* be built as Class I to ensure safety.

These classes of insulation do not merely apply to the appliance, they also apply to the mains flex from the wall socket. Since domestic flexes are not usually sheathed with an earthed conductive sheath, the flex should be double insulated, and a single layer of insulation is insufficient.

Earthing

Earthing is the cause of many problems in amateur constructed equipment, but if thought about logically, there is no need for it to cause any problems whatsoever. Colloquially, the term 'earthing' refers both to the mains earth safety bond to the metal chassis, and also the 0 V signal wiring, but the two are quite distinct.

Earth safety bonding

The three wires leaving a domestic supply are line, neutral and earth. Neutral and earth are commoned together at the substation, or possibly at the electricity supply company's cable head within the house. This means that if line contacts earth, a large fault current flows, determined by the *earth loop resistance*, which is the entire resistance around the loop, including the resistance of the line wires. See Fig. 7.22.

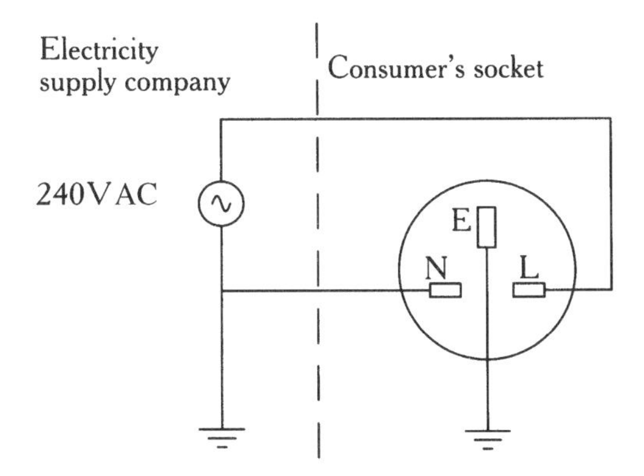

Fig. 7.22 *Earth loop resistance*

The purpose of the safety earth bond is to provide a sufficiently low resistance path to earth that if the line wire of the mains comes into contact with the exposed metalwork (which would then be a shock hazard), the resulting line to earth fault current is sufficiently great to rupture the fuse *quickly*. The time taken for the fuse to rupture is proportional to the earth loop resistance, so there is no such thing as an earth loop resistance that is too low.

Although exposed valves may appear to conform to Class II, because the electrodes are insulated by a vacuum and the glass envelope, if the envelope is broken, the secondary layer of insulation also disappears. Valves on the top of the chassis should therefore be enclosed by a perforated metal cover to ensure that they conform to Class I.

If we build an amplifier on a chassis with exposed metal, then the construction must be to Class I, and all hazardous voltages must be insulated from, and totally enclosed by, earth bonded metalwork. The ideal place to achieve this bonding is near to the entry of the power cable. The bond should be made using a solder tag bolted to the chassis with a shakeproof (star) washer *between* the washer and the chassis because this bites into the metal of the chassis and the tag to provide a gas-tight joint. If the chassis is anodized aluminium, the surface anodizing should be thoroughly scraped away underneath the tag to ensure a good bond.

The nut and bolt should be prevented from loosening using further shakeproof washers or locknuts. The earth bond bolt should *never* pass through plastic, such as the mains input socket, because plastic creeps and will cause the bond to loosen in time.

The ideal earth bond uses an M6 or 0BA bolt passing through the chassis with a shakeproof washer either side of the chassis, another shakeproof washer above the earth tag, followed by a flat washer (to prevent the tag rotating when the bolt is tightened), and secured with a locknut. The earth wire from the mains cable passes directly to the tag, where it should form a mechanical soldered joint; it is much easier to solder this joint before bolting the tag to the chassis. See Fig. 7.23.

Fig. 7.23 *Achieving a low resistance earth bond to chassis*

A thick cable should be used to bring mains earth to the bond point in order to reduce earth resistance. Although safety dictates that 3 A rated equipment may have 0.5 Ω from the pin of the mains plug to the chassis (*not* measured at the bond point), reducing this resistance to 0.1 Ω, or less, by using 2.5 mm^2 mains cable reduces the likelihood of hum and improves safety.

The preceding arguments apply to equipment that is directly powered from the mains, but pre-amplifiers often have remote power supplies. Nevertheless, the same argument should still be applied, and a substantial cable should be used to bring mains earth to the pre-amplifier chassis, and the bonding technique should be the same. Likewise, turntable motors such as the Garrard 301/401 should be firmly earth bonded via their mains cable and not via the pick-up arm lead.

Sometimes there will be metal objects that could come into contact with mains voltages but do not have a guaranteed electrical path to the main earth bond. Examples of this are:

- The suspended motor on a turntable.
- A mains transformer or HT choke that is acoustically isolated because of vibration.
- Any separate anodized aluminium panel supporting a mains connection, such as a front panel mains switch, or mains transformer on the baseplate.

Each of these should have a connection, such as a wire or a screw, via a star washer, to bond them to the main earth bond, but in the first two examples it is important that the wire should not be so stiff that it short circuits the acoustical isolation, and a loop or short helix of wire is ideal.

0 V system earthing

It is the 0 V signal earthing that causes the hum due to hum loops between multiple earths, *not* the safety bond.

Hum loops are circuits within earth paths that can have hum currents induced into them by mains transformers. If there is any resistance in the circuit, a voltage is developed, and this causes the audible hum. See Fig. 7.24.

To remove the hum, the loop must be broken, and this is often done by removing the earth wire from within the mains plug of one of the

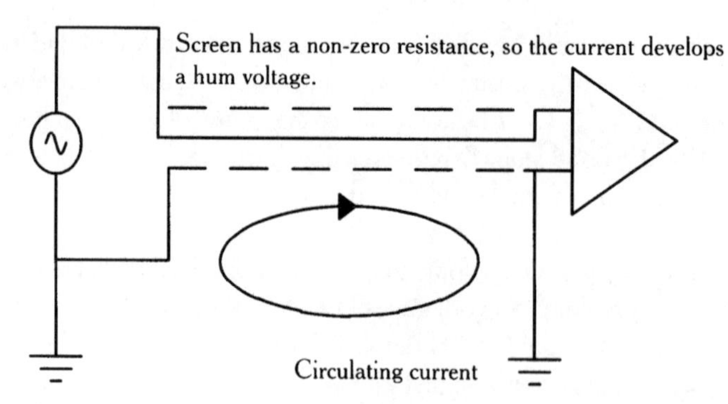

Fig. 7.24 *Why a hum loop causes hum*

affected pieces of equipment, *but this is extremely dangerous*. The loop should be broken by removing the 0 V signal earth bond to chassis from one of the pieces of equipment.

Fortunately, most modern equipment is double insulated, so hum loops will not often occur, but a modern improvement on earthed equipment is to provide a *ground lift* switch or pluggable link that can make, or break, the 0 V signal earth to chassis connection at will on each piece of equipment. This method allows the optimum 0 V system earthing arrangement to be determined quickly and safely.

To avoid ground loops, there should be only one 0 V signal earth bond to chassis, and hence mains earth. The optimum place for this bond is as close as possible to the input amplifier of the RIAA disc input stage. This bond should be made with as short and thick a wire as possible, to reduce its inductance and ensure that it is a good bond even at RF.

Screens and interconnects

The 0 V signal earth is a *signal* wire because it allows the signal current to return to the source. It is therefore most important that we treat it with the same care and consideration that we would apply to the more obvious signal wire.

A particularly useful technique was alluded to in Chapter 6, whereby the signal earth and signal could advantageously be treated as a balanced signal, even if they are unbalanced. The author has used twisted pair/overall screen interconnect cables on unbalanced systems since 1976 because of their superior rejection of external fields, but it is important that the screen is connected to 0 V signal earth at the correct end.

The screen should be earthed at the source end because the source has a low output impedance and is quite capable of defining its output as being the difference between the two output wires. One of these wires is now connected to the screen of the output cable. The screen will pick up RF interference which it will superimpose onto the commoned signal wire. The RF will also be superimposed onto the other signal wire via the output impedance of the source, and if the source has a truly zero output impedance, even at RF, then both output wires now have the full RF superimposed on them. This might seem undesirable, but . . .

At the amplifier end, the input stage responds to the *difference* in signal between the two input wires, and therefore rejects the RF that is identical on both input wires.

If we now connect the cable so that the screen is commoned to one side of the signal at the destination end, the induced RF picked up by the screen now has to travel down the entire length of one inner wire of the cable to the source before it can be coupled via the output impedance of the source to the other inner wire. The coupled RF on this other wire also now has to travel the entire length of the cable before it can arrive at the amplifier input.

The RF signal on one wire has now had to travel twice the length of the inner wires, and the inner wires have *inductance*. This inductance, in combination with the stray shunt impedances, will cause attenuation that rises with frequency. This means that one of the wires at the input to the amplifier has the full RF signal, and the other has an attenuated signal, resulting in a difference signal to which the amplifier is sensitive.

The above argument assumes equal impedances to RF earth (whatever that might be) at the pre-amplifier input. Whilst this is not exactly true for domestic equipment, the quasi-balanced method of interconnecting cables described above is a considerable improvement on coaxial cable or 'screened lead' because the two wires (the inner and screen) have wildly different signals induced into them, and there is no possibility of common-mode rejection.

Internal wiring of amplifiers

Once within an amplifier, the 0 V signal earth path can either travel in the same way as it did between equipment, in which case it is known as *ground follows signal* or it can be *star earthed*.

Ground follows signal is the traditional method of wiring valve amplifiers, and is probably still the best if executed properly. The traditional method uses an *earth busbar* which is a thick (at least 1.6 mm diameter, or 16 swg) tinned copper wire 0 V signal earth connected directly from input sockets to the 0 V source of the power supply (commonly the reservoir capacitor, or if fitted the 0 V output of the regulator).

The repetition of '0 V signal earth' may seem pedantic, but it reminds us that the signal earth also carries power supply currents. This last factor is extremely important, since the power currents are many times larger than their associated signal currents. It is to reduce the possibility of these power supply currents developing spurious signal voltages along the earth busbar that its resistance must be made as low as possible.

Even the precaution of having a low resistance busbar is not sufficient, and connections must be made to the busbar in the correct order such that voltages are not developed in sensitive input circuitry. The author remembers an RIAA disc pre-amplifier that had been constructed to the Mullard two-valve design, but had considerable hum. The hum was cured by moving *one* wire 150 mm along the (1.6 mm) earth busbar.

The correct order from the input socket is: input circuitry (such as shunt capacitors, etc.), grid leak resistor, cathode bypass capacitor (if fitted), cathode resistor, any anode signal circuitry (such as equalization), next valve's grid leak resistor, etc.

If necessary, the earth busbar is then bonded to the chassis via a tag using one of the screws retaining the input valve socket. Centre spigots of the valveholders (if fitted) should also be bonded to the busbar as they help reduce capacitance between pins of the valve socket.

To make a neat earth busbar, the thick wire needs to be straightened, and this is not a trivial task. The traditional way to achieve this is as follows.

Grip one end of the wire in a vice, and then grip the other end in a substantial pair of pliers, such as would be used for working on a car. The wire is then wrapped one turn round the jaws of the pliers, and the pliers are firmly gripped with both hands whilst one foot is braced against the vice. The wire is then pulled until it can be felt to stretch, and *without moving the position of the pliers* is cut at the vice end. A beautifully straight piece of wire is now the result, and this can now be cut away from the pliers.

It should be realized that considerable force is required to achieve this result, and this can apply dangerous forces to your back if you do not position yourself correctly. If you are in any doubt as to how to position yourself, or have back problems, *do not attempt to use this method.* It is far better to have tatty wiring that can be seen, than beautiful wiring that cannot be seen because you are lying down with your back in traction.

The ultimate expression of the earth busbar is the RF ground plane, which is a two-dimensional conducting earthed surface to which earth connections are made (a wire is only one dimensional). This construction is now common on audio printed circuit boards as designers realize how important RF immunity is to audio circuitry. On a PCB, the entire upper surface of the board can be used as a ground plane, which has low inductance because it is so wide, and therefore guarantees a good RF earth at every point that a contact is made.

An intermediate step is to use a strip of 10 mm × 1 mm silver as the busbar, the width lowers the inductance, and the very large cross-sectional area ensures low resistance. Holes can be drilled to suit component wire diameter, or V notches filed at the edge to support the component whilst soldering. The iron should be held to the silver first to bring it up to temperature, then moved so that it touches the component, and more solder applied to form the joint. A 200 W iron is needed, as a smaller iron takes so long to warm the silver to soldering temperature that the component is damaged.

Some old valve amplifiers (such as the Rogers Cadet) approximated to a ground plane by soldering directly to the (steel) chassis, and using this as the 0 V signal earth. Neither this method nor a steel chassis is recommended.

Star earthing is achieved by having a *single* earth point, often bonded directly to chassis, to which all 0 V connections are starred. Ideally, all the connections to this point should be made with short leads to minimize inductance, but building an entire amplifier neatly using this method is extremely difficult. For this reason, many designs use a combination of star and busbar earthing, with the input stage star earthed, and following stages earthed to the busbar.

High current circuitry

Some parts of an amplifier carry high currents. In a capacitor input power supply, the loop from the mains transformer to the reservoir capacitor carries the capacitor ripple current, and it is essential that no connections are made to the 0 V signal earth within this loop. In order to reduce the electromagnetic field caused by the passage of these currents, the length of this loop should be as short as possible, which also reduces its resistance and voltage drop.

The output to the loudspeaker from a power amplifier is also a high current loop, and additional connections to sense this voltage, such as global negative feedback, should be made very carefully. The ideal method is to connect a screened twisted pair to the output terminals, with the screen connected to chassis at the loudspeaker output end only. This cable is then routed to the input stage, where one side is connected to the lower end of the cathode resistor, and the other is connected via a series resistor to the cathode (assuming cathode feedback). This method ensures that the feedback voltage is derived from the correct point, and that the current is applied to the correct point.

Layout of components

The preceding discussions have implicitly assumed *hardwiring*, where components are soldered directly to one another and are self-supporting. In contrast to a printed circuit board, this is a three-dimensional layout, and offers considerable freedom in position of components.

Good layout in a valve amplifier means low stray capacitances (because at typical valve impedances, practical inductances are insignificant).

The best way of reducing capacitance between wires or components is to cross them at right angles. The best way of reducing capacitance to earth is to use short leads. These two requirements result in component layouts that appear to have been thrown together by an avant garde sculptor. We thus arrive at the surprising conclusion that a good layout probably looks untidy, although the converse is not necessarily so.

Tagboards, tagstrips and stand-offs are all useful ways of supporting components and making joints insulated from earth. See Fig. 7.25.

Whether the layout is a 'tidy' PCB or an optimized hardwired layout, good layout requires considerable care, and thermal considerations must also be accommodated.

PCBs have been mentioned several times, and the author uses them frequently, but they are not ideal for the novice. This is because they are really a production method of construction, and it requires considerable confidence to design a theoretical circuit and commit it directly to a PCB. Whilst it cannot be argued that a good PCB gives a thoroughly professional appearance to the finished project, hardwiring will often give superior performance.

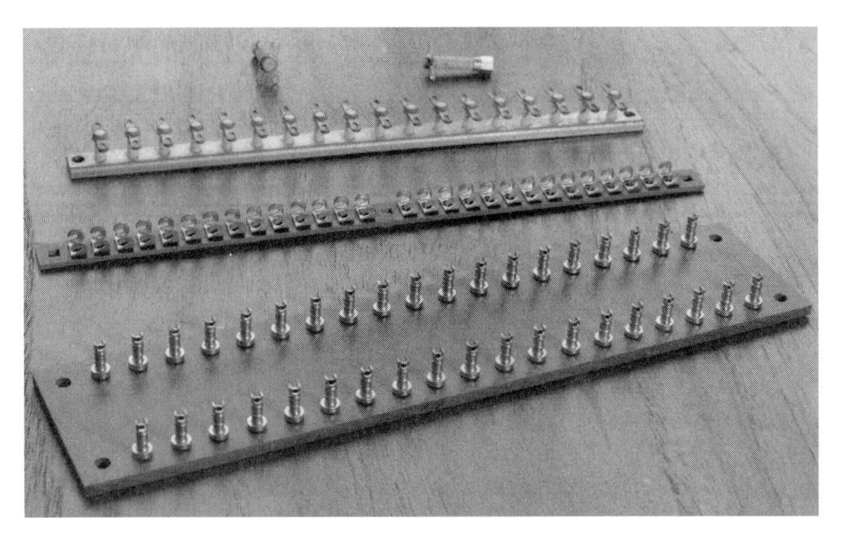

Fig. 7.25 *Front to back: Turret board, tag strip, ceramic stand-off strip, individual stand-offs*

Safety

Valve amplifiers contain mains and HT voltages which can *kill*. Safety is therefore very important, and we should understand the dangers so that we can safeguard against them.

An understanding of the physiological effects of electric shock, whilst somewhat ghoulish, serves to underline why safety is so important.

Electric shock

The best way of avoiding electric shock is to understand how it can be received.

The electricity supply leaving the wall socket in your home is a very good approximation to a pure Thévenin source of zero resistance, with one side of the source connected to earth. In this instance, we do not mean earth in its purely technical sense, the supply really is commoned to the planet Earth. You and I perform most of our activities on the surface of the earth, and we are therefore electrically connected to it, albeit usually by a high resistance path. We can improve our electrical contact to earth by standing barefoot on a damp floor, or by gripping something that is electrically bonded to earth.

Manual activities like hobbies are precisely that, *manual*. They involve our hands going inside objects, and touching them. Humans generally possess two hands, and so what is more natural than to put *both* hands inside a piece of equipment?

The scene has now been set for one hand to be holding the (earthed) chassis of a piece of equipment, whilst the other is moving around and accidentally comes into contact with live mains.

Look at your hands. They are on the ends of your arms, and your arms are joined to your torso. Trace a line from the fingers of one hand to the fingers of your other hand, without the line leaving your body. Note the path.

The easiest path for an electric current to flow from one hand to the other crosses near the heart. Similarly, a shock from one hand to earth passes near to the heart.

The effects of electric shock

The main danger from electric shock is fibrillation of the heart. The heart normally beats at a slow, regular pace on the command of electrical impulses from the brain. If we apply 240 V 50 Hz to the heart, it will attempt to lock to this signal, and will fail miserably. The result is that the heart pulses quickly, and the flow regulating valves do not operate correctly. No blood is pumped and the brain dies of oxygen starvation in about 10 minutes.

A sustained current of 20 mA through the heart is sufficient to cause fibrillation, resulting in the adage, '20 mills kills'.

Sustained currents below 20 mA may not kill, but may cause irreversible injury due to the heating effect of the current. RF burns are notorious for this, and can result in limbs having to be amputated to prevent the spread of gangrene.

A still lower sustained current may cause injury from which the victim will recover, eventually. Skin grafts may be necessary.

A current of 20 mA will result from a 240 V supply connected across a resistance of 12 kΩ. Very few power supplies have a source resistance as high as this, so shock current is determined primarily by skin resistance, and a shock from a small transformer delivering 240 V is just as dangerous as a shock received directly from a 100 A mains feeder. Skin resistance is reduced by damp hands, and standing in the rain in a puddle of water will lower resistance further.

All of the above considerations refer to the direct consequences of electric shock, but does not consider secondary effects. A minor shock that causes the victim to lose their footing and fall over could be fatal if they happen to be standing on a ladder 30 ft above concrete. Another possibility is that the reflex muscle jerk in reaction to the shock could cause the victim to throw themselves through a window and bleed to death.

Even after a minor shock, the victim will be confused and disoriented, and shock in its full medical sense is a possibility. Shock kills.

Burns

Although the primary hazard of electrical equipment is shock from high voltages, it should be realized that low voltages can be just as hazardous. A low voltage/high current DC supply will have a large, low ESR, reservoir capacitor capable of delivering many amps of current into a short circuit.

Even more dangerous than capacitors are lead–acid and NiCad batteries, since these are capable of sourcing enormous currents for an extended time. (Think about it, a car starter motor requires 100 A or more for several seconds.) These batteries are not merely capable of burning, they can vaporize metal bracelets, watchstraps and tools.

Do not wear jewellery whilst working on live equipment.

Avoiding shock and burns

It should now be obvious that electric shock is potentially lethal, that burns can be serious, and both must be avoided at all costs.

Provided that you have made, or modified, your equipment carefully, there will be no exposed voltages, and all metalwork will be earthed, resulting in a very low risk of shock. The danger arises when you *deliberately* remove the safety covers and start testing the equipment with power applied.

It has been suggested that you should always work on live equipment with your left arm behind your back, so that any shock received will not pass from arm to arm across the heart. Whilst it is true that this will reduce the severity of the shock, it also increases the risk of receiving a shock.

The best way of improving safety is to think about safety, and to *think about what you are doing*. It might seem obvious to think about what you are doing, but for most of our lives we think about many things at once. For instance, when driving, are you thinking *only* about driving, or are you thinking about what you are going to say to your boss when you arrive late for work, and when is that idiot in front of you going to turn into the junction, and isn't that a rather attractive male/female/alien over there by the bus stop?

Thinking about what you are doing means not working late. Do not attempt to test a newly completed project at 11.30 at night; you will not be alert and could damage the project and/or yourself.

Test equipment

Before we can test our new amplifier we will need some test equipment. Test equipment may be divided into two categories:

- *Safety testing*. Is the device under test safe? For instance, a multimeter can be used to perform a number of basic safety tests *before* you even switch the equipment on.
- *Functionality testing*. Does the device under test work as it should?

A fully equipped electronic workshop will have the entire gamut of test gear from spectrum analysers and oscilloscopes to insulation testers and variable power supplies. As an amateur, you cannot afford to be this well equipped, and need to pick and choose carefully.

Multimeters

The industry standard meter in the valve era was the AVO Model 8 moving-coil multimeter. You will find many circuit diagrams stating that voltages were measured using a 20 kΩ/voltmeter, which refers to the loading that the AVO 8 imposes on the circuit. Nowadays, everybody wants a digital meter because they have 10 MΩ input impedance, and hardly load the circuit, whilst being cheap and accurate.

Not all digital multimeters are accurate, and their specification should be read very carefully before purchase. Autoranging models may not measure current, but this is not a great loss, since to measure current you must break a wire and later reconnect it. Some models will measure capacitance, but they do not usually have a very wide range, and you may feel that it is better to put the extra money aside towards a second-hand component bridge.

Digital meters divide into pen types or types with flying leads. The types with flying leads are ideal for connecting to equipment before switching on power, but are not so good if you need to test the equipment when live. Poking probes into valve equipment and then looking away to read the meter is courting disaster, and the author usually prefers to use a pen type multimeter where the display is close to the probe tip.

Moving-coil meters still have their uses. Digital multimeters take time to stabilize to a reading, and by then it may be too late. A moving-coil multimeter will flick quickly to its reading, and this may warn you sufficiently quickly to switch off before damage is done. Digital meters are often very poor at measuring one parameter in the presence of another, and some autoranging digital meters refuse to measure the primary resistance of an output transformer because of the large inductance that is also present.

Because of these problems, you will find that it is very useful to have a digital multimeter *and* a very cheap moving-coil multimeter. See Fig. 7.26.

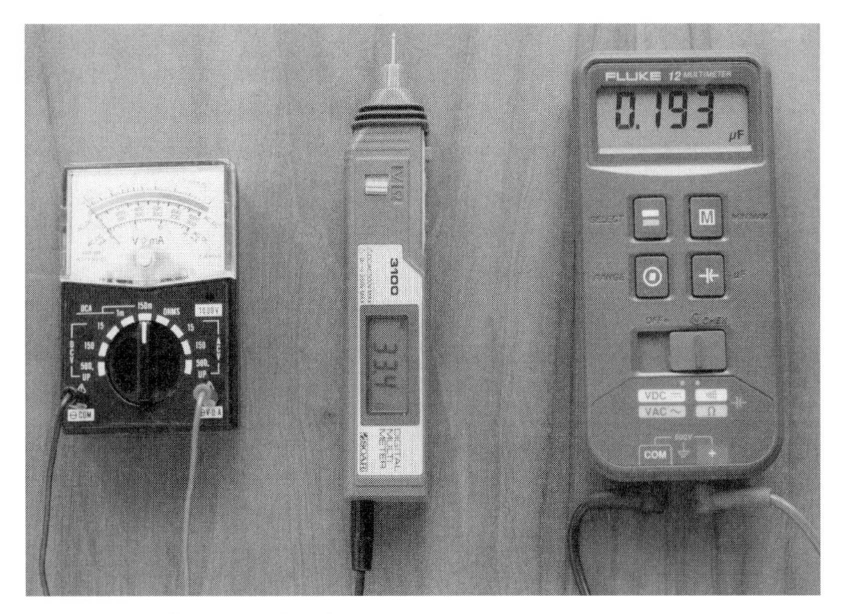

Fig. 7.26 *Selection of multimeters*

'Voltstick'

These have various brand names and are incredibly useful. They usually look like a fat pen with a white tip; if the tip is near to mains, an internal LED lights (runs off 2 × AAA batteries for years). They require no contact and are a lifesaver. Always use them before cutting any cable that, potentially, could carry mains.

The preceding test equipment is the absolute minimum required, and the following can be classed as 'nice to have'.

Component bridges

A component bridge allows you to accurately measure capacitors and inductors. This is particularly useful when building filters or equalization networks, and allows you to remove initial component value as a source of error. Component bridges also measure resistance, often to a rather lower value than a digital multimeter.

The Marconi TF2700 is an excellent instrument, and you will see it advertised in the electronics magazines for a very reasonable price second-hand. It uses a single PP9 battery, and consumption is so low that it is not worth the bother of making a mains adapter. It will measure capacitance (0.5 pF–1100 μF), inductance (0.2 μH–110 H) and resistance (10 mΩ–11 MΩ) to a basic accuracy of 1%, but it is easily tweaked (using itself as a reference, plus external 0.1% resistors) to an accuracy of 0.25%. It can indicate loss factor of capacitors (very useful), and can measure air-cored inductors – digital bridges often can't. The circuit is very simple so it can be fixed easily if it develops a fault.

'Variac'

This is a variable output voltage mains transformer, *not an isolating transformer*, and will allow you to gently bring up the volts on your newly built, mega-expensive amplifier, and check for faults before smoke appears. They are often quite cheap second-hand, but check that the windings and brushes aren't damaged if it is unenclosed, and put it in a box to make it safe.

Oscilloscope

Curiously, this is of limited use in analogue audio, other than for checking the stability of amplifiers and hum in power supplies. For digital audio, an oscilloscope is invaluable but it will then need to be >100 MHz bandwidth, preferably with a dual timebase and two channels. This will be expensive, so you will need a very good reason for buying one.

Oscillator/dedicated audio test set

If you have an oscilloscope, you also need an oscillator; it is also *very* useful for supplying an external source of AC to the component bridge so that components can be tested at different frequencies. Air-cored inductors are more easily measured at 20 kHz than at 1 kHz ($X_L = 2\pi fL$), whereas the primary inductance of an output transformer should be measured at 20 Hz.

If the oscillator has a square wave output, then this can be useful for testing stability of amplifiers by observing the leading edge, which should be gently rounded (depending on frequency), but ideally should not have ringing or coggles.

You could either build a dedicated oscillator (hard work), or buy a cheap one second-hand. Fortunately, a good number of general purpose audio test sets are now appearing on the second-hand market.

Audio test sets contain at the very least:

- Fairly low distortion (<0.05%) sine wave oscillator (typically 20 Hz–20 kHz).
- Wide-band meter calibrated for reading sine waves.
- A simple form of Total Harmonic Distortion (THD) measurement.

Better test sets may include noise and/or wow and flutter (W&F) measurements; however, the main advantage of audio test sets is that they have meters scaled directly in dB, and can therefore measure audio frequency responses quickly and precisely.

In the UK, there is a variety of test sets (mostly ex-BBC) available at prices attractive to the amateur, but it should be remembered that there is almost always a reason for test equipment being cheap:

- *BBC ATM1 plus TS10*. Valve based, and both are usually lethally packaged. Don't even think about them. The Tone Source (TS10) has poor distortion. The Audio Test Meter (ATM1) has superb attenuators and the mean/flat meter very low stiction, but the amplifiers are noisy. It can't even measure THD without external assistance. The PPM has incorrect ballistics, and PPM1 to PPM2 is 6 dB, rather than 4 dB.
- *BBC EP14/1*. This was the first IC-based BBC test set. It is a true piece of laboratory equipment, capable (when calibrated) of making repeatable measurements to an accuracy of 0.05 dB. THD measurements can be made at 100 Hz and 1 kHz. The oscillator section is somewhat poorer distortion than the meter, which includes a Peak Programme Meter (PPM), THD and noise measurement.
- *Ferrograph ATS1*. An idiosyncratic, but very versatile, piece of test gear. Designed (predictably, as Ferrograph made some quite nice tape machines) for comprehensive testing of tape machines, it also includes

W&F measurement, but not a PPM. Oddly, it tends to be quite a lot more expensive than the BBC alternatives.

- *BBC ME2/5*. A 'cooking' piece of test gear designed to replace the EP14/1 in less critical uses. The oscillator is digitally synthesized, with frequency sweep. The meter section is only accurate to 0.1 dB but contains a digital frequency meter. The unit is newer, smaller, lighter, more expensive and less reliable than the EP14/1.
- *Technical Projects MJS401D and Neutrik derivatives*. A splendid piece of equipment with a wide bandwidth meter section, even better than the EP14/1, including THD measurement at *any* frequency, frequency meter, and comprehensive filters, etc. The oscillator section is potentially very good (THD <0.01%) but may be covered in microprocessor noise due to failed local power supply decoupling electrolytics. (This can be checked by invoking the 22 kHz filter whilst measuring oscillator distortion; a noticeable drop in THD shows the problem.) Options may add W&F, IMD. Later versions include a phase meter, so price may be variable; expect it to cost rather more than an EP14/1, but possibly need more attention. Thoroughly recommended, but balance is a little questionable due to electronic rather than transformer balancing.
- *Lindos LA100*. Not (yet) available at amateur prices, this is an excellent piece of semi-automated portable equipment with an LCD display rather than a moving-coil meter. It indicates to 0.01 dB, and under optimum conditions can measure distortion (at only five frequencies) to ≈0.005%. Excellent for fast routine testing of tape machines. Includes everything previously described, and more (except real mechanical PPM and transformer balancing). Can talk to printers and PCs. Super. (Rechargeable batteries included.)

Professional equipment is inevitably balanced, so test equipment is designed to cope with balanced equipment. Traditionally, transformers were used, but electronic balancing is now common, which can cause problems when used improperly. In this context, 'improperly' includes unbalancing the test set by connecting it directly to domestic (unbalanced) equipment, hence the caveats about electronic balancing. Transformer balanced equipment has no such problems.

Valve tester

Not especially recommended unless you have a very large stock of valves or are a keen designer. The AVO VCM163 is the best, but they are relatively expensive and there is a good chance that one of the (many) switches will be irredeemably faulty, rather than just dirty. Check that the code book that lives in a tray at the bottom is not missing, or incorrect for the instrument. Curiously, the construction quality of the oscillator and amplifier PCBs is rather poor, and combined with old components this can mean that a number of faults have to be fixed before the tester is able to correctly measure gm. The tester/valve is critically dependent on main voltage, and so for design work a stabilized AC mains supply is needed. Constant voltage transformers are not suitable, but stabilized mains supplies based on a motor-driven variac are ideal.

Testing

The word 'testing' implies a degree of ambiguity about the results of the test; if we *knew* that our new amplifier was going to work perfectly from the moment that it was completed, we would not need to test it. However, we know that wiring mistakes can be made, and that components could be faulty, so we test our amplifier. Carefully.

Second-hand equipment versus freshly constructed new equipment

Both types of equipment should be treated with a great deal of suspicion and apprehension. The only sensible state of mind when first switching on is controlled fear.

Quite clearly, a piece of newly constructed equipment *will* be switched on at some point, but some old equipment may be so dangerous, or riddled with faults, that it should never be energized, other than applying kinetic energy to throw it into a skip. The state of old equipment can easily be determined by looking at the components.

Things to avoid are:

- Wire insulated with rubber and covered with cotton.
- Enormous resistors marked with tip, body and ring colour codes.
- Electrolytic capacitors with bulges in the surface supporting the tags.
- Previous evidence of fire.
- Insulating tape anywhere!

Vintage radios may incorporate all of these features but still be of value to someone, so try to check with someone else before destroying them.[2]

For a professional, the worst possible sign is previous modification by an amateur. The professional then has to decide whether or not the amateur knew what they were doing. What is the effect of their modification, and was it done competently and safely? For this reason, modified equipment is usually worth *less* than unmodified equipment, so bear this in mind before you modify.

Second-hand equipment can often be dated by the date on the electrolytic capacitors. If it is over 30 years old, it is likely to need major refurbishment even to make it work, so this should be taken into account if you are considering purchase.

The first application of power

Before applying power for the first time, the following 'Ten Commandments' should be thoroughly checked:

- Inspect the earth bonding. Does the chassis appear to be properly earthed?
- When measured, is the resistance from the earth pin of the mains plug to the chassis of the equipment less than $0.5\,\Omega$?
- With the power switch on the equipment (if fitted) switched on, is the resistance from the other pins of the mains plug to the earth pin infinite?
- Does the mains cable look safe (i.e. not frayed, perished, or melted by a soldering iron?)
- Is the mains plug wired correctly?
- Does the plug grip the sheath of the cable correctly?
- Does the plug look safe (i.e. no cracks, chips, dirty pins, etc.)?
- Is a fuse of appropriate rating fitted? (It is most unlikely that the fuse rating should be greater than 3 A.)
- Does all of the internal wiring of the chassis look secure?
- Is the chassis clear of swarf and odd off-cuts of wire? Turn it so that bits can fall out, and give it a really good shake, whilst blowing vigorously into the chassis to free small parts.

If all appears to be well, you can move on to the next stage.

The author *always* assumes at this stage that when power is applied the amplifier will explode, or at the very least, catch fire. It does not, therefore, make sense to stand with your face over the amplifier, or to place it in the middle of a pile of inflammable debris.

If the equipment is a power amplifier, dummy loads or 'disposable' (cheap) loudspeakers should be connected across its outputs. If the amplifier is newly built, there is a 50/50 chance that the global negative feedback taken from the loudspeaker output will turn out to be *positive* feedback, and the amplifier will become a power oscillator. Before oscillation starts, a steadily increasing hum is heard from the loudspeaker, if the amplifier is switched off at this point, only a brief shriek of oscillation will be suffered. The negative loudspeaker terminal of the amplifier is connected to 0 V, and in an amplifier with series applied negative feedback, the other terminal becomes the non-inverted output by definition. Therefore, in order to maintain absolute phase, positive feedback should be cured by swapping the inputs to the output valves, and not by changing the feedback connection.

The equipment should be placed on a cleared bench or table, and a voltmeter set to the appropriate range should be connected across the HT supply. Leave the meter where the display can be clearly seen from a distance.

Retire to a safe distance, and switch on the power, in silence. This way, you will hear any unusual noises, such as the crackles or pops that presage destruction. If nothing untoward happens, move a little closer, and sniff the air, can you smell burning? Are there any little wisps of smoke leaving the chassis? If all still seems to be well, look closely at the heaters; they should be glowing. Listen. Are the loudspeakers making any unusual noises? In this instance, silence really is golden. Is the HT voltage correct? If the HT voltage is correct, then it is highly likely that the circuit is working as it should, and you may breathe a quiet sigh of relief. If, at any point, something untoward happens, switch off immediately at the mains outlet, and unplug the mains plug. Usually, if anything is wrong in an electronic circuit, heat is generated, and components are burnt, so look for charred resistors or wires. Once the burnt parts are found, the fault is often blindingly obvious.

If you switched off hurriedly because of a burning smell, what sort of a smell was it? Old equipment will often be dusty, and so a slight burnt dust smell will be normal. Bacon smells are sometimes produced by burning mains transformers, whereas burning PCBs often smell like underground railway stations with a hint of charcoal, and burning wiring gives off an acrid smell.

If the amplifier appeared to be satisfactory, leave it switched on for a minute or two longer, whilst keeping an eagle eye on everything, particularly output valves, which should not have glowing anodes, or purple and white flashes. Switch off and sniff the internals closely for unusual smells. Some engineers go one step further and touch components with their finger to check temperature, but this is not recommended as some HT may still be present.

If all seems to be well, the amplifier can be switched on again and all the DC voltages checked; if it still looks good, then it probably *is* good, and can be pressed into service.

For the first few weeks of service, a new amplifier should be watched like a hawk for signs of incipient self-immolation, and should not be left unattended when switched on.

Faultfinding

DC conditions

The most common fault is a lack of signal, or gross distortion.

Most faults can be found very quickly by measuring the *DC conditions* of the circuit. For a well-designed circuit to be observably faulty, the DC voltages usually need to be very wrong; consequently, checking the measured voltages against the design voltages will quickly pinpoint the fault. Mark the measured voltages (lightly, in pencil) on the circuit diagram. This will usually have the effect of making the fault appear blindingly obvious.

Sometimes you will not have the circuit diagram of the amplifier, let alone the design voltages. No matter, there are very few variations possible in valve circuitry, and most circuits are so simple that it is not usually too difficult to produce a block diagram of the amplifier. At this point, consider how you would design the amplifier, and look for similarities in the actual amplifier. It should now be possible to obtain a rough idea of what sensible voltages might be, and these can be checked against the faulty circuit.

The most important question you can ask is, 'Did it work once?' If it worked once, then you are looking for a faulty component, and the resistance range of your multimeter will prove invaluable for finding resistors that have 'gone high' in value from age, or capacitors that have gone leaky. If the circuit is freshly built, then you are probably looking for a wiring error.

In all cases, try to break the problem down into blocks. If you are looking at a power amplifier, is the output stage innocent? If it is, then check the driver stage, etc. Check power supplies. If power supply volts are suspiciously high, then something is not drawing current. If they are low, then there ought to be some smoke somewhere.

A calculator is invaluable for calculating currents through resistors, and generally deciding whether measured voltages make sense.

Do not implicitly believe what you digital voltmeter tells you. Even the 10 MΩ input impedance of a digital voltmeter can load some circuits, particularly the grid circuit of cathode followers. The author was once convinced that audible distortion was due to the DC conditions within a valve active crossover, and a digital voltmeter appeared to confirm the theory, but a valve voltmeter with 90 MΩ input impedance measured a more correct value, and the distortion turned out to be due to a faulty loudspeaker drive unit.

Be careful as you poke around within the amplifier. More than one piece of equipment has been blown up by the slip of a probe.

AC conditions

Occasionally, testing DC conditions will not reveal the fault. An oscilloscope to probe around and trace the signal is the obvious approach, but not everybody has an oscilloscope.

The inherent microphony in valves can be useful here, and valves can be gently thumped with a screwdriver handle whilst listening to a 'disposable' loudspeaker on the output. When the 'ting' stops dead, the area of the fault has been found, and components can be replaced or tested until the fault is found.

A more awkward fault is noise, or intermittent operation due to a dodgy joint. One way to deal with this is to gently move each valve in its valve base. The one that significantly changes the fault is the stage with the fault. The next ploy is to thump components and soldered joints with the handle of an insulated screwdriver, the most sensitive part is the faulty part.

Semiconductor circuitry can be tested by heating with a hair dryer to produce the fault, and then selectively cooling parts with a can of aerosol freezer. This can be expensive in freezer, and should be used as a last resort after thumping with a screwdriver handle. Squirting freezer spray near a valve is liable to result in a cracked valve envelope.

HF oscillation can only be satisfactorily investigated with an oscilloscope and an oscillator. A square wave is excellent for provoking oscillations, but the negative feedback loop of a power amplifier may either be the cause of oscillations or conceal the faulty stage, so try removing the loop, and overdriving the amplifier whilst examining each stage.

UHF oscillation can sometimes be deduced by soldering a wire to the 0 V HT rail and touching it to the chassis. If it makes a lot of noise as it is scraped, UHF oscillation is likely, and valves can be successively removed until the problem disappears. This problem is only likely with circuits incorporating E88CC, 5842, etc. without grid stopper resistors.

Note that on DC coupled amplifiers or amplifiers with series heaters, removing valves will not be permissible, and may even cause damage. Check your circuit first.

Motorboating is a low frequency (1 Hz) oscillation that is invariably due to unwanted power supply coupling, so experiment to see if extra capacitance at any point will change the frequency. If you can change something, then you must be near to the source of the fault.

100 Hz hum is usually due to poor power supply smoothing. Magnetically induced hum (hum loops, wires near mains transformers) will have a strong 50 Hz content, whereas electrostatically induced hum (unscreened wires, poor earthing) is more of a buzz due to the additional higher harmonics.

Classic amplifiers: comments

It should be realized that the following remarks relate to the author's personal experience of a few samples of each amplifier, but the comments are included because some guidance is better than none. Various amplifiers, such as Radford, are not included, not because the author has any bias against them, but simply because he has not owned one.

As a very broad generalization, amplifiers using more expensive output valves will be better. So amplifiers using KT66 may be better than EL34, which will be more powerful than EL84, and ECL82 or ECL86 are at the bottom of the heap. Curiously, amplifiers using KT88 *may* be worse than any of these, because they may have been designed as public address amplifiers, purely for their high output power.

The quality of output transformers is crucial. Poor output transformers will be small for their rating, although C-core transformers may be an exception to this rule and are almost a guarantee of good quality.

Quad II

There are still an awful lot of these about, and they generally require very little work to restore them to their original (very good) performance. They are popular with tweakers, and various modifications can be found. Mostly the modifications replace the GZ34 with silicon to increase output power, and others replace the (virtually unobtainable) GEC KT66 with EL34.

Williamson

These were almost all amateur made, so finding a matching pair is unlikely, and build quality may be less than wonderful. However, at the very least, the output transformer is well worth salvaging.

Leak TL12 and BBC LSM/8 derivative

This amplifier has an output stage very similar to the Williamson, and has recently become very fashionable (expensive), but they are likely to be in quite poor condition because of their age.

The BBC version lived in a compartment at the bottom of the LSU/10 loudspeaker (where it became very hot). Designed for a studio environment, the amplifier has a transformer balanced input and a volume control. In common with many BBC loudspeaker amplifiers, some versions included a bass equalizer. Once the BBC modifications are removed, the two amplifiers are identical.

Leak TL12+

This is a completely different beast to the TL12, and is very similar to a Mullard 5–10 using EL84 output valves. They are comparatively recent (typically only 30 years old), and so may not need much remedial work, but by modern standards they are noisy, due to high sensitivity and an EF86 input pentode.

Leak Stereo 20

More common than the TL12+, this is almost a pair of TL12+ on one chassis, but sharing a slightly under-rated mains transformer, and with a few other corners cut. Input valve is ECC83, but they are still noisy. A pair of TL12+ is preferable unless you are only buying the amplifier for the chassis and transformers.

Leak TL10

Similar in design to the Mullard 5–20, but uses 6SN7 phase splitter. Quite a nice conservatively rated amplifier designed for KT61 (irreplaceable), but can be modified for EL34 – not often seen.

Rogers Cadet

Only 6W using ECL82 output valves. There were two versions, an integrated version and a separate chassis version. Ideal for the beginner to cut their teeth on, but not of great intrinsic value. Once component failures begin, they have a habit of continuing catastrophically. Oddly, the disc input stage was very good for its time, but it uses ECC807 (μ = 140, and irreplaceable), now frequently modified to use ECC83 (μ = 100).

BBC amplifiers

Unfortunately, many BBC designed amplifiers were designed for 25 Ω loudspeakers, and cannot be modified for 8 Ω without replacing the output transformer. It is also well worth asking why the BBC disposed of the amplifier, particularly if it is thought to have come from an impoverished local radio station. It is most unlikely that it has spent its time cherished in a protective box in a dry cupboard.

Further reading

Radio Society of Great Britain (1976) *Radio Communication Handbook*, 5th ed. RSGB, Potters Bar, Herts.

Emmerson, Andrew (1998) *Electronic Classics*. Newnes. ISBN 07506 3788 9.

Appendix

Thermionic emission

The Richardson/Dushmann equation for emitted cathode current (I) per unit area (m^2) is:

$$I = RT^2 . \varepsilon^{-q_e \phi / kT}$$

where: $R = \left(\dfrac{4\pi m_e q_e k^2}{h^3} \right) = 1.204 \times 10^6 \text{ Am}^{-2} \text{ K}^{-2}$, see note.

T = absolute temperature
ε = base of natural logarithms
m_e = electron rest mass
q_e = electronic charge
Φ = work function of the cathode surface
k = Boltzmann's constant
h = Planck's constant

Note: Although the theoretical value for $R = 1.204 \times 10^6$, the experimental value is about half this value.

Square wave sag and low frequency $f_{-3\,\text{dB}}$

A square wave with LF sag is a decaying exponential, whose instantaneous voltage at any time 't' may be found from:

$$v = V_0 e^{-t/\tau}$$

Rearranging and solving for τ:

$$\tau = \frac{-t}{\ln\left(\frac{v}{V_0}\right)}$$

't' is the time allowed for the decay across the bar top, but for a square wave with equal positive and negative durations, it is half of the periodic time T:

$$T = 2t$$

But T is the reciprocal of frequency:

$$f = \frac{1}{T}$$

So:

$$t = \frac{1}{2f}$$

Substituting:

$$\tau = \frac{-1}{2f\ln\left(\frac{v}{V_0}\right)}$$

From the frequency domain, a CR filter has a -3 dB cut-off frequency:

$$f_{-3\,\text{dB}} = \frac{1}{2\pi CR}$$

But $CR = \tau$, and $\tau = L/R$, so a universal equation, valid for both CR and LR, is:

$$f_{-3\,\text{dB}} = \frac{1}{2\pi\tau}$$

Rearranging:

$$\tau = \frac{1}{2\pi f_{-3\,\text{dB}}}$$

We now have two formulae for τ, which can be equated:

$$\frac{1}{2\pi f_{-3\,\mathrm{dB}}} = \frac{-1}{2f\ln\left(\dfrac{v}{V_0}\right)}$$

Solving for the ratio $f/f_{-3\,\mathrm{dB}}$:

$$\frac{f}{f_{-3\,\mathrm{dB}}} = \frac{-\pi}{\ln\left(\dfrac{v}{V_0}\right)}$$

Sag is the percentage of peak to peak level by which the horizontal bar has sagged in level, and 10% sag is easily measured on an oscilloscope, and means that the level has fallen from 100% to 90% so:

$$10\%\ \mathrm{sag} \equiv \frac{v}{V_0} = 0.9$$

Applying 10% sag to the $f/f_{-3\,\mathrm{dB}}$ formula:

$$\frac{f}{f_{-3\,\mathrm{dB}}} \approx 30$$

So 10% sag means that the applied square wave frequency is 30 times higher than $f_{-3\,\mathrm{dB}}$.

Sag observed using a square wave of frequency '*f*'	**$f/f_{-3\,\mathrm{dB}}$**
10%	30
5%	60
1%	300

Valve data

This data is reproduced by courtesy of Philips Components Limited.

Characteristics

	ECC81	ECC82	ECC83	E88CC	ECF80 (triode)
V_a	250 V	250 V	250 V	90 V	100 V
I_a	10 mA	10.5 mA	1.2 mA	15 mA	14 mA
V_g	−2 V	−8.5 V	−2 V	−1.2 V	−2 V
gm	5.5 mA/V	2.2 mA/V	1.6 mA/V	12.5 mA/V	5 mA/V
μ	60	17	100	33	20
r_a	11 kΩ	7.7 kΩ	62.5 kΩ	2.65 kΩ	4 kΩ

Capacitances

Unfortunately, because the valves shown in the next table were not all designed for the same use, measurements made for one valve were not necessarily made for another, hence the gaps in the table. For audio use, C_{ag} is by far the most important, whilst C_{in} is the capacitance from the control grid to all other electrodes, and C_{out} is the capacitance from the anode to all other electrodes. Note that the E88CC is the only valve included that incorporates a screen between anodes.*

	ECC81	ECC82	ECC83	E88CC	ECF80 (triode)
C_{ag}	1.6 pF	1.5 pF	1.7 pF	1.4 pF	1.5 pF
C_{gk}			1.6 pF	3.3 pF	2.5 pF
C_{ak}	200 fF			180 fF	1.8 pF
C_{as}				1.3 pF*	
C_{in}	2.3 pF	1.8 pF			
$C_{out'}$		370 fF			
$C_{out''}$		250 fF			
$C_{a'a''}$	<400 fF	<1.1 pF	<1.2 pF	<45 fF	
$C_{a'g''}$	<70 fF	<110 fF	<100 fF	5 fF	
$C_{a''g'}$	<40 fF	<60 fF	<100 fF	5 fF	

Limiting values

Again, not all valves have all parameters measured. Note that although the E88CC specifies the permissible heater to cathode voltages by defining leakage current as <6 μA, the other valves do not mention what limits were applied.

	ECC81	ECC82	ECC83	E88CC	ECF80 (triode)
$V_{a(b)\ max.}$	550 V	550 V	550 V	400 V	550 V
$V_{a\ max.}$	300 V	300 V	300 V	220 V	250 V
$p_{a\ max.}$	2.5 W	2.75 W	1 W	1.5 W	1.5 W
$I_{k\ max.}$	15 mA	20 mA	8 mA	20 mA	14 mA
V_g (I_g = +300 nA)	–1.3 V	–1.3 V	–0.9 V		–1.3 V
$R_{gk\ max.}$ (cathode bias)	1 MΩ		22 MΩ	1 MΩ	
$R_{gk\ max.}$ (grid bias)		1.5 MΩ	2.2 MΩ		
$V_{hk\ max.}$	150 V	180 V	180 V		
$V_{hk\ max.}\ k+$				120 V	150 V
$V_{hk\ max.}\ k-$				60 V	100 V
$R_{hk\ max.}$	20 kΩ	20 kΩ	20 kΩ		

Heaters

		ECC81	ECC82	ECC83	E88CC	ECF80
Current (mA)	Series (12.6 V)	150 mA	150 mA	150 mA		
	Parallel (6.3 V)	300 mA	300 mA	300 mA	300 mA	430 mA

The E88CC/6922 is a special quality version of the ECC88/6DJ8, and all electronic characteristics are identical, except that the heater is specified as 365 mA. *However*, only Mullard/Philips/Amperex seem to have adhered to this specification, and most manufacturers use 300 mA heaters. Brimar specify 365 mA, but the author tested 10 Brimar ECC88/6DJ8, and found they were 300 mA.

Pin connections

Unlike ICs, B9A and International Octal valves are numbered *viewed from underneath* counting clockwise.

ECF80 pin connections

a_t	g_t	k_t	a_p	g_{2p}	g_{1p}	k_p	h	h
1	9	8	6	3	2	7	4	5

Triode pin connections

<table>
<tr><th></th><th>ECC81</th><th>ECC82</th><th>ECC83</th><th>E88CC</th></tr>
<tr><td>a_1</td><td colspan="4">1</td></tr>
<tr><td>g_1</td><td colspan="4">2</td></tr>
<tr><td>k_1</td><td colspan="4">3</td></tr>
<tr><td>a_2</td><td colspan="4">6</td></tr>
<tr><td>g_2</td><td colspan="4">7</td></tr>
<tr><td>k_2</td><td colspan="4">8</td></tr>
<tr><td>h</td><td colspan="3">4</td><td>4</td></tr>
<tr><td>h</td><td colspan="3">5</td><td>5</td></tr>
<tr><td>$h(ct)$</td><td colspan="3">9</td><td>none</td></tr>
<tr><td>screen</td><td colspan="3">none</td><td>9</td></tr>
</table>

Graphical valve data

The following six pages contain the anode characteristic (I_a against V_a) curves for each of the above valves.

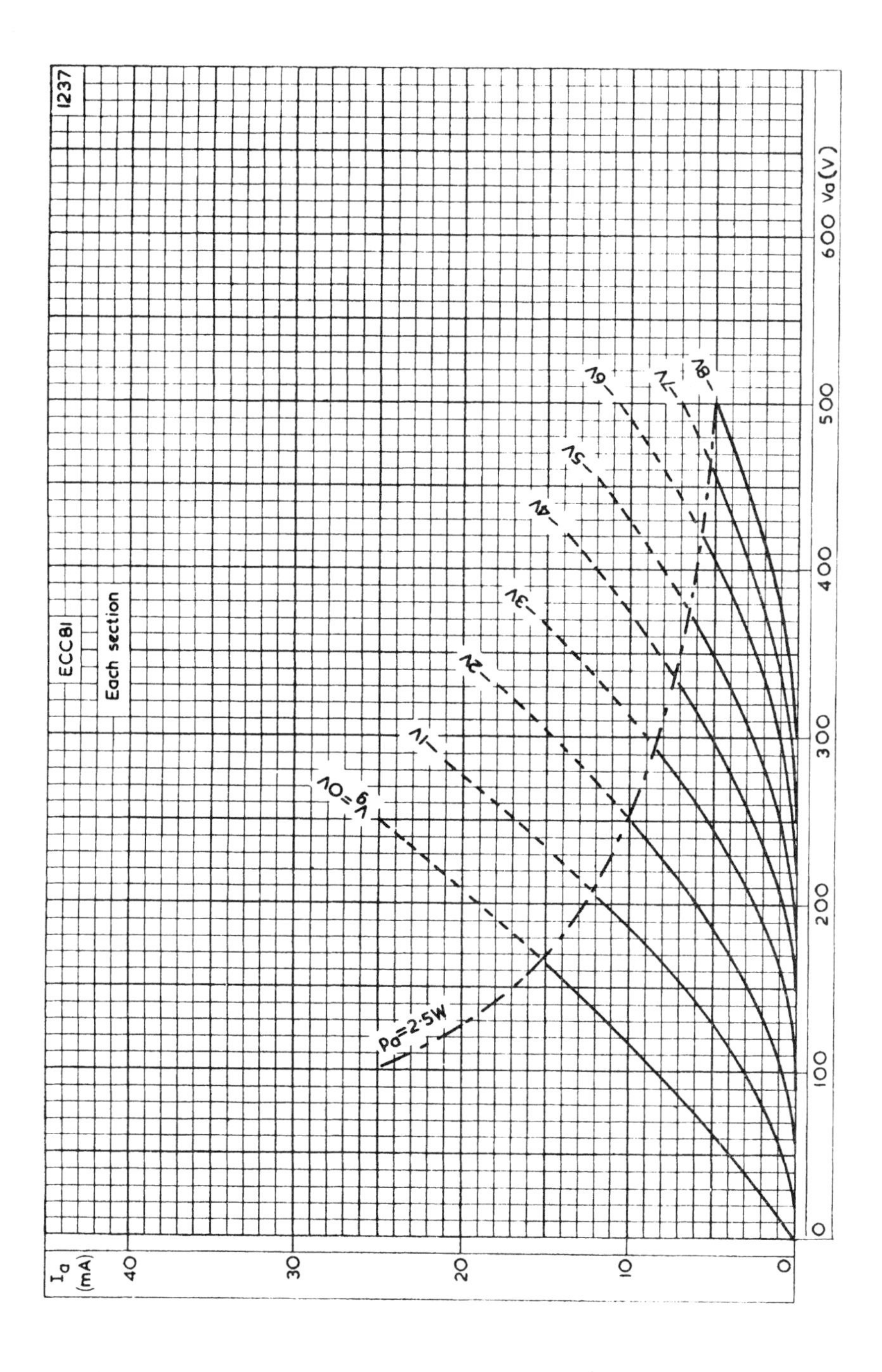

1237
ECC81
Each section
I_a (mA)
V_a (V)
P_a=2·5W
V_g=0V
-1V
-2V
-3V
-4V
-5V
-6V
-7V
-8V
0
10
20
30
40
0
100
200
300
400
500
600

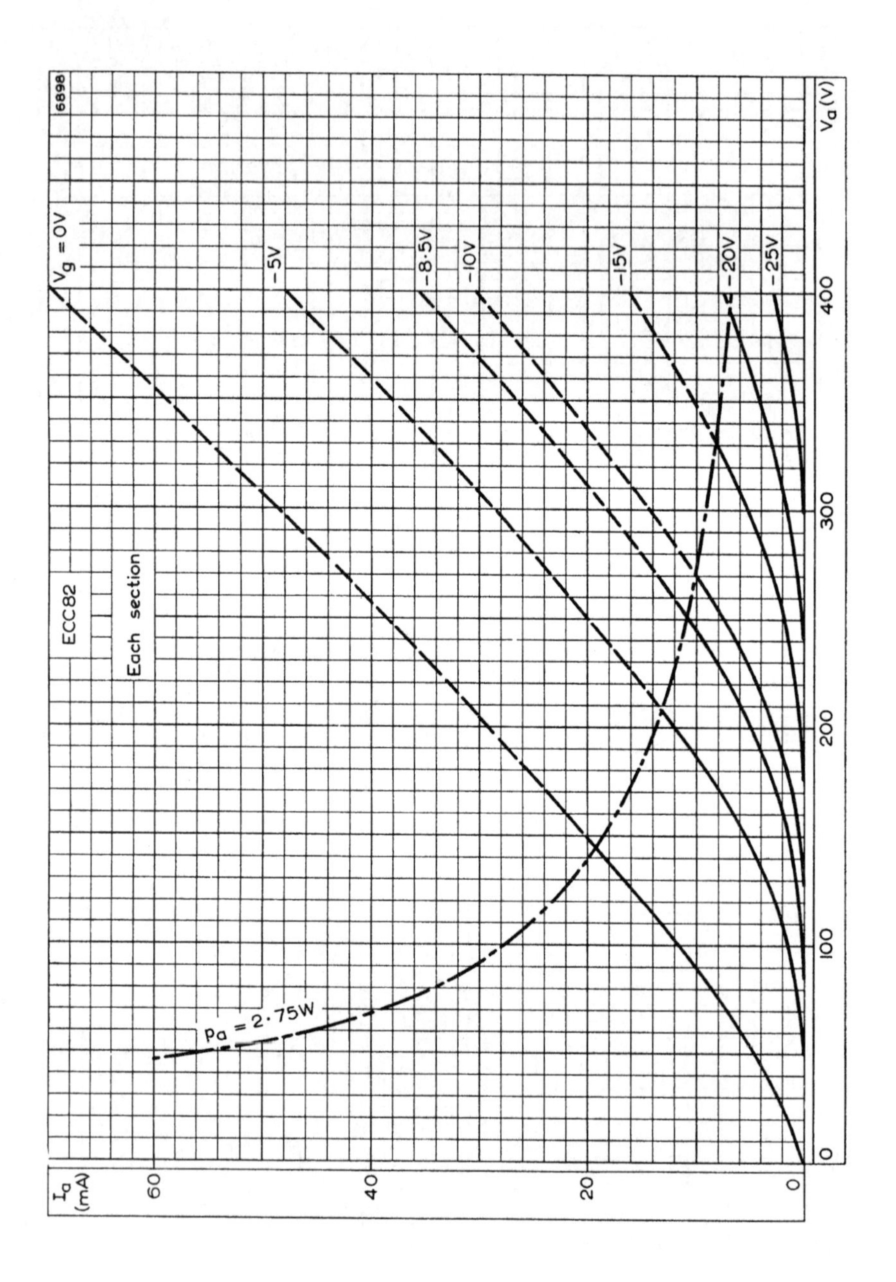
ECC82
Each section
6898
V_g = 0V
-5V
-8·5V
-10V
-15V
-20V
-25V
P_a = 2·75W
I_a (mA)
V_a (V)
0
20
40
60
0
100
200
300
400

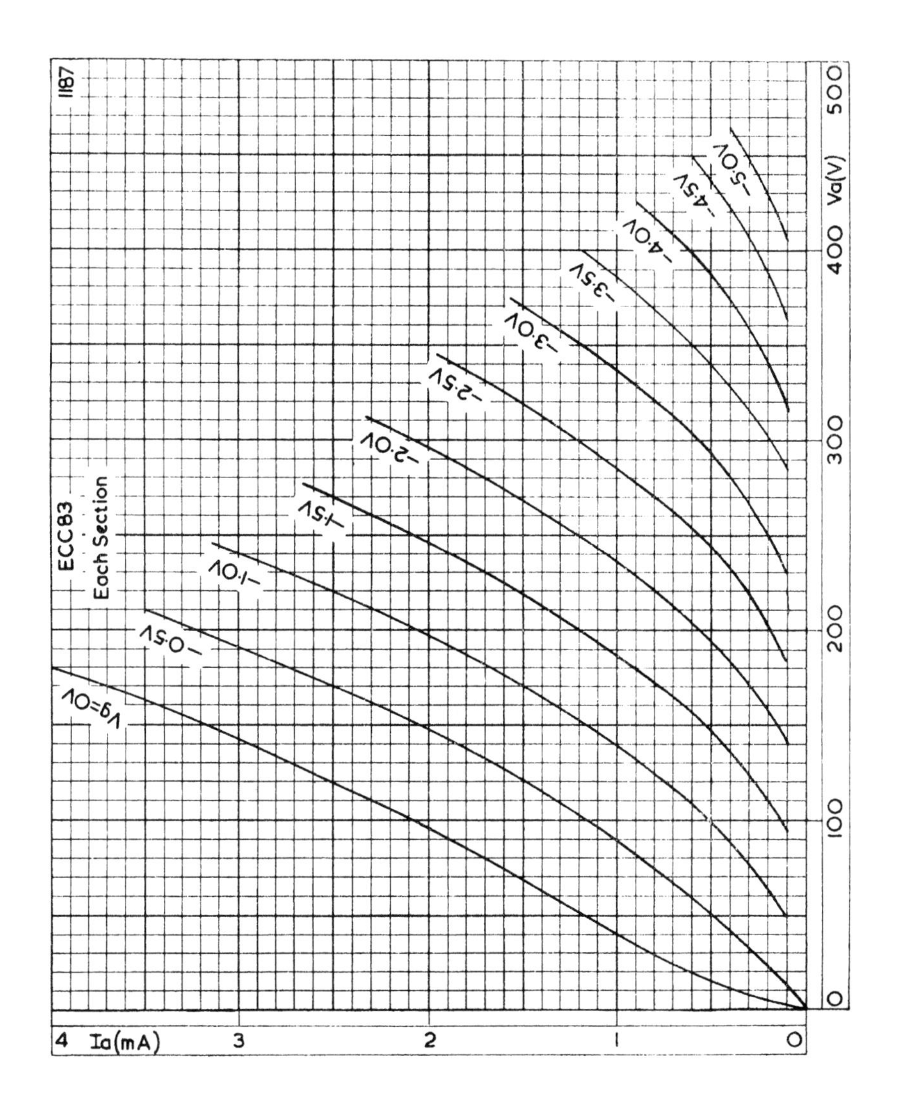

ECC83
Each Section
1187
Vg=0V
-0.5V
-1.0V
-1.5V
-2.0V
-2.5V
-3.0V
-3.5V
-4.0V
-4.5V
-5.0V
Va(V)
0
100
200
300
400
500
4
Ia(mA)
3
2
1
0

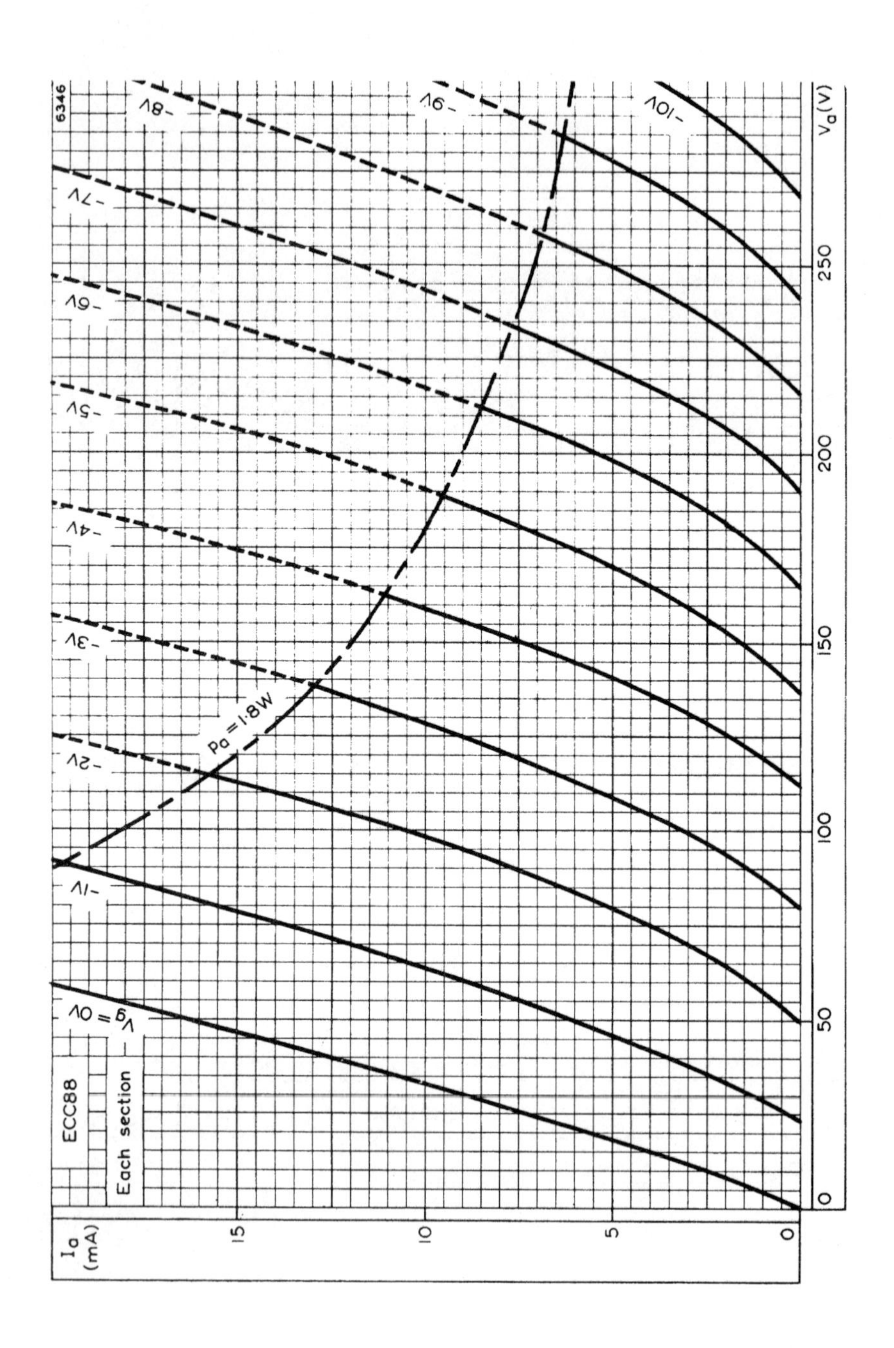

ECC88
Each section
6346
I_a (mA)
V_a (V)
V_g = 0V
-1V
-2V
-3V
-4V
-5V
-6V
-7V
-8V
-9V
-10V
Pa = 1·8W
0
5
10
15
0
50
100
150
200
250

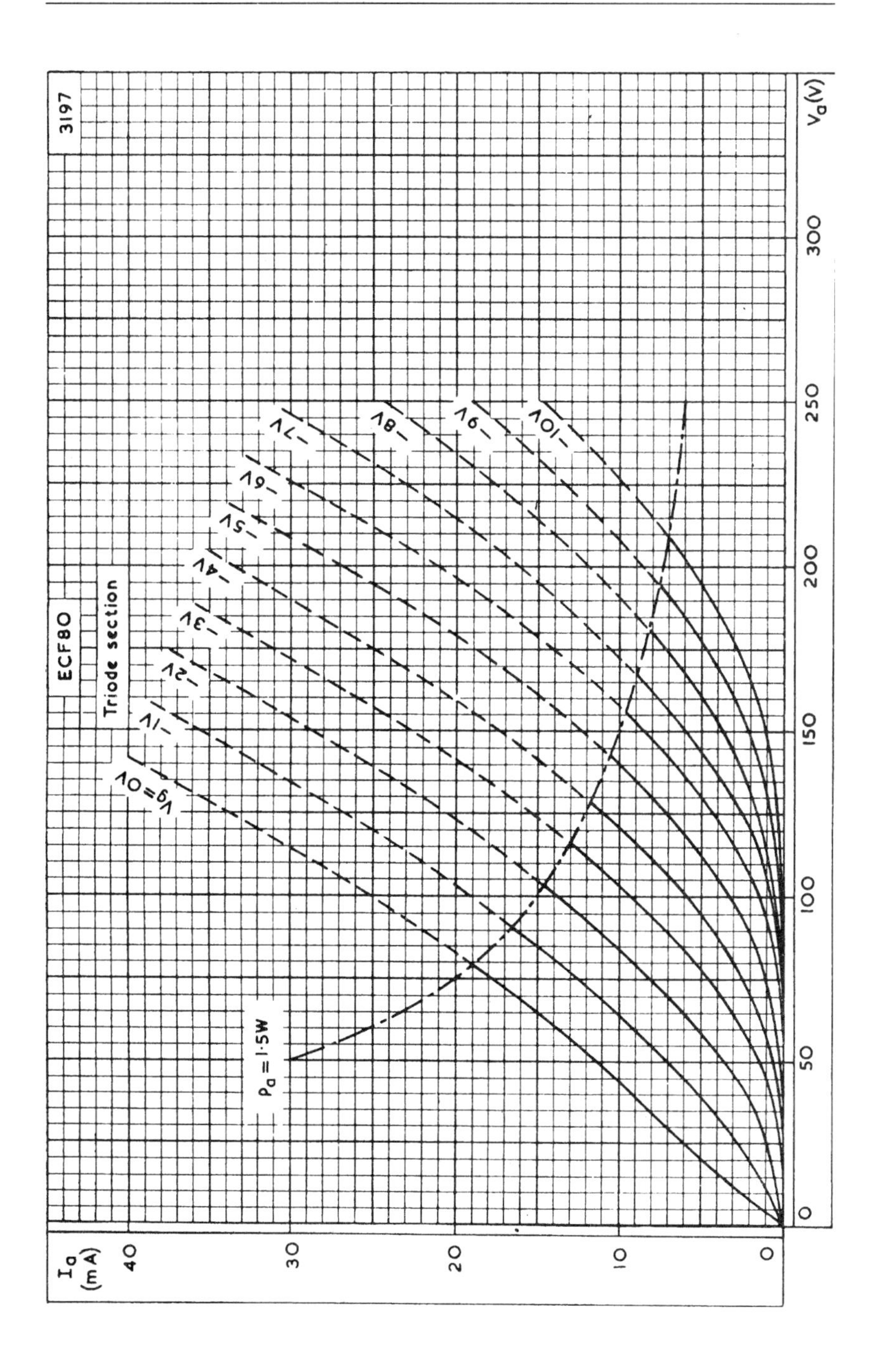
ECF80
Triode section
3197
$V_g = 0V$
−1V
−2V
−3V
−4V
−5V
−6V
−7V
−8V
−9V
−10V
$P_a = 1{\cdot}5W$
I_a (mA)
V_a(V)
0
10
20
30
40
0
50
100
150
200
250
300

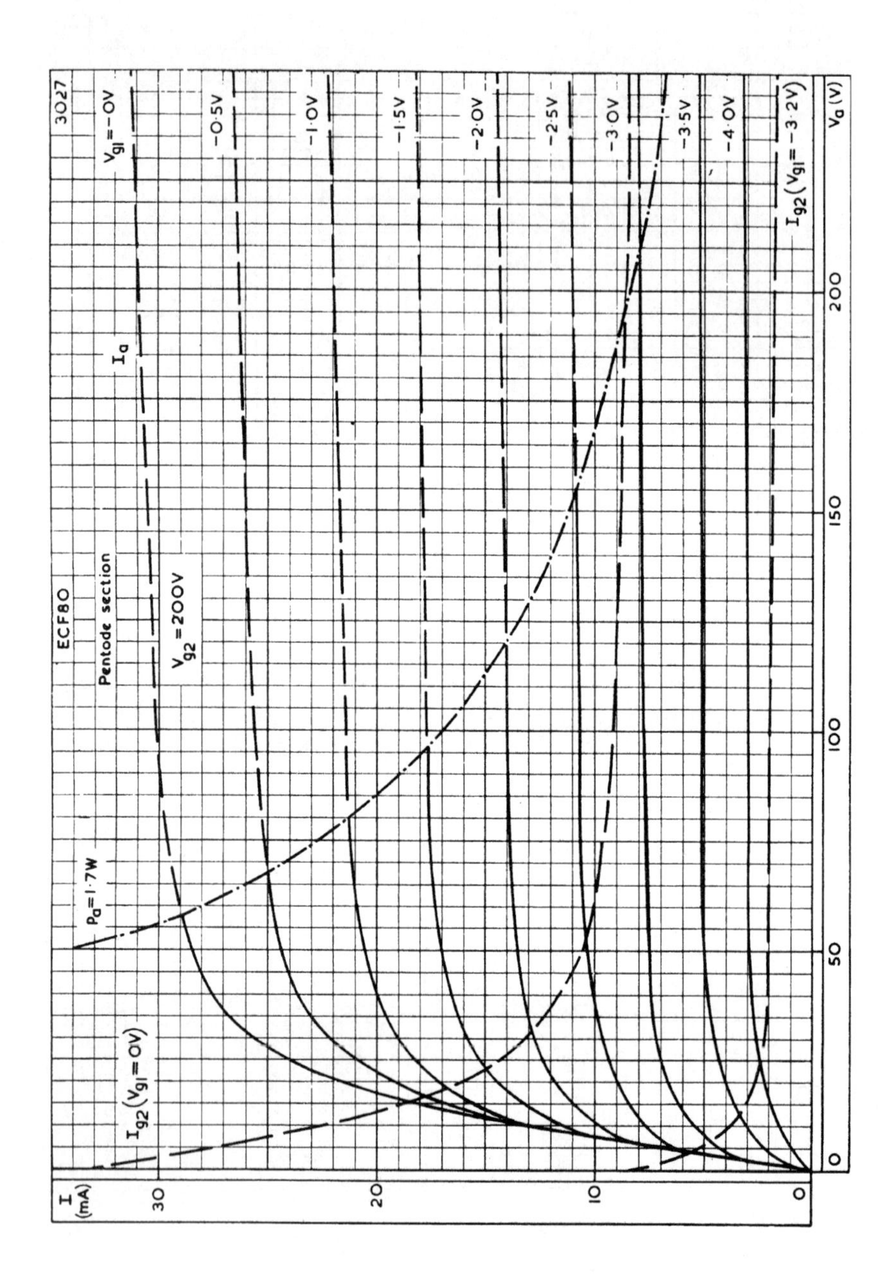
3027
ECF80
Pentode section
V_{g2} = 200V
V_{g1} = −0V
I_a
−0·5V
−1·0V
−1·5V
−2·0V
−2·5V
−3·0V
−3·5V
−4·0V
P_a = 1·7W
I_{g2} (V_{g1} = 0V)
I_{g2} (V_{g1} = −3·2V)
I (mA)
30
20
10
0
V_a (V)
0
50
100
150
200

Playing 78s

You have just inherited a collection of 78s that appear to be in superb condition, some of them are original recordings of legendary performers, and you are desperate to play them.

There are four main problems.

Correct speed

Although colloquially known as '78s', referring to their speed, very early 78s were recorded on rather crude lathes and their recorded speed was somewhat variable. The first requirement for replay is therefore a turntable that will not only rotate at 78 rpm, but also has varispeed, so the Garrard 301/401, and Thorens TD124 are obvious contenders. The BBC modified the Technics SP10 direct drive turntable to give varispeed, added a pick-up arm/cartridge plus elaborate electronics, and called the whole confection an RP2/10 (ReProducer 2, version 10). The Technics SL1200 turntable is also capable of playing 78s.

Groove size

The 78 has a coarse groove, and was traditionally played with a crude steel 'needle'. The LP stylus of a modern cartridge is far too small, and a dedicated large diameter stylus is required. Since a dedicated 78 cartridge would be commercial suicide, a 78 stylus is usually only offered by original equipment manufacturers as a replaceable stylus for moving magnet cartridges, and only then if the primary market for that cartridge is likely to use 78s. Broadcasters need robust cartridges, and are likely to play 78s, so the Shure SC35 and the Ortofon OM Pro were both offered with 78 styli.

The OM Pro was designed a decade later, and superseded the SC35 in 1985/86 – just in time to be eclipsed by CD. Fortunately, clubs prefer to use vinyl, partly for the sound, but mostly because turntables (such as the Technics SL1200) have varispeed, and allow the beat of one record to be synchronized to the beat of the next, allowing a seamless crossfade. The OM Pro and Concorde variants are now the universal professional cartridges and are readily available, so the 78 stylus is also obtainable, although it may need to be specially ordered. Interestingly, Shure have just introduced a new mono cartridge, the M78S, designed specifically for playing 78s. It is a great advantage for the cartridge to be mono, since it simplifies cartridge construction, and the dedicated 78 pre-amplifier now only needs one channel –

thus avoiding the technically dubious 'mono' switch (which usually forces one low resistance source to drive another).

However, a specialist re-tipping concern may be prepared to fit a 78 rondel to your cartridge.

78s *must* be played with a 78 stylus. Because the groove of a 78 is so much larger, an expensive microgroove stylus will founder on the bottom of the groove, producing unnecessary noise, and wearing itself on the abrasive shellac as it does so.

Pick-up arm mechanics

Playing a 78 drives a lot of vibration into the pick-up arm, and loose bearings will cause rattles and mistracking. At a more subtle level, a stylus traversing an imperfection, or speck of dust, produces an mechanical impulse and excites arm resonances which greatly magnify the subjective nuisance. Paradoxically, the inferior medium needs a good arm to replay it adequately – a modern arm such as the Rega RB250 arm and derivatives seems a minimum requirement.

Equalization

It took some time before the manufacturers of 78s and LPs standardized on their equalization. The following table gives the electrical time constants used by major organizations, and therefore an indication of the likely equalization required.[1]

Time constants (μs)		t_3	t_4	t_5
78	'Standard'	—	636	—
	Decca 'ffrr'/European	—	636	25
	AES	—	400	63.6
	Pre-1954 DG	—	450	50
	International	**3180**	**450**	**50**
LP	Pre-1954 DG	1590	450	50
	Pre-1954 Decca	1590	318	50
	Columbia/EMI	1590	318	100
	European	2230	318	50
	NAB	3180	318	100
	RCA New Orthophonic	3180	318	75
	RIAA	**3180**	**318**	**75**

(All time constants are specified according to Lipshitz' notation.[2])

Since t_3 for most 78 standards was not specified, and a practical design cannot continue bass boost indefinitely, 3180 μs should be used unless otherwise specified. Errors in t_4 and t_5 cause peaks and troughs in the critical mid-band, and despite popular belief cannot possibly be corrected using tone controls.

If only the later 78s (when an international standard had been fixed) are to be played in addition to modern LPs, then t_4 need only be switchable between 318 μs and 450 μs, and t_5 between 75 μs and 50 μs.

Analogue 'microgroove'

RIAA: 3180 μs, 318 μs, 75 μs.
IEC: 7950 μs, 3180 μs, 318 μs, 75 μs.

Note that the 7950 μs time constant for IEC is *replay only*, and is a 20 Hz high-pass filter intended to remove rumble produced by turntables. See Chapter 6 for detail on implementation.

CD

50 μs, 15 μs.

This equalization is only very rarely used on CDs, and is accompanied by a sub-code flag to enable the player to switch in the equalization. Many oversampling filters implement this equalization digitally, but older players need analogue equalization.

NICAM

Various countries now have terrestrial broadcast TV with digital sound, encoded as NICAM (Near Instantaneous Companding And Modulation). This uses a CCITT equalization characteristic known as J17, whose response curve can be calculated from the following formula:

$$\text{Loss}_{(\text{pre-emphasis})} = 10 \log_{10} \frac{75 + \left(\frac{\omega}{3000}\right)^2}{1 + \left(\frac{\omega}{3000}\right)^2} \text{ (dB)}$$

This is not a useful equation for determining component values for de-emphasis, so the following equations will be found to be far more useful:

$$C = \frac{1}{3000x}$$

$$x = y\left(\sqrt{75} - 1\right)$$

$$y = \frac{x}{\sqrt{75} - 1}$$

where x, y, and C are the components in the diagram. See Fig. A.1.

Standard resistor values of 36 and 4.7 for x and y result in a ratio error of <0.01%.

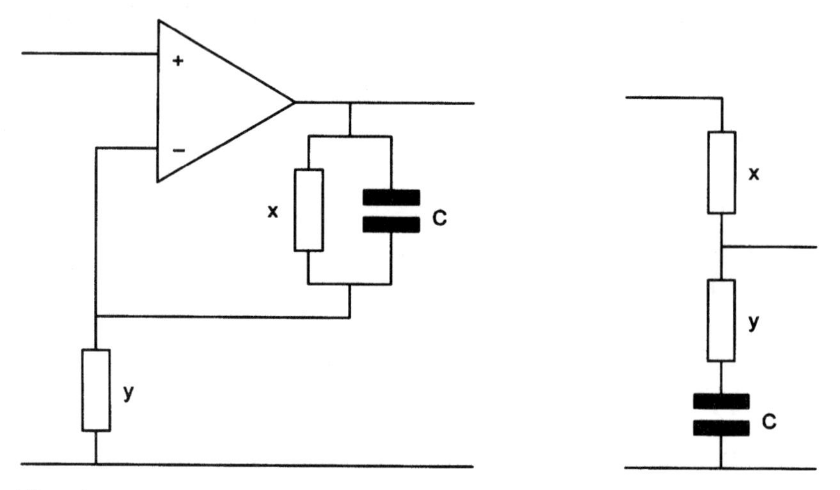

Fig. A.1.

Standard component values

The following series of components covers one decade; other values are obtained by multiplying or dividing by factors of ten.

E6

1	1.5	2.2	3.3	4.7	6.8

E12

1	1.2	1.5	1.8	2.2	2.7	3.3	3.9	4.7	5.6	6.8	8.2

E24

1	1.1	1.2	1.3	1.5	1.6	1.8	2	2.2	2.4	2.7	3
3.3	3.6	3.9	4.3	4.7	5.1	5.6	6.2	6.8	7.5	8.2	9.1

E96

1	1.02	1.05	1.07	**1.1**	1.13	1.15	1.18	1.21	1.24	1.27	**1.3**
1.33	1.37	1.4	1.43	1.47	**1.5**	1.54	1.58	1.62	1.65	1.69	1.74
1.78	1.82	1.87	1.91	1.96	**2**	2.05	2.1	2.15	2.21	2.26	2.32
2.37	2.43	2.49	2.55	2.61	2.67	2.74	2.8	2.87	2.94	3.01	3.09
3.16	3.24	3.32	3.4	3.48	3.57	3.65	3.74	3.83	3.92	4.02	4.12
4.22	4.32	4.42	4.53	4.64	4.75	4.87	4.99	5.11	5.23	5.36	5.49
5.62	5.76	5.9	6.04	6.19	6.34	6.49	6.65	6.81	6.98	7.15	7.32
7.5	7.68	7.87	8.06	8.25	8.45	8.66	8.87	9.09	9.31	9.53	9.76

Values in bold are common to E24 series.

Resistor colour code

Most resistors are marked with their value in the form of a colour code consisting of four or six concentric bands of paint on the body of the component which are read from left to right. See Fig. A.2.

Four-band resistors

The first two bands denote the two significant digits of the value.

The third band is the *multiplier*, whose value is 10^x, where x is the value of the band. Gold used as a multiplier means $10^{-1} = 0.1$, and silver means $10^{-2} = 0.01$.

The fourth band is the tolerance, which will commonly be 1% (brown) or 2% (red). On older equipment, you will see gold (5%) and silver (10%); the use of these colours as tolerances dates from the days when 5% was considered to be close tolerance! If there is no fourth band, the tolerance is 20%.

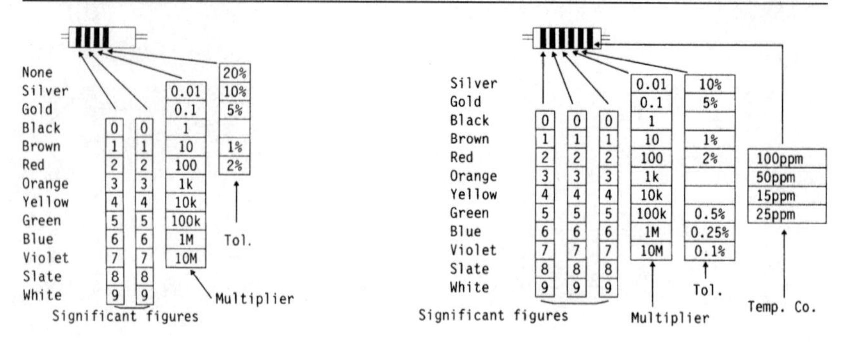

Fig. A.2.

Six band resistors

The first three bands denote the significant digits, and the fourth band is the multiplier.

The fifth band is the tolerance, note that six-band resistors imply greater precision and so 5%, or worse, tolerance will not be seen.

Examples

Yellow, violet, yellow, red = 470 kΩ 2%
Yellow, violet, black, orange, brown, red = 470 kΩ 1% 50 ppm

Red, red, red, red = 2.2 kΩ 2%
Red, red, black, brown, brown, red = 2.2 kΩ 1% 50 ppm

Brown, black, black, red = 10 kΩ 2%
Brown, black, black, gold, brown, red = 10 kΩ 1% 50 ppm

Note that because the six-band resistors have an extra significant digit, their multiplier is always one level lower than for the same value in a four-band component.

Sometimes it can be difficult to decide which end of the resistor is which, and the value makes sense either way round:

Brown, orange, yellow, red = 130 kΩ 2%, but read the other way round = 24 kΩ 1%

If in doubt, measure the resistor with a digital multimeter; it is far easier to change the component now than when it has been soldered into place.

UK sources of components

It used to be very difficult to obtain components for valve amplifiers, but there is now a plethora of vendors eager for your money. Suppliers of specialist components can be found in the hi-fi magazines, particularly those that tend to have constructional articles. Standard components can be bought from the following suppliers.

Maplin Electronics

They stock valves. See adverts in electronics magazines for shop addresses. W. H. Smith stock catalogues.

RS Components

RS do not deal directly with the public, but their retail arm Electromail stocks all of the RS range.

Farnell Electronic Components

Nominally a trade distributor, but FEC will accept orders placed on a personal credit card, so all you need to obtain is a current catalogue.

Farnell Electronic Services

A trade distributor once known as STC or ESD, they stock an unusual range of components that other stockists do not. They will accept a credit card order.

Bargains and dealing directly

The above companies distribute general electronic components, but do not necessarily stock specialist audio components such as Vishay resistors, etc.; it is well worth shopping around for specialist components, as some stockists have imaginative pricing policies.

If you and your friends are able to club together to generate a large order it can be worth approaching manufacturers directly. After all, the worst they can do is to laugh at your order.

Companies at hi-fi shows often give a 'show' discount on their goods, so it is often worth timing your order to coincide with a show. They may even be prepared to negotiate a quantity discount on the last day of the show.

Components such as transformers can often be made specially to order, and the author has had a number of transformers made by Sowter Transformers of Ipswich. If you choose to follow this course, specify as completely as possible what it is you need, and remember that every additional complication will add to the finished price.

Electronics surplus shops are excellent places for picking up bargains, provided that you know what to look for. It is a good idea to take a digital multimeter, tape measure and calculator with you.

When large companies close down a site, they often auction equipment, and this can be a fruitful source of second-hand test gear. It is your responsibility to check the condition of what you buy, and you have no comeback afterwards. Don't get carried away at an auction, and remember that VAT will be added to your bid price.

When buying second-hand equipment privately, remember to add the cost of your journey to inspect the goods, and whatever cost is required to refurbish. This can make second-hand goods quite expensive, so bargains are often best found through your circle of friends, who are aware of your hobby.

Sallen & Key filters for active crossovers

In 1955, Sallen & Key[3] wrote a seminal paper on *RC* active filter design, and all component notation and page numbers hereon refer to this paper.

Cathode followers can be used to implement Sallen & Key filters, but they are by no means perfect. A low-pass filter has a pair of series resistors (R_1, R_2) at the input, and a capacitor is connected from the centre tap of these to the output of the cathode follower. Remembering that for the cathode follower, $r_{out} \approx 1/gm$, then typically $r_{out} \approx 500\,\Omega$, if $gm = 2$ mA/V. The output resistance forms a potential divider in conjunction with the feedback resistor or capacitor, which determines the maximum attenuation possible from the input *RC* pair, so a high *gm* valve maximizes stop band attenuation, which is $\approx R_1/R_{out}$.

As an example, a 12 dB/octave $f_{-3\,\text{dB}} = 860$ Hz 6J5 section was computer simulated as a possible contender for an active crossover. Practical values for the circuit resulted in a response falling to –54 dB at 18 kHz, at which point the response began to rise again at 6 dB/octave. Fortunately, two sections in series were proposed, and deviations at –108 dB almost out of the audio band were not felt to be very significant.

Because Sallen & Key filters rely on positive feedback to operate, they considerably worsen the distortion of their gain stage,[4] so it is crucial to minimize the distortion of the amplifier before it is enveloped by the filter. Clearly, the valve and operating point should be chosen carefully, but a slightly less obvious approach is to use a high μ valve which allows plenty of (distortion reducing) negative feedback before the positive feedback is applied, and further lower distortion with a constant current load, perhaps an EF184 pentode. At a typical operating point of the single triode EC91, $\mu = 80$ and $gm = 5.7$ mA/V, making this an excellent theoretical choice, although the intrinsic linearity of the 6J5 or 6SN7 could well prove superior in the final analysis. See Fig. A.3.

Using the Sallen & Key paper

Unfortunately, the paper is somewhat tortuous, with much important information buried in obscure footnotes. The following example aims to guide you through the maze.

1 Determine the common cathode gain of the valve under chosen operating conditions, this is A_0. If we use a valve with a constant current load, then its gain is simply μ.

2 A cathode follower has 100% negative feedback, so the gain of the cathode follower is:

$$K = \frac{A_0}{A_0 + 1}$$

If we choose a 6J5 as the audio device, with a constant current load:

$$K = \frac{20}{21} = 0.9524$$

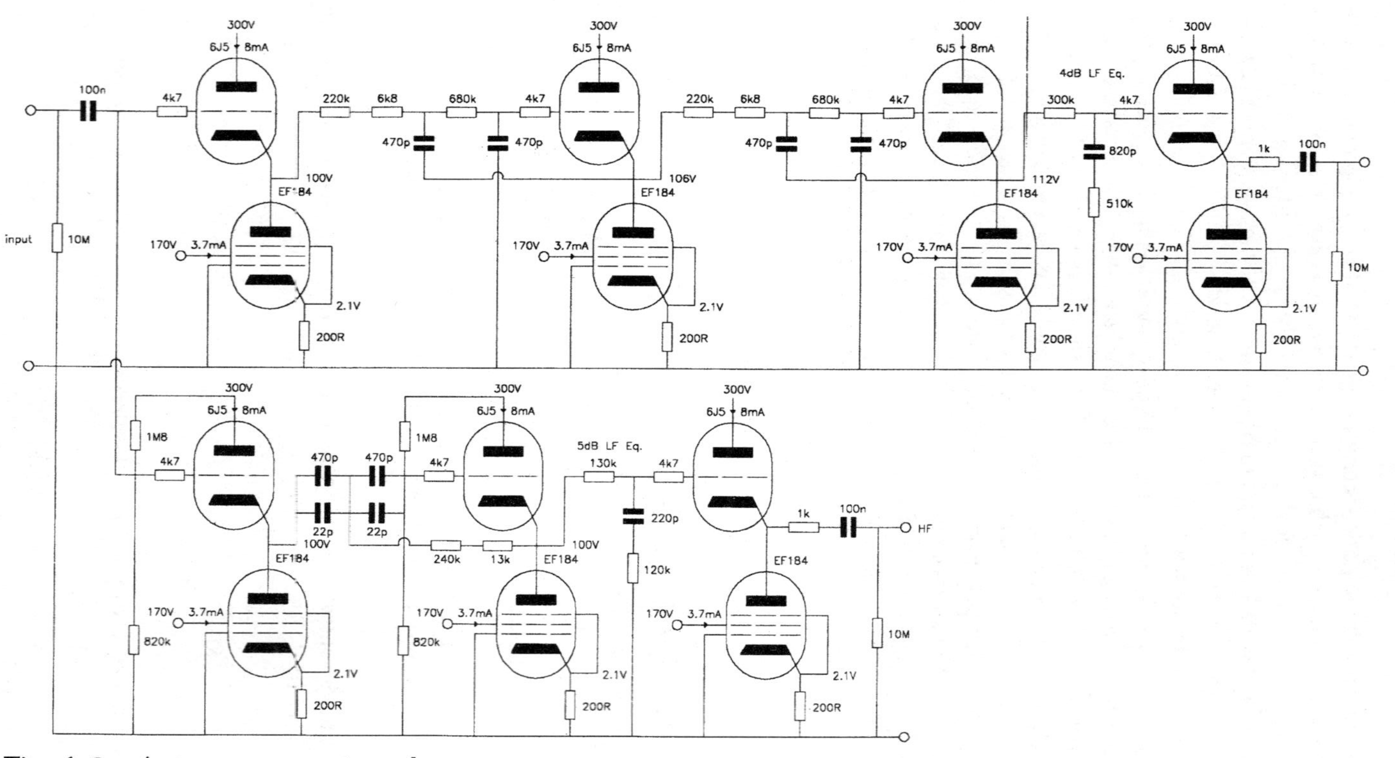

Fig. A.3 *Active crossover using valves*

3 24 dB/octave Linkwitz–Riley[5] crossovers are made by cascading a pair of 12 dB/octave Butterworth filters, for which:

$$Q = \frac{1}{\sqrt{2}}$$

Alternatively, we can specify *d*:

$$d = \frac{1}{Q} = \sqrt{2}$$

4 It is easy to change resistor values, but capacitors are only available in a restricted range, so the two capacitor values should be equal:

$$\gamma = \frac{C_1}{C_2} = 1$$

5 The high-pass filter (which is the more awkward one to design) has a transfer function:

$$\frac{hs^2}{s^2 + ds + 1}$$

We know that we will use a cathode follower (which is non-inverting), so we look at diagram 3 from the table on p. 76.

R_2 may well need to be a potential divider chain, such as on p. 75, which means that we are restricted in our choice of R_2, and because:

$$T_2 = R_2C_2$$

Our choice of T_2 has now been restricted.

(A better method would be to provide a simple bias regulator that could be shared by all the valves, since this would be quieter and not restrict R_2. A version of the circuit used in Chapter 4 for elevating heater supplies would be ideal.)

6 We now know that both γ and T_2 are restricted, and we should therefore refer to Formulas Group III. This table contains general purpose formulae, which are explained by their footnote. We want to find T_2, so we look to column b:

$$\frac{d}{2(1-K)} \cdot \left(1 \pm \sqrt{1 - \frac{4(1+x)(1-K)}{d^2}}\right)$$

We substitute our values: $\gamma = x = 1$, $K = 1$, $D = \sqrt{2}$

$$T_2 = \frac{\sqrt{2}}{2(1-0.9524)} \cdot \left(1 \pm \sqrt{1 - \frac{4(1+1)(1-0.9524)}{2}}\right)$$

The lower value of T_2 is more stable (as explained on p. 81), so we only evaluate the equation for the case where we subtract the large square root term. We are calculating a time constant, so it must have units (seconds), giving:

$$T_2 = 1.493\,\text{s}$$

7 Referring back to the diagram on p. 76 to find T_1, we see that:

$$T_1 T_2 = 1, \text{ or, } T_1 = \frac{1}{T_2} = \frac{1}{1.493} = 0.6698$$

8 The filter designs have all been designed for a frequency (ω) of 1 radian/second, a procedure known as normalization. We want to be able to scale to our chosen frequency f (in Hz), so:

$$CR_2 = \frac{T_2}{2\pi f}, \quad CR_1 = \frac{T_1}{2\pi f}$$

We can now choose practical resistor and capacitor values for our chosen frequency:

(a) Remember that a high-pass filter biased by a potential divider has a value for R_2 that is the Thévenin resistance seen looking back into the divider chain (the two resistors in parallel). We choose resistor values to set our DC bias voltage first, then see if they produce a convenient capacitor value. Since the exact bias potential is unlikely to be critical, we can adjust the resistor values until a convenient capacitor value emerges.

(b) Having set the value of C, we can now calculate the value of R_1.

To find values for the low-pass filter, we repeat steps (6) to (9) using diagram 1 on p. 76, again setting $\gamma = 1$, so that we can use the same value of capacitor as used in the high-pass section. Not only is it cheaper to buy many capacitors of the same value, but it is much easier to change the frequency of the crossover later by changing standard capacitor values than by changing (and re-calculating) non-standard resistor values.

References

1. Acoustical Manufacturing Company (1953–1967) *QC11 Owner's Manual.*
2. Lipshitz, Stanley P. (1977) On RIAA Equalisation Networks. *JAES*, June.
3. Sallen & Key. (1955) A Practical Method of Designing RC Active Filters. *IRE Transactions*, March.
4. Billiam, Peter J. (1978) Harmonic Distortion in a Class of Linear Active Filter Networks. *JAES*, June, Vol. 26, No. 6.
5. Linkwitz, Siegfried (1976) Active Crossover Networks for Non-coincident Drivers. *JAES* Jan./Feb., Vol. 24, No. 1.

Index